Improve Your Grade!

Access included with any new book.

Get help with exam prep! REGISTER NOW!

Registration will let you:

- **Prepare for exams using online quizzes**

- **Master difficult concepts with a detailed PowerPoint™ lecture outline for each chapter**

- **Access GIS exercise data and a variety of GIS internet resources**

- **Read current events about GIS topics with RSS feeds that are updated daily**

www.mygeoscienceplace.com

TO REGISTER

1. Go to **www.mygeoscienceplace.com**
2. Click "Register."
3. Follow the on-screen instructions to create your login name and password.

Your Access Code is:

Note: If there is no silver foil covering the access code, it may already have been redeemed, and therefore may no longer be valid. In that case, you can purchase access online using a major credit card or PayPal account. To do so, go to www.mygeoscienceplace.com, click on "Buy Access," and follow the on-screen instructions.

TO LOG IN

1. Go to **www.mygeoscienceplace.com**
2. Click "Log In."
3. Pick your book cover.
4. Enter your login name and password.
5. Click "Log In."

> **Hint:**
> Remember to bookmark the site after you log in.
>
> **Technical Support:**
> http://247pearsoned.custhelp.com

Pearson Prentice Hall Series in
Geographic Information Science
Keith C. Clarke, Series Editor

About our Sustainability Initiatives

This book is carefully crafted to minimize environmental impact. The materials used to manufacture this book originated from sources committed to responsible forestry practices. The paper is Forest Stewardship Council (FSC) certified.

The printing, binding, cover, and paper come from facilities that minimize waste, energy consumption, and the use of harmful chemicals.

Pearson closes the loop by recycling every out-of-date text returned to our warehouse. We pulp the books, and the pulp is used to produce items such as paper coffee cups and shopping bags. In addition, Pearson aims to become the first climate neutral educational publishing company.

The future holds great promise for reducing our impact on Earth's environment, and Pearson is proud to be leading the way. We strive to publish the best books with the most up-to-date and accurate content, and to do so in ways that minimize our impact on Earth.

Mixed Sources
Product group from well-managed forests, controlled sources and recycled wood or fibre

Cert no. SW-COC-002985
www.fsc.org
© 1996 Forest Stewardship Council

Prentice Hall
is an imprint of

Pearson Prentice Hall Series in
Geographic Information Science
Keith C. Clarke, Series Editor

Getting Started with Geographic Information Systems

Fifth Edition

Keith C. Clarke
University of California, Santa Barbara

Prentice Hall

Boston Columbus Indianapolis New York San Francisco Upper Saddle River
Amsterdam Cape Town Dubai London Madrid Milan Munich Paris Montréal Toronto
Delhi Mexico City São Paulo Sydney Hong Kong Seoul Singapore Taipei Tokyo

Library of Congress Cataloging-in-Publication Data

Clarke, Keith C.
Getting started with geographic information systems / Keith C. Clarke. -- 5th ed.
    p. cm. -- (Prentice Hall series in geographic information science)
Includes bibliographical references and index.
ISBN-13: 978-0-13-149498-5
ISBN-10: 0-13-149498-8
1. Geographic information systems. I. Title.
G70.212.C57 2011
910.285--dc22

                          2009041842

Geography Editor: *Christian Botting*
Editor in Chief, Geosciences and Chemistry: *Nicole Folchetti*
Marketing Manager: *Maureen McLaughlin*
Project Manager, Editorial: *Tim Flem*
Assistant Editor: *Jennifer Aranda*
Editorial Assistant: *Christina Ferraro*
Managing Editor, Geosciences and Chemistry: *Gina M. Cheselka*
Senior Project Manager, Science: *Beth Sweeten*
Media Producer: *Lee Ann Doctor*
Art Director: *Jayne Conte*
Cover Design: *Bruce Kenselaar*
Senior Manufacturing and Operations Manager: *Nick Sklitsis*
Operations Specialist: *Maura Zaldivar*
Production Supervision/Composition: *GGS Higher Education Resources, A Division of Premedia Global, Inc.*
Cover Credit: *Keith Clarke*

Prentice Hall
is an imprint of

**This book is not for sale or
distribution in the U.S.A. or Canada**

1221069

www.pearsonhighered.com

For Margot and Lila

Brief Contents

Contents

New to the Fifth Edition

Instructors who have used the book in the past will notice that this is a substantial rewrite and update. Since the fourth edition in 2003, much has happened in Geographic Information Science (GIS), and room had to be found to include many new parts of the field, while other content was showing its age. New features include:

- Chapter 6 on Spatial Analysis has been revised to follow publicly-available data sets through the reasoning and logic of spatial analysis, rather than the details of calculations and statistics.

- Chapter 7 on Terrain Analysis is completely new. Too few GIS students become familiar with the tools for transforming terrain, and for too long GIS has lived in a two-dimensional world.

- Chapter 9 on GIS Functionality includes an updated survey of the growing number of software and server-based tools available, including open-source GIS.

- Chapter 11 has been revised for currency and examines current and future trends in the field.

- The end-of-chapter Study Guides have been reworked, making them more student-friendly.

- Updated information sources and websites ensure that students are getting the most current and accurate data available.

- Each chapter concludes with new "People in GIS" interviews with individuals currently working in the field of GIS, with themes matched to the chapter and student interest.

- A dedicated website (www.mygeoscienceplace.com) offers a variety of resources for students and instructors, including lab data files, quizzes, PowerPoint™ lecture outlines, RSS feeds, links to Web resources, and Class Manager & GradeTracker Gradebook functionality for instructors.

- The book now has an accompanying *GIS Exercise Student Workbook*, including exercises using ESRI's ArcGIS software and Google Earth™, carefully matched with the structure of the book.

Foreword

The other day I was standing in line at the supermarket and overheard the following conversation.

"So you're going to college in the fall, what are you going to study?"
"I want to do environmental studies, and specialize in GIS. Do you know what that is?"
"Sure, I know all about it. Cool."

I quietly smiled to myself, because in the fourteen years since I wrote the first edition of this book, GIS has gone from an obscure technical concentration for nerds in Geography to a central part of everyday life. When I crossed the country during my move to California in 1996, there was no Google Maps™ or Google Earth™, GPS had reached operational status just the previous year, and most GIS software was buggy and expensive. How much things have changed. GIS has even become cool.

Yet each Fall brings another generation of students into the classroom, eager for knowledge. It was for this 1996 group that I originally wrote this book, largely because their needs were being ignored by the more advanced GIS books that initially catered to a higher level educational arena. I am delighted that the book still has a life, among a new generation of scholars with talents and interest in righting the ills of the world through science. I wish you happy studies, and challenge you to make your mark in the world. I hope the new edition, and the added *GIS Exercise Student Workbook*, continue to get people started with GIS without too much pain and suffering.

The fifth edition includes a completely new chapter on terrain analysis. This area is of critical importance, and new instruments such as LiDAR and datasets such as Aster GDEM and SRTM have made global scale terrain analysis possible. Too few GIS students become familiar with the tools for transforming terrain, and for too long GIS has lived in a two-dimensional world. I have also revamped the end-of-chapter Study Guides, making them more student-friendly. The addition of Indy Hurt's accompanying GIS Workbook also formalizes years of work in creating challenging but interesting exercises to reinforce the topical coverage of the book. Thanks for all the hard work, Indy, and the several generations of teaching assistants who have taught the class at UCSB (Geography 176A) along with me. After carefully reading the evaluations of my fourth edition, I have traded the cartoons that introduced the chapters with geographically-themed artwork. Call it my 2% for art, but I also strongly believe that *art* is p*art* of c*art*ography. As for the quotes, well, I'm afraid Brits can adapt to the American way of life, but they don't lose their British sense of humor. At least *I* had fun putting them together!

Chapter 1 introduces the field of GIS and a little of its history, then covers organizations and resources where additional information and help can be found. I call this the self-help approach to learning, something infinitely easier with the World Wide Web in place. Chapter 2 covers basic cartography: geodesy, map projections, scale and coordinate systems. Without this material much of GIS will remain a mystery, so I believe in putting these topics up front. Chapter 3 introduces GIS data structures, understanding of which is essential before a student attempts to seek out his or her own data. Chapter 4 covers data acquisition and input, both the vast amounts of ready-for-use data on the servers of the Internet and how to get your own maps into a GIS. Chapter 5 touches on the data management part of GIS, introducing the principles of database management and covering some of the issues involved. Chapter 6 introduces spatial analysis, and has been revised to follow one or two publicly-available data sets through the reasoning and logic of spatial analysis, rather than the details of calculations and statistics. Chapter 7 is new, and introduces into basic GIS coverage the geographical surface along with its particular issues, display options, and analyses. Chapter 8 returns to cartography and covers some of the methods for cartographic display and the means by which maps can be made both more effective and more aesthetic. Chapter 9 looks at GIS functionality and surveys the large number of software and server-based tools available for GIS. Chapter 10 covers four GIS case studies and examines what each is able to teach about how GIS is actually used in the workplace. Finally, Chapter 11 is more of an essay, examining some of the current and future trends in the field and the issues that will arise from them. Each chapter is followed by an interview with people working in the field of GIS whose experience both supplements the topic of the chapter, and shares experiences likely to help students. To the ten willing interviewees, I give my collective thanks, and add thanks to Bill Norrington, who transcribed the audio.

So many people have now been involved with this book that to thank them by name would require another chapter. I have tried to work as many names into credits as possible. If I have left you out, you know who you are. Thanks very much. Among the many reviews of the book that I have used for guidance was the observation that this book is more of a conversation with students than a standard text. That this idea, which drove the first edition, has survived fourteen years and countless changes is an achievement I view with pride. Let's keep the conversation going.

Lastly, no one deserves more thanks than my wife of 30 years today, Margot, and our new daughter Lila. It is they who have put up with most. Sitting at a play computer in a toy store recently, when asked what she was doing, 2-year-old Lila said to be quiet, she was working on her book. I know the feeling! It is their love and support, and that of my whole family, that keep me going.

Keith C. Clarke
Santa Barbara, CA

CHAPTER 1

What Is a GIS?

GISs are simultaneously the telescope, the microscope, the computer, and the Xerox machine of regional analysis and synthesis of spatial data.

Ron Abler

1.1 GETTING STARTED

Every thing, every person, and every event on earth exists or happens somewhere. If the location of this somewhere can be fixed, then information about almost anything can be placed on a map, and this map then used to organize, search, and analyze the information. There is already an extraordinary amount of mapped, remotely sensed, and measured information that is geographically intelligent, i.e., it knows where it is, and much of it is easily available. There is also a technology and suite of methods, geographic information systems and science, that can digest the information, and give us the power of seeing meaning and structure, perhaps gaining understanding and knowledge. The power that comes from using GIS can win elections, feed the hungry, wage wars, protect the environment, save lives, and make the world sustainable. In this book, you will learn how this technology and these methods work, what they are capable of, and why GIS is changing the world and almost everything in it. The goal is to understand and improve the world through careful analysis of geography through the lens of GIS. We may not be able to solve the world's problems immediately, but every solution has to get started somewhere!

Getting started with GIS can be a long, slow, expensive, and sometimes painful climb up the learning curve. On the other hand, in the last decade the software side of GIS has improved by leaps and bounds, resolving many of these old problems. As your first book in GIS, this one will set the foundation for a more breadth-first tour through the discipline than the more advanced books can offer. A goal is to remain lean, mean, and relevant, and not to engage in the "feature creep" and time-sensitive content that other books use. By keeping the text up to date, the author and editors are still working hard to ensure that your first GIS experience is timely, pleasant, and constructive.

The introduction gets started with GIS definitions, outlines the development of the field, and maps out some of the sources of information that can teach you more about GIS. It should be clear at the outset that GIS is not really a "killer app," namely a must-have innovative and essential computer application like a spreadsheet, a word processor, or a database manager, although some GIS spin-offs like Google Earth and MapQuest are headed that way. GIS is partly a killer app, but the upward shift in capability that its users receive is not due to computer software alone. Instead, GIS has built on the collective knowledge of the scholarly fields of geography and cartography, with some geodesy, database theory, computer science, and mathematics thrown in for good measure. As Ron Abler's definition above shows, GIS is not just one, but many simultaneous technological revolutions. *Getting Started with Geographic Information Systems* introduces a distilled version of the theory and content from the fields of these disciplines—the minimum necessary to get you started—and then offers some signposts pointing toward where the revolutions will lead next. If you choose to go further, there are an infinite number of paths forward.

Using GIS requires you to think like a geographic information scientist. Geographic information science was born in the 1990s by merging skills and theory across many different disciplines, and it has now reached maturity after years of development. Like all fields when first encountered, geographic information science requires some mental readjustment. The purpose of this book is to gently guide you, the reader, through this process. Since you are reading this book, the chances are that you are already used to thinking graphically, mapping out information, and building analytical solutions around these maps and graphics. If not, I hope that this text will serve both to get you started and to unleash a part of the brain you may never have used before—the spatial part—that holds real power as a new way of problem solving.

1.1.1   WHAT IS SPATIAL INFORMATION?

We live in an information society. Few of us can imagine life without cellular telephones, streaming video, Facebook, 24/7 television news, or on-line news services. Yet most of the information that reaches us contains only a few elements: text, numbers, images, video, and animations. Among the most abundant and easily searched information is text, and indeed if we look at the source file for a web page, all we see is text, and a set of special codes, in an ordered sequence. There are many analog equivalents of digital information flows: tables, lists, indexes, catalogs, cross-references, and so forth. Information is often unintelligible without some form of ordering. A random listing of people in the

Figure 1.1: Forms of information organization in everyday life—books, catalogs, and lists.

telephone book, for example, would be meaningless because to find any particular person's phone number, you would have to read the names one by one until you found the correct one. The simplest form of order is sequence, and things can be arranged alphabetically, chronologically, or in any one of dozens of clever ways. Another way to impose order is to use an index. Each piece of information or "record" can have a number or index, through which it can be found. At the back of this book, for example, key words that you may wish to find references to later are listed in alphabetical sequence in an "index," and the information is retrievable by giving a page number where the term was used. Lists, tables, indexes, and their equivalents are everywhere: in books, catalogs, lists, and even television listings (Figure 1.1).

One extremely versatile way of organizing information is in two dimensions, by geographic location. We do this by creating lists of points, ordered in not one but two dimensions, such that they make a sequence that "traces" the geographical extent of something in the real world. When data are listed by location, then they can be placed on a map to show the order that must be imposed by some other means when the data are not geographical. Such data are called spatial data. A GIS is the ordering sequence, the list, the index, and the catalog for spatial data all put together. As such it is the key to understanding geographic and spatial data, and in addition, a way to organize virtually all other sorts of information, simply because almost everywhere is somewhere. Geography is a very powerful organizational tool indeed. For example, imagine the vast quantity of information, photographs, video, text, and discussion about Martin Luther King's "I Have A Dream" speech at the Lincoln Memorial in Washington, D.C., on August 28, 1963. All of this data could be attached to the single location on the Memorial's steps shown in Figure 1.2, at 38°53'21.5"N 77°02'59.4"W (WGS84), where the speech is commemorated.

Figure 1.2: Inscription at the spatial location 38°53'21.5"N 77°02'59.4"W (WGS84).

1.2   SOME DEFINITIONS OF GIS

Good science starts with clear definitions. In the case of geographic information systems, however, different definitions have evolved over the years as they were needed, or as GIS was rediscovered. Searching the Internet using Google and the term "geographic information system" today yields 1,640,000 pages of information or hits. It is no surprise, then, that GIS can be defined in many different ways. Wikipedia, for example, defines GIS as "any information system capable of integrating, storing, editing, analyzing, sharing, and displaying geographically referenced information." Which definition works for you depends on what questions you are trying to answer. Common to all GIS-related questions is that they relate to one type of data, spatial data, that is unique because it can be linked to a geographic map.

Spatial means related to the space around us, in which we live and function. Our own definition of GIS can start with a simple description of the three parts of a GIS, which are (1) the database (2) the spatial or map information, and (3) some way create and use links between the two. Necessary parts are a computer, some software, and people to use the system. We also need an underlying problem or task that the GIS will be used to solve or perform, such as choosing a site for a nature preserve, routing an ambulance to a house, determining which neighborhoods in a city are susceptible to flooding, or maintaining a map guide that citizens of a town can use to find public services. Then, of course, we also need understanding and experience, both of the system and of the problem. As you will quickly learn, the last two items may be the hardest to come by.

1.2.1   A GIS Is a Toolbox

A GIS can be seen as a set of tools for analyzing spatial data. A toolbox is a handy container in which to store tools and bring them to the place where they are to be used. GIS tools are, of course, computer tools, and a GIS can be thought of as a software package containing the elements necessary for working with spatial data. Just as the toolbox can hold a hammer, a screwdriver, pliers, and a wrench, so too a GIS can contain basic functions that allow generic map and database operations (Figure 1.3). A saw can "cut out" geographic areas, for example, or a glue gun merge them together.

Figure 1.3: The GIS-as-toolbox analogy. A GIS can be thought of as a collection of tools for spatial tasks, each with different specialist functions.

Two decades ago, Peter Burrough defined GIS as "a powerful set of tools for storing and retrieving at will, transforming and displaying spatial data from the real world for a particular set of purposes" (Burrough, 1986, p. 6). The key word in this definition is "powerful." Burrough's definition implies that GIS is a tool for geographic analysis. This is often called the *toolbox definition* of GIS, because it stresses a set of tools each designed to solve specific problems. Most GIS software today contains hundreds or even thousands of both generic and special purpose tools; indeed some GIS research has been aimed at identifying which tools all GISs share, a sort of minimum set of only about twenty (Albrecht, 1998).

If a GIS is a toolbox, a logical question is: What types of tools does the box contain? Several authors have tried to define a GIS in terms of what it does, offering a functional definition of GIS. Most agree that the functions fall into categories and that the categories are sub-tasks that are arranged sequentially as data move from the information source to a map and then to the GIS user and decision maker. Another GIS definition, for example, states that GISs are "automated systems for the capture, storage, retrieval, analysis, and display of spatial data" (Clarke, 1995, p. 13). This has been called a *process definition*, because we start with the tasks closest to the collection of data, and end with tasks that analyze and interpret the geographical information. The chapters in this book are structured around this sequence of functions—each will be discussed in detail as the book progresses.

1.2.2 A GIS Is an Information System

Santa Barbara geographers Jack Estes and Jeffrey Star once defined a GIS as: "An information system that is designed to work with data referenced by spatial or geographic coordinates. In other words, a GIS is both a database system with specific capabilities for spatially-referenced data, as well as a set of operations for working with the data" (Star and Estes, 1990, p. 2).

This definition stresses that a GIS is a system for delivering answers to questions or queries, what might be called an information system definition. This means that a GIS collects data, sifts and sorts them, and selects and rebuilds them to create precisely the

Figure 1.4: GIS as an information system. An analogy to a map library. Map librarians can search for and find specific maps in answer to your queries.

right organization of the information to answer a specific question. The reference to geographic coordinates is an important one, because the coordinates are literally how we are able to link data with the map. This theme is taken up further in Chapter 2. An information system is designed to re-organize information in such a way as to make it useful, i.e., to convert raw data into more valuable pure information.

A simple example of an analog information system for geographical data is a map library (Figure 1.4). The maps are stored in drawers contained in cabinets. These have to be indexed for any particular map to be found. A search system might be a map index or catalog that gives us the indexes. When we have the maps we need, we can examine then together and use them for problem solving. To do so, we might need to enlarge them, copy them, annotate them, or view multiple maps together.

Another information-system definition of a GIS is one that has stood the test of time remarkably well. As such, this definition is worth considerable thought. In 1979, during the stone-age of GIS technology, Ken Dueker defined a GIS as "a special case of information systems where the database consists of observations on spatially distributed features, activities or events, which are definable in space as points, lines, or areas. A geographic information system manipulates data about these points, lines, and areas to retrieve data for ad hoc queries and analyses" (Dueker, 1979, p. 106). Like a fine wine, this thoughtful definition has actually improved with age.

The phrase "special case of information systems" implies that GIS has a heritage in information systems technology, which it indeed does. GIS did not invent database management and there exists in computer science a 50-year tradition in this field all the way from the earliest mainframe databases and spreadsheet programs, through relational database management, to the object-oriented database management and data warehouses of today. Information systems are used extensively in library science (including in map librarianship), in business, and around the Internet. In fact, today it is through the Internet that we usually meet these versatile and powerful tools that are at the heart of every GIS.

In Dueker's definition of GIS, the database consists of a set of observations, which implies a scientific approach to measurement. Scientists take measurements and record those measurements in some kind of system to help them analyze the data. The observations are spatially distributed; that is, they occur over space at different times and at different locations at the same time. For example, the National Oceanic and Atmospheric Administration (NOAA) collects weather data from a large number of weather stations around the country, and integrates their information on temperature, rainfall, humidity and so forth to give us the weather maps we might see on the weather channel or the weather website as a single map.

The observations are those of features, activities, and events. A feature is a term from cartography meaning an item to be placed on a map. Point features, such as an elevation bench mark, location of a transmission tower, or public survey section corner (Figure 1.5), have only a location. Line features have several locations strung out along the line in sequence like a bead necklace, an example being a road or a stream. Area features consist of one or more lines that form a loop, such as the shoreline of the ocean or a wide river, or the edge of a patch of vegetation. Traditionally, the source of geographic information is the map, and the information on a map consists of a set of graphic symbols, such as colors, lines, patterns, and shades.

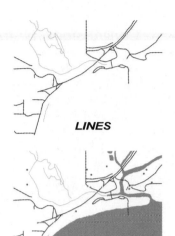

POINTS

LINES

AREAS

Figure 1.5: Point, line and area features extracted from a USGS topographic map. Any map consists of these features, plus text.

"Activities" implies a link to the social sciences. Human activities create geographical patterns and distributions, and these activities are linked to points, lines and areas too.

For example, take the act of voting. At 18 years old in the U.S., we gain the right to register to vote and must do so in a voting district. Every ten years, the constitution says that we must redistrict voting areas to represent equal numbers of people, so that one vote means the same across the whole country. As a result of constant revision, the congressional districts begin to look highly spatially irregular, in spite of their equal representation (Figure 1.6). In most cases, the redrawing has been done using a GIS. On election night, the resultant voting behavior is obvious for all to see as we watch the tallies come in on the nightly news. In all cases, each person's vote has been tallied up and recorded in a congressional or other voting district which can be represented as a polygon.

Many social science elements of life are related to polygons, regions delimited on the ground with boundaries, visible or invisible. Nations, states, provinces, counties, health districts, neighborhoods, school districts, and police precincts are all examples that affect us all. There are also the more intangible districts, eco-zones, climatic types, hazard and disease areas, ecosystems and watersheds. Even point and line features can be counted by region, for example the number of oil wells in each US state.

Human activities lead to the population map, census map, distribution of disease incidences, retailing patterns, location of schools and bus routes, and so on—all related to how people live their daily lives. The "event" part of GIS implies that geographic data fall

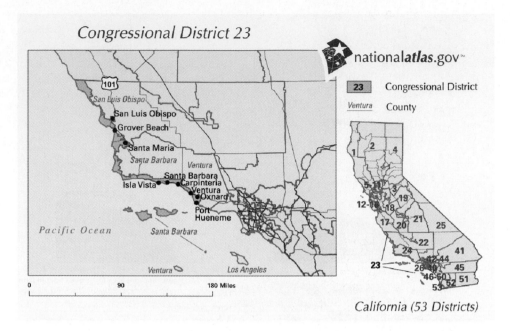

Figure 1.6: Congressional District 23 in California. Polygons, or areas, of equal population rather than area relating to an "event," an election.

not only into space but also into time. Time gives us a fourth dimension and becomes a part of the data because events happen in time and features exist over a duration. The coastline in Figure 1.5, for example, has been eroding away, and so is not shown in the same place on a map made 100 years ago. So an important use of GIS is to track not only the distribution of human activities over space, but also changes in them.

Dueker's GIS definition assumes that events also have expressions as points, lines, or areas in space and on the map. An example of a point event is the location of a traffic accident. A line activity could be the flow of electricity along a segment of a power cable. An area event could be the freezing of a body of water, such as the Central Park reservoir in New York City. The information element becomes useful to the GIS user because it exists, it has data associated with it, and it has cartographic reality as a feature on a map.

We use the information mapped in the GIS for doing exactly what an information system should do: solve problems, do queries, come up with the answer, or try out a possible solution. So we manipulate the data, not by hand, but digitally. We manipulate data about events or activities by using the digital map features that represent them as "handles." In other words, the points, lines, and areas in this map database are used to manage the data. Another key part of Dueker's GIS definition is that the queries must be ad hoc or context-specific queries. We do not have to know in advance when building a GIS exactly what we want to use it for. This means that GIS is a generic problem-solving tool—it is not something built just for a specific project or to get this week's GIS class assignment done. The value of GIS comes from its ability to apply general geographic methods to specific geographic regions. The GIS supplies the methods; you as a user supply the region.

Finally, in Dueker's definition a GIS can also do analysis. Usually, the purpose of having data in GIS form is so that an analyst can extract what is necessary to make predictions and explanations about geographic phenomena. If we have data, for example, on traffic flow and the location of accidents in Los Angeles, how can we use that information to lessen traffic congestion? A focus on GIS technology ignores the fact that the ultimate purpose of the system is to solve problems. Geographic information science goes beyond description, to include analysis, modeling, and prediction. The information systems definition, then, leads back to the role of a GIS as a problem solver. It begs the question: Is this just one more scientific method, or is this a new scientific approach?

1.2.3   GIS Is an Approach to Science

As a tool or as an information system, GIS technology has changed the entire approach to spatial data analysis. GIS has already been compared to not one but several simultaneous revolutionary changes in the way that data can be managed. The convergence of GIS with other geospatial technologies, those of surveying, remote sensing, air photography, the global positioning system (GPS), and mobile computing and communications, has fed a spectacular growth of these technologies. This is a revolution that has changed much already, and that is still almost daily changing how we live our lives. One consequence of the emergence of GIS as science has been the use of GIS methods in many other disciplines, from archaeology to zoology. GIS has also spearheaded a convergence of mostly Internet-based tools for mapping and made them easy to use and very practical. Thirdly, this mutation has led to a culling of the body of knowledge that constitutes geography so that it is suitable for use in these parallel fields as a new approach to science.

Goodchild called this "geographical information science" (Goodchild, 1992). In the United States the preferred term is geographic information science. This renaming of the field has had a major influence, including the renaming of journals, professional societies and conferences (Figure 1.7).

Goodchild defined geographic information science as "the generic issues that surround the use of GIS technology, impede its successful implementation, or emerge from

 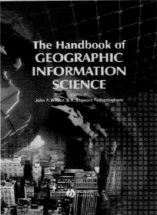 

Figure 1.7: GIS as geographic information science, not systems.

an understanding of its potential capabilities." He also noted that this involved both research on GIS and research with GIS. Geographic information science (GISc or GISci) is the academic theory behind the development, use, and application of geographic information and is concerned with GIS hardware, software, and geospatial data. Geographic information science addresses fundamental issues raised by the use of GIS and related technologies (Wilson and Fotheringham, 2007; Kemp, 2007). Supporting the science are the uniqueness of geographic data, a distinct set of questions about the world that can only be asked geographically, the commonality of interest of GIS meetings, and a supply of books and journals. On the other hand, Goodchild noted that the level of interest depends on innovation, that it is hard to sustain a multidisciplinary (rather than interdisciplinary) science, and that at the core of the science, in geography, a social science tradition has to some extent an antipathy toward technological approaches.

The body of knowledge of GIScience is based on geographical data, the rules for its use, the means by which evidence can be extracted from the data by analysis, and the combination of results to form new maps and new geographical knowledge. Key sources of variation and uncertainty in this process are also studied by geographical information science, including scale and resolution, classification and abstraction, use of methods, and the testing of these methods.

This book is an effort to distill from the discipline of geography exactly those components that are derived from the areas of research outlined by Goodchild. As such, this book adopts Goodchild's GIScience approach. The chapters that compress the principles of cartography are Chapters 2 and 7; analytical cartography's contributions fall into Chapters 3–5; and spatial analysis is discussed in Chapter 6. Added to these are doses of general geography, database management, and applied GIS. This knowledge base constitutes the new and strengthening field of geographic information science.

1.2.4   GIS Is a Multibillion-Dollar Business

Groups monitoring the GIS industry estimate the total value of the hardware, software, and services conducted by the private, governmental, educational, and other sectors that handle spatial data to be billions of dollars a year. Furthermore, for the last two decades, the industry has seen double-digit annual growth. Anyone who attends a national or international conference in the field can feel an overwhelming sense of rapid growth, sophistication, and the sheer magnitude of the transformation that GIS has led in the worlds of business and industry (Pick, 2005; 2008).

Largely responsible for this situation were the massive cost reductions in technology dating from about 1982, when computers moved out from behind glass windows as tended by people in white coats—and onto the desktop. This decline in cost, aided by the success of the workstation as a tool in engineering settings, has led to a rapid increase in what is usually called the "installed base" of GIS. Just about every major academic institution in the United States and in many other countries now teaches at least one class in GIS, and many teach several courses, specialties or even entire degrees. Most local, state, and federal government agencies use GIS, as do businesses, planners, architects, foresters, geologists, archeologists, and so on. This growth in pure numbers, added to the increase in sophistication of the systems, is what has led to the big business aspect of GIS.

However, other steps have been critical to the booming (and blooming) of GIS. First, the industry was founded on vast amounts of inexpensive or free federal government data, mostly data of the U.S. Census Bureau and the U.S. Geological Survey (USGS). Second, the community has been a successful advocate of the field and has rapidly developed an infrastructure for self-support, user groups, network conference groups, and so on. Third, the addition of graphical user interfaces and the addition of extremely useful features such as help screens and automatic installation routines have played an essential role. Fourth, GIS has merged successfully with parallel technologies and has benefited from the resultant multiplier effect.

One view of the impact that GIS has on business can be gained by examining the field of Business Geodemographics. In this GIS specialty, small areas the size of neighborhoods to counties are delineated characterized by the demographic properties of the population, often from census data, and by any available purchasing data. This allows specialized marketing or promotions aimed at highly specialized sub-markets. For example, districts with large numbers of young couples with children for baby products, or districts with retired but active adults who often travel, for marketing cruise ships. Examples are ESRI's Community Tapestry (Figure 1.8) and in the United Kingdom, Experian's Mosaic. This approach is used very effectively also in canvassing for votes.

The growth of GIS has been a marketing phenomenon of amazing breadth and depth and will remain so for many years to come. Clearly, GIS will continue to integrate its way into our everyday life to such an extent that it will soon be impossible to imagine how we functioned before. The Internet with tools like map servers and Google Earth, and mobile systems that use location based services have brought GIS to the public as a commodity. GIS operations have become so transparent to the public that we often do not even realize

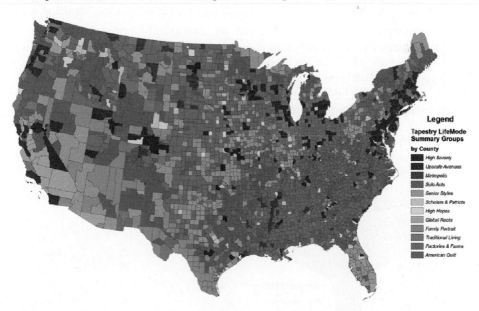

Legend

**Tapestry LifeMode
Summary Groups
by County**

- High Society
- Upscale Avenues
- Metropolis
- Solo Acts
- Senior Styles
- Scholars & Patriots
- High Hopes
- Global Roots
- Family Portrait
- Traditional Living
- Factories & Farms
- American Quilt

Figure 1.8: Geodemographics. ESRI's community tapestry, US lower 48 states by county with 12 "lifestyles" created with ArcGIS Business Analyst Segmentation Module. Copyright © 2009 ESRI and its licensors. All rights reserved. Used by permission. HYPERLINK "http://www.esri.com/"www.esri.com

that GIS is there, just as we give no thought to the microprocessor that calculates our change at the cash register, or the GIS that controlled the van delivery of FedEx packages we received, right down to the signature on the driver's handheld computer.

1.2.5 GIS Plays a Role in Society

Many people doing research on GIS have argued that defining GIS narrowly, as a technology, as software or as a science, ignores the role that GIS plays in changing the way people live and work. Not only has GIS radically changed how we do day-to-day business, but also how we operate within human organizations. Nick Chrisman (1999) has defined GIS as "organized activity by which people measure and represent geographic phenomena then transform these representations into other forms while interacting with social structures." This definition has emerged from an area of GIS research that has examined how GIS fits into society as a whole, including its institutions and organizations, and how GIS can be used in decision making, especially in a public setting such as a town meeting, or on a community group website (Figure 1.9). This latter field is termed PPGIS, for Public Participation GIS (Corbett and Keller, 2006).

Few people would doubt that when GIS has become part of the way of doing business within many organizations, such as planning offices or state planning agencies, the result has been a shifting in the work assignments, job descriptions, responsibilities, and even the power relationships of the organization. For example, when GIS is first introduced into a work environment, it has been found that it is very important to have a "champion," someone who is a GIS advocate from within the group. Many in the study of GIS have focused on describing and analyzing these impacts rather than looking technically at GIS, or at GIS in its application. So far, the field has generated a history of the discipline (Foresman, 1997) and an increasingly popular set of meetings and conferences. Several books have stimulated interest in this approach, including *Ground Truth* (Pickles, 1995), which introduced the somewhat more humanistic and social science dimension to GIS research work.

Nick Chrisman's definition of GIS includes all of the social processes of GIS functions. For example, a GIS may be used to capture data about land holdings as

Figure 1.9: Chrisman's diagrammatic view of how GIS fits into a broader context of people, organizations, and goals.

Social and Cultural Context

Institutional Context

Transformations
Operations

Representation
Measurement

ownership parcels. However, the use and purpose of the data and its dissemination will vary according to the philosophy and traditions of the community in which the data are being used. In a growth-oriented community, for example, a GIS might be seen as a mechanism for expediting building permits and increasing land sales. In a more conservation-oriented community, a GIS might be seen and used as a vehicle for raising public awareness about environmental issues, supporting community planning, or enforcing pollution controls. Although essentially the same GIS software, hardware, and data may be in place in the two settings, the staff, their work assignments, and the degree of administrative control might be very different. It is the human factors involved that determine much about the GIS, rather than the technical capabilities.

Another component that Chrisman's definition recognizes is the importance of a basis in measurement. In the abstract sense, a GIS supports measurements about the land with many different levels of accuracy and reliability. In most cases, the GIS is based on the "best available data," but virtually always some of the data are incomplete, outdated, or missing. How GIS users come to terms with this problem is very often as large a factor in the GIS's capabilities and effective use as are the software, hardware and processes involved. As we state later, a GIS, like a map, is often a set of errors that have been agreed upon. This definition notes that not just the errors and the system supporting them define GIS, but also those critical agreements about the data that result among the people involved.

1.3   A BRIEF HISTORY OF GIS

Many of the principles of the new geographic information science have been around for quite some time. General-purpose maps date back centuries and usually focused on topography, the lay of the land, and transportation features such as roads and rivers. In the last century, thematic maps came into use. Thematic maps contain information about a specific subject or a theme, such as surface geology, land use, soils, political units, and data collection areas. Although both types of maps are used in GIS, it is the thematic map that led cartography toward GIS. Some themes on maps are clearly linked. For example, a map of vegetation is closely tied to a map of soils.

It was the field of planning that first began to exploit thematic maps by extracting data from one map to place it onto another. As an early example, the geographic extent of the German city of Dusseldorf was mapped at different time periods in this way in 1912, and a set of four maps of Billerica, Massachusetts, were prepared as part of a traffic circulation and land-use plan in the same year (Steinitz et al., 1976). By 1922, these concepts had been refined to the extent that a series of regional maps were prepared for Doncaster, England, which showed general land use and included contours or isolines of traffic accessibility. Similarly, the 1929 "Survey of New York and Its Environs" clearly shows that overlaying maps on top of each other was an integral part of the analysis, in this case of population and land value.

In 1950, the publication of the Town and Country Planning Textbook in Britain included a landmark chapter, "Surveys for Planning" by Jacqueline Tyrwhitt (Steinitz et al., 1976). Various data themes, including land elevation, surface geology, hydrology/soil

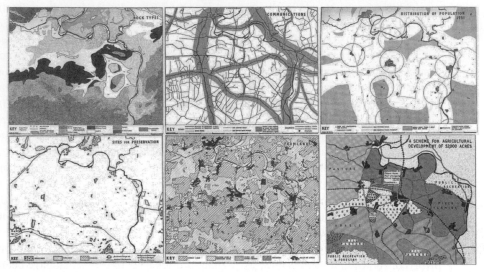

Figure 1.10: "Surveys for Planning," by Tyrwhitt introduced map overlay methods for integrative planning

drainage, and farmland, were brought together and combined into a single map of "land characteristics" (Figure 1.10)

The author described how the maps were drafted at the same scale, and how some map features were duplicated so that the maps could be superimposed precisely, using these features as a guide. Although the methods were manual they provided a basis for later methods by computer. Just as many others had "discovered" America, it was Columbus who is remembered because he was the first to write about it (and, incidentally, to draw a map!)..

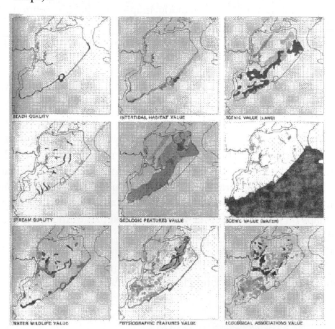

Figure 1.11: McHarg's composite overlay method used for siting Staten Island, New York's Richmond Parkway as detailed in *Design with Nature.*

In 1950, then, the technique of map overlay, now so common in GIS, was "invented" by Tyrwhitt, although it is likely that there were earlier precedents. Nevertheless, it is clear that by 1950, maps were regularly being traced onto transparent overlays for use in land analysis and presentation. Twenty years later, Ian McHarg in his 1969 book *Design with Nature* described using blacked-out transparent overlays to assist in finding locations for Richmond Parkway on New York's Staten Island that were solutions to multiple siting control factors (Figure 1.11).

As early as 1962, two planners at the Massachusetts Institute of Technology had evolved the map overlay idea to include weighting, by making the overlays different in their importance with respect to each other. The plan involved 26 maps showing the desirability of highways. Maps were ordered in a "procedural tree" and different combinations were made by reordering the map layers photographically.

During the 1960s, many new types of thematic maps were becoming available in standardized scales, such as topographic and land cover maps from the U. S. Geological Survey and soil maps from the U.S. Department of Agriculture's Soil Conservation Service (now the Natural Resource Conservation Service). It became fairly straightforward to select the right maps, trace off a layer, or photographically build a "separation" for one type of feature on the map, and then to combine the layers mechanically. GIS became known for providing the means by which map overlays could be made, giving us the layer model for GIS data, in which thematic overlays were held separately for composite analysis. The "layer cake" diagram rapidly became the preferred means to explain what a GIS was and could do (Figure 1.12).

The scene was set for the arrival of the computer. In 1959, Waldo Tobler, then a graduate student, published a paper in *Geographical Review* outlining a simple model for applying the computer to cartography (Tobler, 1959). His model, often referred to as a MIMO (map in-map out) system, had three elements: a map input, map "manipulation," and a map output stage. These three simple steps were the distant origins of the geocoding and data capture, data management and analysis, and data display modules now part of every GIS package.

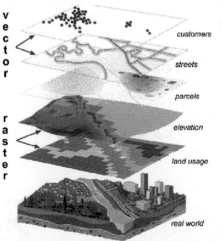

Figure 1.12: The layer cake model of GIS as a means to combine geographical themes. Source: NOAA.

Within just a few years, many people were busy writing computer programs using programming languages such as FORTRAN to draw maps using primitive printers and plotters. The new demands on computing led to the development of the first digitizer by the New Haven group planning the 1960 census and to the development of many other new devices. As new capabilities for mapping came along, the first experiments with entirely new mapping methods such as animation and automated hill-shading took place. Nevertheless, none of these early systems could be described as a GIS. During the early years, development of computer mapping resulted in less and less dependence on individual computer programs and more and more on software packages, sets of linked computer programs that had common formats, structures, and files. When modular computer programming languages came along during the 1960s, the process of writing integrated software became easier. Among the early computer mapping packages were SURFACE II, IMGRID, CALFORM, CAM, MOSS, and SYMAP.

Most of these programs were sets of modules for the analysis and manipulation of data and the production of choropleth (shaded area) and isoline (contour) maps. With these packages it was possible to overlay data sets, reducing the hard work of doing this only with transparencies. Closely related to the mapping software was the development of the first systematic map databases. First came the Central Intelligence Agency's (CIA) World Data Bank, a global map of coastlines, rivers, and national boundaries still in use today, along with the CAM software that projected it onto maps at different scales (Figure 1.13).

After many prototype systems, the DIME (dual independent map encoding) coding system was devised by the U.S. Census Bureau as an experiment in digital mapping and data handling. The DIME and the resultant files, called geographic base files (GBFs), were a major breakthrough in the history of geographic information representation. The GBF/ DIME recognized that attribute information, in this case all the data collected by the census, and the computer maps used in planning the census could be integrated not just for mapping but also to search for geographic patterns and distributions. Some landmark early systems were the Canada Geographic Information System (CGIS) in 1964, the Minnesota Land Management System (MLMIS) in 1969, and the Land Use and Natural Resources Inventory System in New York (LUNR) in 1967. Both MLMIS and LUNR were derivatives of the GRID system that replaced SYMAP at Harvard University.

During the mid– and late 1960s a cluster of faculty and students at Harvard University's Laboratory for Computer Graphics and Spatial Analysis made some major theoretical contributions and developed and implemented several new systems (Chrisman, 2006). Most influential among these was the GIS program Odyssey. With program modules named for sections of Homer's *The Odyssey*, the team pioneered a set of data

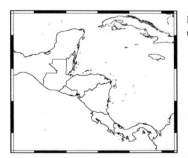

Figure 1.13: Section of the coastline and political boundaries from the CIA's World Data Bank.

structures that came into common use after their publication in 1975 (Peucker and Chrisman, 1975) called the arc/node or vector data structure. The computer routines that sorted digitized chains and lines and assembled topologically connected polygons, for example, was called the Whirlpool. Odyssey was a highly influential arc/node-based GIS and influenced much of the software that followed.

In Chapter 4 we will examine this structure in some detail, but what was different then was that the data structure captured polygon information using a series of nodes; there was a beginning node and an ending node with an arc between them. The arcs could be assembled to construct a polygon because the structure contained information about adjacency and connectivity between features. Many GIS packages, including Arc/Info, have been based on this simple model of geographic features.

In 1974 the International Geographical Union surveyed software in the mapping sciences and found enough GIS software to publish an entire inventory volume entitled "Complete Geographical Information Systems." While in the early days many different terms were used to describe a GIS, this report began the convergence on the term GIS as a generic name for this new application and research field. Reporting on the results of the survey, Kurt Brassel noted that "we understand that a mapping system is mainly designed for display purposes, even though it may fulfill some secondary functions that are not graphical. A geographical information system is designed for a broader range of applications, even though mapping functions may represent an important subset of its activities" (Brassel, 1977, p. 71). Both GIS and computer mapping continue to have this significant and constructive overlap in their content.

Development of GIS persisted into the 1980s, with large computers and FORTRAN continuing to dominate. In 1982, IBM introduced its PC, or personal computer, following from the Apple II microcomputer of a few years earlier. The impact of this single advance cannot be understated. Within just a few years, some of the large GIS packages, such as Arc/Info, had made the difficult transition to the microcomputer. Others, such as IDRISI, owe their origin to the low cost and high degree of efficiency that characterized the first generation of PCs. Other packages migrated instead to the new workstation platform that had developed from the minicomputer and networking trend. Again, other packages, such as GRASS, owe their origins to this transition.

The 1980s and early 1990s saw GIS mature as a technology. Many older packages that failed to move to the new languages and platforms died out, to be replaced by newer systems that could exploit the capabilities of the more powerful equipment. Costs of storage fell remarkably, computer power increased many-fold, and the first generation of GUIs, or graphical user interfaces, among them X-Windows, Microsoft Windows, and Apple's Macintosh, made the software considerably easier to use, adding features such as menus, on-line manuals, and context-sensitive help. During the 1980s, the Internet arose out of the collection of early networks such as Arpanet and NSFNet that were beginning to link scientists and became a significant new component of computing.

The 1980s also saw the origins of the infrastructure for GIS: the books, journals, conferences, and other resources that are so critical to finding out about GIS. During this era, the National Science Foundation created the National Center for Geographic Information and Analysis (NCGIA), which devised a national college curriculum and developed broad research agendas for academic research on GIS.

The decade of the 1990s saw remarkable growth in the GIS world. Several new factors emerged. First GIS spread far beyond its origins in the mapping sciences to encompass developments in new fields such as geology, archeology, epidemiology, and criminal justice. Also, the cost of GIS fell markedly after a series of "desktop" GIS products emerged. The increasing market penetration of personal computers, and the more mobile laptop and portable digital assistants, took GIS into many new work environments. Object-oriented programming approaches made radical improvements in the software engineering that could be applied to GIS software, and allowed the portability of programs across many computer platforms. In addition, GIS became fully integrated with the global positioning system, greatly enhancing the systems data capture capability. High-resolution imagery became common as a reference base for GIS data. Finally, the emergence of the Internet and e-commerce placed GIS onto the World Wide Web as Web-GIS.

During the first decade of the 21st century, leading factors influencing GIS have been focused on the Internet. GIS software developed connections to webservers, including the ability to easily post GIS maps on the Internet, to make them interactive, and to support interaction through a large number of "geobrowsers," applications interfaces connected to tools such as Google Earth, Googlemaps, MapQuest and NASA's World-wind. The Internet has also led to a rise in open source software and shared tools, including some GIS packages, that often get patched together to perform a GIS function, in what has become called a map mash-up. At the same time, vast depositories of geographic information have found their way onto the Internet, meaning that maps and imagery are broadly available at many scales, and at high levels of detail for some areas. At the same time, GIS has been influenced by the new generation of mobile devices, such as PDAs, tablet PCs, and palmtop computers. Even the ubiquitous cellular telephone has become a GIS platform, with many phones incorporating positioning from GPS and Internet search capability. Such mobile technologies incorporating spatial positioning and search have become known as location-based services.

And so we arrive at the present. While GIS's lineage dates back to the roots of cartography, and although thematic cartography and map overlay date from the nineteenth century, what is today known as GIS owes its birth to a cluster of interrelated events and human interactions in the 1960s, and its spectacular growth to the microcomputer, the workstation, and the Internet. It is, indeed, a rather short history and one that is still being written.

1.4 SOURCES OF INFORMATION ON GIS

The amount of information available about GIS is somewhat overwhelming. An excellent place to begin one's search is at a library, or from home by connecting to the Internet and using one of the World Wide Web search tools. A visit to the library may be more productive. Some libraries have facilities to connect to network search systems and even specialized staff with training in geographic information.

As in our definition of geographic information science, the information sources on GIS fall into the broad categories of research with GIS and research on GIS. As a beginner, try restricting your search to basic material rather than going straight to the research frontier. This can come later. A good way to research a topic is to find publications that came out at about the time a new idea was being introduced. In the older papers, articles, or book chapters, the authors had to write for an audience that would be unfamiliar with the language and concepts under discussion. This is the case in several "classic" papers in the GIS arena. The writing remains today as a good first step toward understanding and an excellent place to get started with GIS.

1.4.1   The Internet and the World Wide Web

An extraordinary, indeed overwhelming, amount of information about GIS can be found on the Internet using the World Wide Web (WWW). Everything from newsgroup FAQs to commercial GIS software vendor's websites, to entire on-line and downloadable GIS packages, such as GRASS, are available. A Google search for "Geographic Information System," for example, in early 2009 gave 1,640,000 hits.

The best way to search is to load a suitable Web browser such as Internet Explorer or Mozilla Firefox, and then follow your own interest. The bibliography at the end of this chapter lists some of the places where GIS information resides, but there are many, many more, and even more are added every day. The network news group GIS-L (comp.infosystems.gis) is a long-standing source of technical information on GIS. Users post questions to the list, and people answer back. Replies are archived, and when common threads emerge, they are compiled into an FAQ (frequently asked questions) list, sometimes echoed and hosted on sites across the World Wide Web. GIS-L is currently hosted by the URISA professional organization at http://www.hdm.com/urisa3.htm. Following discussions on GIS-L is an excellent way to get an introduction to the software and environment of GIS applications. There are also a very large number of GIS and cartography-oriented web logs or blogs, each with snippets of interesting details and sometimes breaking GIS news. Some on-line services can be configured to send a daily GIS-related alert, with stories and links to GIS issues.

One very useful GIS online resource is the network-accessible copy of the U.S. Geological Survey's brochure Geographic Information Systems (Figure 1.14) accessible at

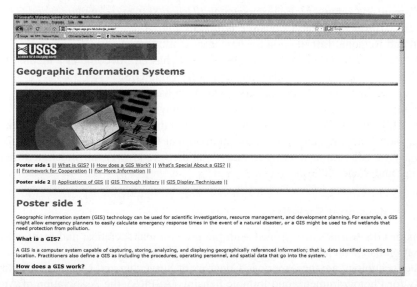

Figure 1.14: USGS poster/website devoted to basic information about GIS. Source: http://erg.usgs.gov/isb/pubs/

http://erg.usgs.gov/isb/pubs/gis_poster. This Web document was originally a wall-size poster, also available free, and contain all sorts of GIS samples, examples, and definitions. A recent addition to the Internet GIS information sources are news services that update frequently, some daily, information about GIS. Among these are GIS Monitor (www.gismonitor.com), Spatial News (www.spatialnews.com), Geospatial Solutions (gismap.geospatial-solutions.com) the GIS Café (www.giscafe.com), and Geoplace, home of several of the news and information journals (www.geoplace.com.) Some of these will send you daily updates on the website's contents by e-mail.

More information about the GIS data available on the Web is included in Chapter 3. Also available are the various data clearinghouses now forming part of the Spatial Data Infrastructure, an online "library catalog" of available GIS format data available free or at cost. Individual cities, counties, and states also maintain their own online GIS, many with search and some with data download capabilities. In addition to acting as a library, the WWW also serves as an information source, a software source, a data source, and even as a place to publish your results. In Chapter 11, the future role of the WWW and the Internet are discussed from a GIS viewpoint.

1.4.2 Books, Journals, and Magazines

Today many journals and magazines publish articles and papers on GIS, and a large number occasionally publish a few papers or a special issue. Journals that publish exclusively on GIS include the academic research journals *International Journal of Geographical Information Science*, *Geographical Systems*, *Transactions in GIS*, and applications periodicals *Geospatial Solutions*, *Geoinformatics*, and *Georeport*. The *International Journal of GIS* has published a set of its most cited papers as a book, along with follow-up papers by the same authors (Fisher, 2006).

Among the scholarly journals that publish academic work about GIS and its uses are the *Annals of the Association of American Geographers*; *Cartographica*; *Cartography and GIS*; *Computer*; *Computers, Environment, and Urban Systems*; *Computers and Geosciences*; *IEEE Transactions on Computer Graphics and Applications*; the *URISA Journal* (Urban and Regional Information Systems Association); and *Photogrammetric Engineering and Remote Sensing*. However, so predominant has GIS become as a method that few science or social science journals have no articles using GIS.

For finding books and papers, two Web-based projects are the GIS Master Bibliography (http://liinwww.ira.uka.de/bibliography/Database/GIS/index.html) and the Spatial Odyssey (http://wwwsgi.ursus.maine.edu/biblio/). There are also a handbook (Wilson and Fotheringham, 2007), an international GIS encyclopedia (Kemp, 2007), a dictionary of GIS (http://www.geo.ed.ac.uk/agidict/), and even an illustrated dictionary (Sommer and Wade, 2006).

1.4.3 Professional Societies

The major GIS journals are aligned with the professional societies, and many have book distribution lists with member and student discounts. The professional societies associated with the technology are the American Congress of Surveying and Mapping, the American Society for Photogrammetry and Remote Sensing (ASPRS), the Association of American

Geographers, the Geospatial Information and Technology Association, and the Urban and Regional Information Systems Association (URISA).

The ACSM (American Congress of Surveying and Mapping) has member organizations, each of which has an interest in GIS, including the Cartographic and Geographic Information Society. Journals produced are *Cartography and Geographic Information Systems* and *Surveying and Land Information Systems*. The American Society for Photogrammetry and Remote Sensing covers the mapping science fields quite broadly. Its journal, *Photogrammetric Engineering and Remote Sensing*, is monthly and has taken on a very strong GIS theme over the last few years. The journal itself publishes GIS articles in an even balance with traditional mapping and remote sensing. The Association of American Geographers (AAG) has a GIS specialty group, which constitutes the largest specialty group within the organization. The association has regular regional and annual national meetings, and the organization supports a newsletter with job listings.

URISA is a large organization aimed primarily at professionals in planning, government, infrastructure, and utilities. The organization holds an annual national conference and hosts many activities, including job listings. It also publishes a journal, and distributes newsletters. Another professional organization is the Geospatial Information and Technology Association. This group hosts an annual national conference, publishes conference proceedings and other publications, issues a newsletter, and provides scholarships and internships for college students to work in GIS firms.

1.4.4   Conferences

As a growing industry, especially in the early days when there was as yet no major journal where research and applications were published, the various professional conferences for GIS served as "literature." As a result, some of the key papers in GIS technology and theory appear, at least in their early and most readable form, as papers in conference proceedings. Unfortunately, these papers are often hard to find.

The earliest conference in GIS was probably the original Harvard Conference on Topological Data Structures. Very soon, the AutoCarto (International Symposium on Automated Cartography) took over as a key place for the publishing of papers. This series now has a CD with all past proceedings, with the most recent conference in 2008. During the 1980s, the GIS/LIS conference became a leading focus of GIS activity but was terminated in 1998, having completed its major task. The proceedings remain as a valuable GIS resource.

Other major conferences have been the URISA annual conference, which has more of a GIS application focus, the ACSM/ASPRS technical meetings, with both a research and applications orientation, and the GITA conference, which is the one many municipalities and industrial GIS users attend. The biannual spatial data handling conference, held alternately in the United States and internationally, has become a major concentration of people working on GIS research and development. The Association for Computing Machinery has a specialty group, called SIGSPATIAL, which hosts an annual International Conference on "Advances in Geographic Information Systems." The bi-annual GIScience conference is another important venue for GIS research. Lastly, several states, including New York, Texas, Minnesota, New Mexico, California, and North Carolina, also hold annual meetings. In addition, the various GIS packages or local areas hold their own user group meetings, some of which even approach the professional conferences in size. Largest of them all is the ESRI User Conference, held annually in San Diego, California, with over 12,000 attendees (Figure 1.15).

Outside of the United States there are many professional organizations that work with GIS. The United Kingdom has the Association for Geographic Information, Canada has the Canadian Institute of Geomatics, South Asia hosts the Centre for Spatial Database Management and Solutions, and Europe has the EUROGI (European Umbrella Organisation for Geographic Information). Other international professional societies include IRLOGI in Ireland, LISA in Iceland, and the Pakistan Society of GIS. China has several institutions that specialize in GIS, and groups such as the International Association of Chinese Professionals in GIS. There are also international bodies that cover the field, including the International Cartographic Association, and the International Society of Photogrammetry and Remote Sensing.

1.4.5   Educational Organizations and Universities

Many colleges and universities teach classes in GIS, and some offer complete programs with course sequences, BA and BSc emphases, and certificates. Some vendors offer certification as instructors, and national certification is monitored by the GIS certification institute, who offer the GISP (GIS Professional) certificate (see: www.gisci.org). Some universities and extension services offer short courses, and most of the major GIS vendors offer short training programs lasting anywhere from a few hours at a national or regional conference to several days or weeks. Every bit as important, URISA manages a Code of Ethics for GIS (see: www.urisa.org/about/). All students of GIS are advised to consult this code, which is both simple and useful.

Within universities and colleges, GIS classes are taught in many departments. Most are in geography, but many are also in departments and programs in geology, environmental science, forestry, civil engineering, computer and information science, and many

Figure 1.15: Views of the exhibits area at the 2008 ESRI User Conference in San Diego, California.

others. Exactly what should be covered in a GIS class is covered in detail by the GIS&T Body of Knowledge, a systematic catalog of what is important to GIS practitioners compiled by the Association of American Geographers and the University Consortium for GIS. The printed version is available at www.aag.org/bok and includes ten knowledge areas, 73 units, 329 topics, and over 1,600 formal educational objectives (DiBiase et al. 2006). Many programs around the country offer just a single class, structured in much the same way as this book. Others use the national GIS curriculum of the National Center for Geographic Information and Analysis (NCGIA). This center is a National Science Foundation-funded program designed to channel GIS research and learning toward improvement of the discipline of geographic information science. The center, a consortium of three universities, maintains a website at http://www.ncgia.ucsb.edu. The group has conducted a comprehensive set of "research initiatives" in GIS covering many different areas. Publications, research reports, outreach activities, and sponsorship of conferences and visitors to the center have been the main activities of the NCGIA. One particular NCGIA project created the Center for Spatially Integrated Social Science, whose website (www.csiss.org) has a wealth of GIS "cookbooks," "classics," and more.

The primary mission of the NCGIA is to conduct basic research, but the organization also coordinated the formation of a far broader Geographic Information Science community in the United States. In 1994, a total of thirty-three universities, research institutions, and the Association of American Geographers met to establish the University Consortium for Geographic Information Science (UCGIS) (Figure 1.16). UCGIS is a nonprofit organization of universities and other research institutions dedicated to advancing the understanding of geographic processes and spatial relationships through improved theory, methods, technology, and data. As of July 2008, there were 84 members of the consortium. Several meetings and resource collections have proven very useful as GIS information

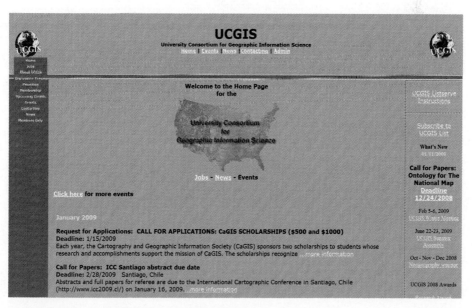

Figure 1.16: Website for the University Consortium for Geographic Information Science (www.ucsgis.org).

sources, and the updated list of UCGIS initiatives gives an indication of research directions in GIS.

A college or university near you may be able to provide information about GIS courses or help you to find out more. University libraries hold many GIS publications and conference proceedings, and these are also a good starting point. Perhaps, after reading this book, you will be tempted to take a college course, or maybe you are using this book as part of one. If so, don't forget that learning never ends and that increasing your GIS education also increases your effectiveness as a GIS user, your ability as a geographic information scientist, and your employability as a GIS specialist.

Several websites provide useful information on GIS as a career. These include: AGI Guide to Geosciences Careers and Employers (guide.agiweb.org/employer/index.html); Earth Science World-Gateway to the Geosciences (www.earthscienceworld.org/careers/ links); Geography Careers listed by the Association of American Geographers (www.aag.org/Careers/What_can_you_do.html); GIS Career Info and Listings from ESRI (careers.esri.com); Profiles of Cool Geography Graduates (AAG) (www.aag.org/Careers/ Geogwork/Intro.html); and Geography Jobs listed by Occupational Outlook Quarterly http://www.bls.gov/opub/ooq/2005/spring/art01.htm. Several salary surveys and a Department of Labor high growth industry profile exist (http://www.doleta.gov/Brg/Indprof/ geospatial_profile.cfm). There are also sites devoted to employment in GIS at all levels of expertise, for example at giscareers.com and geosearch.com (Figure 1.17).

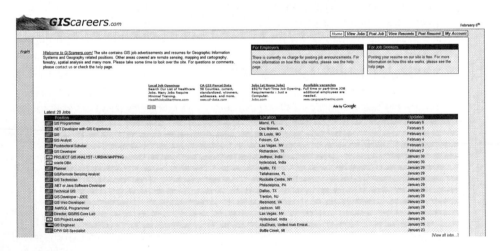

Figure 1.17: Listing of GIS jobs available at www.giscareers.com

1.5   STUDY GUIDE

1.5.1   Quick Study

CHAPTER 1: WHAT IS A GIS?

O *GIS is many simultaneous technological revolutions* O *Geographic information science merges skills and theory across geography, cartography, geodesy, database theory, computer science and mathematics* O *Information in general can be text, lists, tables, indexes, catalogs, and cross references. Simplest way to order information is in sequence, e.g., alphabetically* O *Geographic information is sequenced by location, and so can organize almost any other information type, because everything exists or happens somewhere* O *A GIS is any information system capable of integrating, storing, editing, analyzing, sharing, and displaying geographically referenced information GIS has a database, a map, and links between them* O *A GIS is a set of tools for analyzing spatial data* O *A GIS has separable functions that can be applied in sequence* O *A GIS is an information system for delivering answers to questions or queries* O *It often divides the world into points, lines, and areas called features* O *Human activities create patterns and distributions, which link to the points, lines, and areas* O *Time gives another dimension; events happen at times and features exist over a duration* O *The purpose of GIS is to help an analyst make predictions and explanations about geographical phenomena* O *GIScience is the academic theory behind the development and use of geographic information* O *GIScience also deals with fundamental issues raised by the use of GIS* O *GIS is also a multi-billion dollar a year business, a marketing phenomenon that will remain for years* O *GIS plays a role in changing society, including workplace, organizations, and institutions* O *PPGIS studies how GIS is used in public decision-making, and recognizes that not all GIS data are perfect* O *GIS's historical roots lie in thematic cartography and planning* O *Layers of information were combined first mechanically, then digitally* O *Tobler's model of computer mapping involves map-in, map manipulation, and map out stages* O *Over time, stand-alone software packages became unified, making modern GIS* O *Early examples were SURFACE II, IMGRID, CALFORM, SYMAP, and Odyssey* O *Early data sets were GBF/DIME and the World Data Banks* O *Odyssey was highly influential, supported at Harvard and using the Arc/Node structure* O *In the 1980s GISs evolved graphical user interfaces, and the infrastructure of science—books, journals, and conferences* O *The 1990s saw GIS merge with the Internet, both as data provider and as software interface* O *The 2000s have seen mobile GIS, geobrowsers with GIS functions, and complete integration with the global positioning system* O *To find out about GIS: use the web; consult books, journals, and magazines; and join professional societies* O *Many universities now offer GIS* O *You can become certified as a GIS Professional and can learn the scope of GIS knowledge in the UCGIS's GIS&T Body of Knowledge* O

1.5.2 Study Questions and Activities

Getting Started

1. Use a web browser to examine the actual content of a web page. For example, in Mozilla Firefox, under the View menu, select Page Source. Make a list of the different forms of information contained in the HTML, and note any important sequencing in the data.

2. Find a magazine or newspaper, and cut out all of the tables, images, maps, and text. Sort them by type, and then measure how many square inches or centimeters of space is devoted to each. How much information does each type convey per square unit?

GIS Definitions

3. Look at the various different GIS definitions covered in the chapter. Why are they different? What tradition or discipline does each definition come from?

4. Using your GIS documentation, make a table that compares each function of the software with a real tool in a toolbox, e.g., saw and glue gun with Clip and Merge.

5. Find a photograph of a landscape. Make a list of what features on the image fall into the classes of points, lines, and area. Any exceptions?

The History of GIS

6. Using one of the GIS timelines on the web, choose no more than six milestones in GIS history that account for its success.

7. What was the role of the Harvard Laboratory for Computer Graphics and Spatial Analysis in the history of GIS?

8. How did Ian McHarg's "Design with nature" principles translate into GIS methods?

Sources of Information on GIS

9. Make a list of the key GIS information sources and find as many as you can in your local public library. If your library offers data services or the Internet, use these to find out as much information about GIS information sources (GIS metadata, or "data about data") as possible.

10. What local, regional, or national GIS meetings or conferences are taking place in your area in the near future?

11. Sign up for one of the various daily e-mail distributions offered by the GIS on-line services. After a month, select the three most interesting or useful items of information you received, and share them with your friends or classmates.

12. Which colleges, university or other educational establishments in your area offer GIS classes?

13. Take a look at the UCGIS GIS&T Body of Knowledge. What items in the BoK would you most like to learn, or consider the most important?

1.6   REFERENCES

1.6.1   Chapter References

Abler, R.F. (1988) "Awards, rewards and excellence: keeping geography alive and well," *Professional Geographer*, vol. 40, pp. 135-40.

Albrecht J. (1998) Universal Analytical GIS Operations–A task-oriented systematisation of data-structure-independent GIS functionality. In Craglia M and H. Onsrud (eds.) *Geographic Information Research: transatlantic perspectives.* pp. 577-591. London: Taylor & Francis.

Brassel, K. E. (1977) "A survey of cartographic display software," *International Yearbook of Cartography*, vol. 17, pp. 60-76.

Burrough, P. A. (1986) *Principles of Geographical Information Systems for Land Resources Assessment.* Oxford: Clarendon Press.

Clarke, K. C. (1995) *Analytical and Computer Cartography.* 2nd ed. Upper Saddle River, NJ: Prentice Hall.

Chrisman, N.R. (1999) "What does GIS mean?" *Transactions in GIS* vol. 3, no. 2, pp. 175-186.

Chrisman, N. R. (2006) *Mapping The Unknown: How Computer Mapping Became GIS at Harvard.* ESRI Press, Redlands, CA.

Corbett, J. and Keller, P. (2006) An analytical framework to examine empowerment associated with participatory geographic information systems (PGIS). *Cartographica*, vol 40, no. 4, pp. 91-102.

DiBiase, D., DeMers, M., Johnson, A., Kemp, K., Luck, A., Plewe, B., et al. (eds.) (2006) *Geographic Information Science and Technology Body of Knowledge* (1st ed.). Washington, DC: Association of American Geographers.

Ducker, K. J. (1979) "Land resource information systems: a review of fifteen years' experience," *Geo-Processing*, vol. 1, no. 2, pp. 105-128.

Foresman, T. W. (ed.) (1997) *The History of Geographic Information Systems: Perspectives from the Pioneers.* Upper Saddle River, NJ: Prentice Hall

Fisher, P. (ed.) (2006) *Classics from IJGIS. Twenty Years of the International Journal of Geographical Information Systems and Science.* Taylor and Francis, CRC. Boca Raton, FL.

Goodchild, M. F. (1992), "Geographical information science," *International Journal of Geographical Information Systems*, vol. 6, no. 1, pp. 31-45.

Kemp, K. K. (ed.) (2007) *Encyclopedia of Geographic Information Science.* Thousand Oaks, CA: Sage Publications.

McHarg, I. L. (1969) *Design with Nature.* New York: Wiley.

Peucker, T. K. and Chrisman, N. (1975) "Cartographic data structures," *American Cartographer*, vol. 2, no. 1, pp. 55-69.

Pickles, J. (1995) *Ground Truth: The Social Implications of Geographic Information Systems.* New York: Guilford Press.

Pick, J. B., (ed.) (2005) *Geographic Information Systems in Business.* Hershey, PA: Idea Group Publishing.

Pick, J. B. (2008). *Geo-Business: GIS in the Digital Organization.* New York, NY: John Wiley and Sons.

Sommer, S. and Wade, T. (2006) *A to Z GISs: An Illustrated Dictionary of Geographic Information Systems.* ESRI Press.

Star, J. and Estes J. E. (1990) *Geographic Information Systems: An Introduction.* Upper Saddle River, NJ: Prentice Hall.

Steinitz, C., Parker, P. and Jordan, L. (1976) "Hand-drawn overlays: their history and prospective uses," *Landscape Architecture*, vol. 66, no. 5, pp. 444-455.

Tobler, W. R. (1959) "Automation and cartography," *Geographical Review*, vol. 49, pp. 526-534.

Wilson, J. P. and Fotheringham, A. S. (2007) *The Handbook of Geographic Information Science.* Malden, MA: Blackwell Publishing.

1.6.2 Recent Bibliography

Alibrandi, M. (2003) *GIS in the classroom: using geographic information systems in social studies and environmental science.* Portsmouth, NH: Heinemann.

Belussi, A. (2007) *Spatial data on the Web: modeling and management.* Berlin: Springer.

Bossler, J. D., Jensen, J. R., McMaster, R. B., & Rizos, C. (2002) *Manual of geospatial science and technology.* London: Taylor & Francis.

Breman, J. (2002) *Marine geography: GIS for the oceans and seas.* Redlands, Calif: ESRI Press.

Campagna, M. (2006) *GIS for sustainable development.* Boca Raton: CRC Press.

Chang, K.-T. (2002) *Introduction to geographic information systems.* Boston: McGraw-Hill.

Clarke, K. C., Parks, B. O. and Crane, M. P. (eds.) (2002) *Geographic Information Systems and Environmental Modeling*, Prentice Hall, Upper Saddle River, NJ.

Cromley, E. K. and McLafferty, S. L. (2002) *GIS and Public Health.* New York, NY: Guilford.

Czerniak, R. J., & Genrich, R. L. (2002) *Collecting, processing, and integrating GPS data into GIS.* Washington, D.C.: National Academy Press.

Davis, D. E. (2000) *GIS for everyone: exploring your neighborhood and your world with a geographic information system.* Redlands, Calif: ESRI Press.

DeMers, M. N. (2005) *Fundamentals of geographic information systems.* New York: John Wiley & Sons.

Falconer, A., Foresman, J., & Shrestha, B. R. (2002) *A system for survival: GIS and sustainable development.* Redlands, CA: ESRI Press.

Flynn, J. J., and Pitts, T. (2000) *Inside ArcInfo.* Albany, NY: OnWord Press.

Foresman, T. W. (1998) *The history of geographic information systems: perspectives from the pioneers.* Prentice Hall series in geographic information science. Upper Saddle River, NJ: Prentice Hall PTR.

Fox, T. J. (2003) *Geographic information system tools for conservation planning user's manual.* Reston, Va: U.S. Dept. of the Interior, U.S. Geological Survey.

Goldsmith, V. (2000) *Analyzing crime patterns: frontiers of practice.* Thousand Oaks, Calif: Sage Publications.

Goodchild, M. F., & Janelle, D. G. (2004) *Spatially integrated social science. Spatial information systems.* Oxford [England]: Oxford University Press.

Hilton, B. N. (2007) *Emerging spatial information systems and applications.* Hershey, PA: Idea Group Pub.

Hutchinson, S., & Daniel, L. (2000) *Inside ArcView GIS.* Albany, N.Y.: OnWord Press.

Huxhold, W. E., Fowler, E. M., & Parr, B. (2004) *ArcGIS and the digital city: a hands-on approach for local government.* Redlands, Calif: ESRI Press.

Kanevski, M., & Maignan, M. (2004) *Analysis and modelling of spatial environmental data.* Lausanne, Switzerland: EPFL Press.

Kennedy, M. (2002) *The global positioning system and GIS: An introduction.* London: Taylor & Francis.

Knowles, A. K. (ed.) (2002) *Past Time, Past Place: GIS for History.* Redlands, CA: ESRI Press.

Lang, L. (2000) GIS for health organizations. Redlands, Calif: ESRI Press.

Leuven, R. S. E. W., Poudevigne, I., & Teeuw, R. M. (2002) *Application of geographic information systems and remote sensing in river studies.* Leiden: Backhuys.

Lo, C. P. and Yeung, A. K. W. (2002) *Concepts and Techniques in Geographic Information Systems,* Upper Saddle River, NJ: Prentice Hall.

Longley, P. (2005) *Geographical information systems: principles, techniques, management, and applications.* New York: Wiley.

Longley, P. A., Goodchild, M. F., Maguire, D. J., and Rhind, D. W. (2005) *Geographic Information Systems and Science.* New York, NY: J. Wiley. 2ed.

Lyon, J. G., & McCarthy, J. (1995) *Wetland and environmental applications of GIS.* Mapping sciences series. Boca Raton: CRC Press.

National Academies Press (U.S.) (2006) *Learning to think spatially.* Washington, D.C.: National Academies Press.

National Risk Management Research Laboratory (U.S.) (2000) *Environmental planning for communities: a guide to the environmental visioning process utilizing a geographic information system (GIS)* Cincinnati, OH: Technology Transfer and Support Division, Office of Research and Development, U.S. Environmental Protection Agency.

Neteler, M., & Mitasova, H. (2002) *Open source GIS: A GRASS GIS approach.* Boston: Kluwer Academic.

Okabe, A. (2006) *GIS-based studies in the humanities and social sciences.* Boca Raton, FL: CRC/ Taylor & Francis.

Ormsby, T. (2001) *Getting to know ArcGIS desktop: basics of ArcView, ArcEditor, and ArcInfo.* Redlands, Calif: ESRI Press.

Ott, T., & Swiaczny, F. (2001) *Time-integrative geographic information systems: management and analysis of spatio-temporal data.* Berlin: Springer.

Pinder, G. F. (2002) *Groundwater modeling using geographical information systems.* New York: Wiley.

Price, M. H. (2006) *Mastering ArcGIS.* Dubuque, IA: McGraw-Hill.

Ralston, B. A. (2002) *Developing GIS solutions with MapObjects and Visual Basic.* Albany, N.Y.: OnWord Press.

Shamsi, U. M. (2002) *GIS tools for water, wastewater, and stormwater systems.* Reston, Va: ASCE Press.

Sinha, A. K. (2006) *Geoinformatics: data to knowledge.* Boulder, Colo: Geological Society of America.

Spencer, J. (2003) *Global Positioning System: a field guide for the social sciences.* Malden, MA: Blackwell Pub.

Steede-Terry, K. (2000) *Integrating GIS and the Global Positioning System.* Redlands, Calif: ESRI Press.

Stewart, M. E. (2005) *Exploring environmental science with GIS: an introduction to environmental mapping and analysis.* New York, N.Y.: McGraw Hill Higher Education.

Thill, J.-C. (2000) *Geographic information systems in transportation research.* Amsterdam [Netherlands]. Pergamon.

Thurston, J., Moore, J. P., & Poiker, T. K. (2003) *Integrated geospatial technologies: a guide to GPS, GIS, and data logging.* Hoboken, N.J.: John Wiley & Sons.

Tomlinson, R. F. (2003) *Thinking about GIS: geographic information system planning for managers.* Redlands, Calif: ESRI Press.

Van Sickle, J. (2004) *Basic GIS coordinates.* Boca Raton, Fla: CRC Press.

Walsh, S. J., & Crews-Meyer, K. A. (2002) *Linking people, place, and policy: a GIScience approach.* Boston: Kluwer Academic.

Williams, J. (2001) *GIS processing of geocoded satellite data.* Springer-Praxis books in geophysical sciences. London: Springer.

1.6.3   Professional Organizations

AAG: The Association of American Geographers, 1710 Sixteenth St. NW, Washington, DC 20009-3198. Also publishes AAG Newsletter. E-Mail: gaia@aag.org. Web: www.aag.org

ACSM: American Congress on Surveying and Mapping, 5410 Grosvenor Lane, Suite 100, Bethesda, MD. 20814-2122. Web: http://www.acsm.net

ASPRS: American Society for Photogrammetry and Remote Sensing. 5410 Grosvenor Lane, Suite 210, Bethesda, MD 20814-2162. E-mail: asprs@asprs.org. Web: www.asprs.org

GITA: Geospatial Information and Technology Associations. 14456 East Evans Avenue, Aurora, CO 80014, E-Mail info@gita.org. Web: www.gita.org

NACIS: North American Cartographic Information Society, AGS Collection, P.O. Box 399, Milwaukee, WI 53201. E-mail: nacis@nacis.org. Web: http://www.nacis.org.

URISA: Urban and Regional Information Systems Association. 1460 Renaissance Drive, Suite 305. Park Ridge, IL 60068. E-mail: urisa@macc.wisc.edu. Web: www.urisa.org

1.7   KEY TERMS AND DEFINITIONS

analysis: The stage in science when measurements are sorted, tested, and examined visually for patterns and predictability.

arc/node: Early name for the vector GIS data structure.

arc: A line represented as a set of sequential points.

area feature: A geographic feature recorded on a map as a sequence of locations or lines that, taken together, trace out an enclosed area or ring that represents the feature. Example: a lake shoreline.

attribute: A characteristic of a feature that contains a measurement or value for the feature. Attributes can be labels, categories, or numbers; they can be dates, standardized values, field measurements or other data. An item for which data are collected and organized. A column in a table or data file.

AUTOCARTO (International Symposium on Automated Cartography): A sequence of computer cartography and GIS conferences.

cartography: The science, art, and technology of making, using, and studying maps.

CGIS (Canadian Geographic Information System): An early national land inventory system in Canada that evolved into a full GIS.

choropleth map: A map showing numerical data (but not simply "counts") for a group of regions by (1) grouping the data into classes and (2) shading each class on the map.

computer mapping: Producing maps using the computer as the primary or only tool.

data structure: The logical and physical means by which a map feature or an attribute is digitally encoded.

database: The body of data that can be used in a database management system. A GIS has both a map and an attribute database.

database manager: A computer program or set of programs allowing a user to define the structure and organization of a database, to enter and maintain records in the database, to perform sorting, data reorganization, and searching, and to generate useful products such as reports and graphs.

digitizing tablet: A device for geocoding by semi-automated digitizing. A digitizing tablet looks like a drafting table but is sensitized so that as a map is traced with a cursor on the tablet, the locations are picked up, converted to numbers, and sent to the computer.

FAQ: A list of "frequently asked questions," usually posted on a network news group or conference group to save new users the trouble of asking old questions over again.

feature: A single entity that makes up part of a landscape.

file: Data logically stored together at one location on the storage mechanism of a computer.

format: The specific organization of a digital record.

FORTRAN: An early computer programming language, initially for converting mathematical formulas into computer instructions.

functional definition: Definition of a system by what it does rather than what it is.

GBF (Geographic Base File): A database of DIME records.

general-purpose map: A map designed primarily for reference and navigation use.

geographic information science: Research on the generic issues that surround the use of GIS technology, impede its implementation, or emerge from an understanding of its capabilities.

geographic(al) information system: (1) A set of computer tools for analyzing spatial data; (2) a special case of an information system designed for spatial data; (3) an approach to the scientific analysis and use of spatial data; (4) a multibillion-dollar industry and business; (5) a technology that plays a role in society.

geographic pattern: A spatial distribution explainable as a repetitive distribution.

geography: The science concerned with all aspects of the earth's surface, including natural and human divisions, the distribution and differentiation of regions, and the role of humankind in changing the face of the earth.

GUI (graphical user interface): The set of visual and mechanical tools (such as window, icons, menus, and toolbars, plus a pointing device such as a mouse) through which a user interacts with a computer.

information: The part of a message placed there by a sender and not known by the receiver.

information system: A system designed to allow the user to be delivered the answer to a query from a database.

Internet: A network of many computer networks. Any computer connected to the Internet can access any of the computers accessible through the network.

killer app: A computer program or "application" that, by providing a superior method for accomplishing a task in a new way, becomes indispensable to computer users. Examples are word processors and spreadsheets.

learning curve: The relationship between learning and time. A steep learning curve means that much is learned quickly (usually thought to be the opposite). A difficult learning curve is one where learning takes place slowly, over a long time period.

line feature: A geographic feature recorded on a map as a sequence of locations tracing out a line. An example is a stream.

location: A position on the earth's surface or in geographic space definable by coordinates or some other referencing system, such as a street address or space indexing system.

LUNR (Land Use and Natural Resources Inventory System): An early GIS in New York.

map: A depiction of all or part of the earth or other geographic phenomenon as a set of symbols and at a scale whose representative fraction is less than 1:1. A digital map has had the symbols geocoded and stored as a data structure within the map database.

map overlay: Placing multiple thematic maps in precise registration, with the same scale, projections, and extent, so that a compound view is possible.

measurement: A quantitative assessment of a phenomenon.

menu: A component of a user interface that allows the user to make selections and choices from a preset list.

MIMO system: A term used to describe a first-generation computer mapping system designed to capture the map by computer and reproduce it (map in-map out).

MLMIS (Minnesota Land Management System): An early statewide GIS for the state of Minnesota.

NCGIA (National Science Foundation's National Center for Geographic Information and Analysis): A three-university consortium funded to assist in GIS education, research, outreach, and information generation.

node: At first, any significant point in a map data structure. Later, only those points with topological significance, such as the ends of lines.

observation: The process of recording an objective measurement.

Odyssey: A first generation GIS developed at Harvard to implement the original arc/node vector data structure.

overlay weighting: Any system for map overlay in which the separate thematic map layers are assigned unequal importance.

PC (Personal Computer): A self-contained microcomputer, providing the necessary components for computing, including hardware, software, and a user interface.

point feature: A geographic feature recorded on a map as a location. Example: a single house.

proceedings: The formal record of the papers and other prepared presentations at a conference. Usually available to conference attendees and later distributed as a soft-cover book or CD-ROM.

professional publication: Books, journals, or other information designed primarily for those using GIS technology as part of their job.

query: A question, especially if asked of a database by the user via a database management system or GIS.

record: A set of values for all attributes in a database. Equivalent to a row in a data table.

scientific approach: A method for rationally explaining observations about the natural and human world.

search engine: A software tool designed to search the Internet and the WWW for documents meeting the user's query. Examples: Yahoo and Alta Vista.

software package: A computer program application.

spatial data: Data that can be linked to locations in geographic space, usually via features on a map.

spatial distribution: The locations of features or measurements observed in geographic space.

spreadsheet: A computer program that allows the user to enter numbers and text into a table with rows and columns and then maintain and manipulate those numbers using the table structure.

thematic map: A map designed primarily to show a "theme," a single spatial distribution or pattern, using a specific map type.

topographic map: A map type showing a limited set of features but including, at the minimum, information about elevations or landforms. Example: contour maps. Topographic maps are common for navigation and for use as reference maps.

topology: The numerical description of the relationships between geographic features, as encoded by adjacency, linkage, inclusion, or proximity. Thus a point can be inside a region, a line can connect to others, and a region can have neighbors.

transparent overlay: An analog method for map overlay, where maps are traced or photographed onto transparent paper or film and then overlain mechanically.

U.S. Census Bureau: An agency of the Department of Commerce that provides maps in support of the decennial (every 10 years) census of the United States, especially the census of population.

user group: Any formal or informal organization of users of a system who share experiences, information, news, or help among themselves.

USGS (U. S. Geological Survey): A part of the Department of the Interior and a major provider of digital map data for the United States.

vector: A map data structure using the point or node and the connecting segment as the basic building block for representing geographic features.

workstation: A computing device that includes, at a minimum, a microprocessor, input and output devices, a display, and hardware and software for connecting to a network. Workstations are designed to be used together on local area networks, and to share data, software, and so on.

World Data Bank: One of the first digital maps of the world, published in two versions by the Central Intelligence Agency in the 1960s.

World Wide Web (WWW or W3): A distributed database of information stored on servers connected by the Internet.

1.8 PEOPLE IN GIS

Shoreh Elhami, GIS Director for Delaware County, Ohio Auditors Office

KC: I see that you are located in Ohio.

SE: Central Ohio; actually part of Columbus is in Delaware County. The county is in fact the 10th fastest growing county in the nation. My office is mainly responsible for collection of all data that's related to land parcels.

KC: So how many people work in your division, the GIS Division for a county?

SE: Eight, including myself, and two of them are part time employees. The office not only produces data sets for the Appraisers Office, but also others such as E911 and Sanitary Engineering. So it's kind of the hub of all GIS for the county; we produce a variety of data sets both off and on the Web, and we produce lots of maps.

KC: The people you supervise—are some of them at the beginning level or are they interns?

SE: I have programmers who mostly program dot net technology and Visual Basic, which we use for INS applications, and I have cadastral specialists who are experts in encoding parcel data or transforming CAD data sets we receive from surveyors. Four of my interns are cadastral specialists, and we have student interns that we use to help us with quality control—from address quality control to aerial photo quality control and LIDAR data. We are the main custodian of address files for the entire county, and especially for the 911, being the recipient of all the address files, it's incredibly important that we provide the best address file to the entire county.

KC: What prepared you for your career?

SE: I have actually been with Delaware County for 17 years, 12 of which have been in my current position. I was the GIS coordinator for the planning department where I built their GIS system. I'm a planner by training, with a degree in Architecture from National University of Iran, which is where I'm originally from. I came to the United States in 1985 where I went to graduate school at The Ohio State University, and that's where I got my masters degree in City and Regional Planning. I teach a graduate level course there, Applications of GIS in Professional Planning, and that's how I keep in touch with the planning profession.

KC: I notice you're wearing a URISA pin. Can you tell us about your role in URISA?

SE: I'm on the Board of Directors of URISA, the Urban and Regional Information System Association, and it's an organization that I've been a member of since 1989. I've been very active in many of their subcommittees. One of the main initiatives that I've spearheaded under the URISA umbrella is GISCorps. This program started a few years ago. It was an idea that I had, I started talking to different colleagues, mainly at URISA, and the idea was that we as GIS specialists should be able to share our knowledge on a voluntary basis—and why is it that we aren't doing that? Whenever I spoke they said, "This is a great idea, why not?" When I started asking people, "Would you volunteer?" many of them would say, "Of course, just show me where to go." And so it really started as an idea, and then I had colleagues of mine who helped me form this program under URISA. Its premise is to send GIS specialists for a short period of time to do volunteer work and provide their GIS specialties in developing communities. Now those developing communities mainly are in developing countries, but we also do provide assistance in the United States and Canada in event of a disaster.

KC: Can you give us an example of a disaster where GISCorps has helped?

SE: Actually, there are two of them so far, one international and one a US experience. In early 2005 after the tsunami in Indonesia and India, a total of 13 volunteers helped with that project. Six of them were deployed to Indonesia and seven of them stayed home, because as a GIS volunteer you don't necessarily have to travel, especially if you can't. You can do a lot of work from your home, from your office, and I think that's become a very intriguing part of what GISCorps is, because a lot of people just want to help but they can't travel because of family and many other reasons. So that was the first disaster-related mission. And then Katrina of course: following August 29, 2005, on September 1st we started deploying volunteers to Mississippi mainly. The reason was that the request for volunteers came from the Mississippi Emergency Operations Center (EOC).

KC: What sort of things did the volunteers do in Mississippi?

SE: They actually asked for a very specific expertise. They asked at the very least for 20 people; 10 people they wanted were GPS experts, people who could go out with a laptop and collect GPS data, either damage-related or roads, infrastructure-related and bring it back and kind of plug it in and create different feature classes for other geospatial specialists. They also asked for GIS special-ists mainly in general terms—they empha-sized the maturity of the group, because the conditions, we all knew, were going to be very harsh. We looked in our database, which at the time, actually was about 300 volunteers, but it shot up to over 900 in a matter of a week. Just the first week I had more than 500 people signing up as volunteers, so it was quite chaotic. We weren't ready for that, and we never thought that such a response would come from our community, it was just quite amazing.

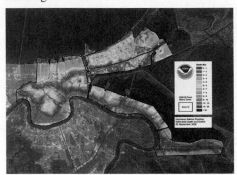

KC: How could a reader of this book find out more information about GISCorps?

SE: The best place is our website which is www.giscorps.org.

KC: Anything you'd like to add?

SE: I would like to encourage anyone who's in the GIS profession to volunteer their expertise with the feeling that the payback is enormous. I think that the effect that one has on others as a volunteer is ultimately the most rewarding. I have had my own experience doing it, and I know that it does make a huge difference.

KC: Thank you very much.

CHAPTER 2

GIS's Roots in Cartography

"A map is the greatest of all epic poems. Its lines and colors show the realization of great dreams."

G. H. Grosvenor

2.1 MAP AND ATTRIBUTE INFORMATION

In Chapter One, we saw that a GIS derives its power by allowing information to be assigned locations in geographic space. Take a look at Figure 2.1, which is a screen snap of Asian cities shown in the uDig open source GIS. On the map, we have zoomed in to Asia, and the map is centered on China, showing major cities as red squares. By selecting the information tool on the toolbar, we have clicked on a particular city (Wuhan in China, home of the major Chinese university for mapping and GIS). When we did so, the table to the right showed up, zoomed in more detail in the inset. These are the two major ways of organizing the data we are using, on the map (as points) and in a table, as records, also known as tuples. Each city has the same information fields, such as the latitude and longitude, the population, the population rank, the name and codes for the country and province, etc. In this most simple of GIS operations, a query to request the attributes for a single place, we have been presented with a table showing the information that is relevant

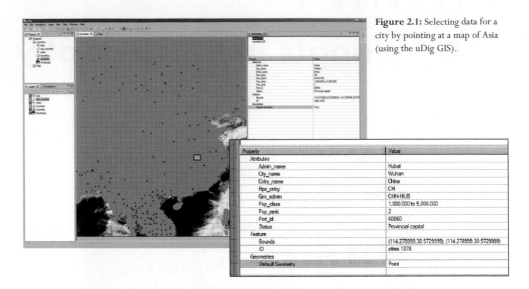

Figure 2.1: Selecting data for a city by pointing at a map of Asia (using the uDig GIS).

to our request. In this case, we clicked with a mouse on one city on the map, but we could have asked for many, or all cities inside a search box. This is the spatial search done by the GIS. Alternatively, we could have asked the database for the city of Wuhan, and then highlighted it to see where it was on the map. This is the database query option. Quite simply put, the GIS stores both the map and the attributes associated with the features on the map. Cities are points (at least at the scale shown), but we could have been searching for line features such as rivers, or area features such as counties. How the data are stored is up to the GIS, and ideally we need never know how the GIS works its magic. It is sufficient to know that the GIS stores the map and the data in the files and folders (or directories) familiar to PC users. Some of these folders and files are shown in the figure in the left panel. If the map and the attribute data are to be used together, then obviously the GIS creates a way to link the two together.

Central to this map and attribute data use is finding a way to link the map with the attributes. As we are using a computer, obviously the link should be in the form of numbers. When we locate people and houses, we usually use street addresses rather than numbers. Later in the book, we will see that a GIS gives us the power to move from one to the other of these descriptions of location with numbers. For now, however, we need a simple number description for a location. In the example here, we used latitude and longitude. Many GIS packages use latitude and longitude, so this is quite appropriate.

Before we move on, however, it is important to get a feel for what these geographic numbers mean and how they correspond to places on both the earth and the map. It is a little more complex than it first seems, but with a little digression, we can quickly come up to speed, and even be experts. This means that to understand GISs, we need to know a little cartography, and even some geodesy. Cartography is the science that deals with the construction, use, and principles behind maps and map use. The basics here go back a long way, to the work of the ancient Greek Ptolemy, the father of latitude and longitude and of map projections. Geodesy is the science that deals with the size and shape of the earth, and its gravity field.

2.2   MAP SCALE AND PROJECTIONS

2.2.1   The Shape of the Earth

Stories about a belief in a flat earth throughout history are apparently so much myth. Even to someone who believes that the lunar landings took place in a Hollywood studio, it would be hard to prove that the earth is flat, appearances aside. A trip to the beach to observe a ship sailing into the distance reveals not a visible dot that gets smaller and smaller, but a dot that eventually just disappears over the curve of the earth. When high in a plane at cruising altitude, hold a ruler at arm's length up to the horizon and judge for yourself whether the horizon line is straight. Three issues are important if the idea is to "contain" the earth, or parts of its surface, within a map inside a GIS. First, how big is the earth? We need to know this because we need to calculate by how much to reduce the drawing of it to fit onto the computer screen. Second, what shape is the earth? We need to know this because the way that we scale the map is not even worldwide due to the fact that the earth's shape is not a perfect sphere. Thirdly, how can a flat map (and simple pairs of numbers or coordinates) be used to describe locations on the earth's surface?

First of all, how big? This question becomes one of what shape we use as a description of the earth. Although for many mapping applications the earth can be assumed to be a perfect sphere, there is a small but significant difference between the distance around the earth pole to pole versus the distance around the equator. This is because the earth resembles more closely than the sphere a figure called an oblate ellipsoid or spheroid, the three-dimensional shape you get by rotating an ellipse about its shorter axis (Figure 2.2). This discovery was a landmark in early scientific discovery, and one of the most important scientific questions of the 18th century.

Figure 2.2: Sphere and ellipsoid (or spheroid). Earth's ellipsoid is actually only about 1/300 off from the sphere.

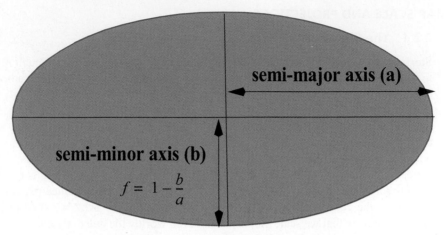

For the WGS84 $a = 6{,}378{,}137$ $b = 6{,}356{,}752.3$, so $f = 1/298.257$

Figure 2.3: The ellipsoid. The long axis is the major axis, the short the minor axis. Half of each of these lengths is used to calculate the flattening of the ellipsoid.

There have been many attempts to measure the size and shape of the earth's ellipsoid. In 1866 the mapping of the United States was based on an ellipsoid measured by Sir Alexander Ross Clarke, which had a basis in measurements taken in Europe, Russia, India, South Africa, and Peru. Spheroids or ellipsoids are determined by two of three numbers, the semi-major and semi-minor radii or the flattening (Figure 2.2). The Clarke 1866 ellipsoid had an equatorial radius of 6,378,206.4 meters and a polar radius of 6,356,538.8 meters. This gave a "flattening" of the ellipsoid of 1/294.9787, where the flattening is one minus the ratio of the semi-axis lengths. In 1924, a simpler measure of 1/297 with a longer radius of 6,378,388 was adopted as an international standard. As mapping had already begun in the United States, the older values were used, adopted as the North American Datum of 1927 (NAD27). The NAD was tied to a particular point on the ground, at Meades Ranch, Kansas. This was the single point, or "datum" from which all other positions and heights were reckoned.

The satellite era has brought with it more accurate means of measurement, including the global positioning system (GPS). An estimate of the ellipsoid allows calculation of the elevation of every point on earth, including sea level, and is often called a datum. The choice of datum is important because the ellipsoid surface is an abstract model or artificial sea level, onto which all points on the earth's surface can be projected up or down. Since up and down are relative to the earth's gravity field, and not the earth's center, this means that as we go from one datum to another, not only elevations change but so also do geographical positions, by small amounts but amounts that can be of importance for mapping (Figure 2.4). It took measurements of the tiny differences in the pull of gravity on satellites to measure the "down" direction (called the deflection of the vertical) precisely. Mapping that used the "down" vector of the plumb bob in surveying gradually noticed that not all degrees of latitude, or even longitude, accounted for the same ground distance. Since gravity is affected by local lumps and bumps on the earth's surface and sub-surface the geodetic latitude differs from the earth-centered latitude, with the lengths of a degree getting longer at the poles as a consequence of the earth's ellipsoidal shape.

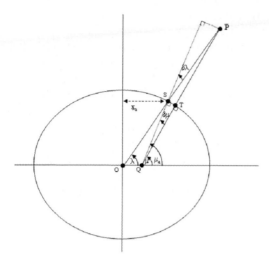

Figure 2.4: Latitude depends on whether we use a sphere or ellipsoid. On the ellipsoid, the "down" vector does not point to the earth's center, resulting in geodetic latitudes. Latitudes to the earth center are called geocentric.

Recent datums have been calculated using the center of the earth as a reference point instead of a point on the ground, as was the case before. It is critical to note that changing data between datums, because of the three dimensions involved, changed both heights and positions. Since often the distances shifted are small, the effect is sometimes barely noticeable. At London's Greenwich Observatory, for example, home of the famous prime meridian, the actual WGS84 meridian is about 104m east of the observatory (Figure 2.5). WGS84 uses the zero meridian as defined by the Bureau International de l'Heure, which was defined by compilation of star observations in different countries. ,

In 1983 a new datum was adopted for the United States, called the North American Datum of 1983 (NAD83), based on measurements taken in 1980 and accepted internationally as the geodetic reference system (GRS80). Efforts have been under way since then to make the slight necessary corrections to maps of the United States, which amount to about 300 meters in places. Many maps in the USA are still based on the older ellipsoids,

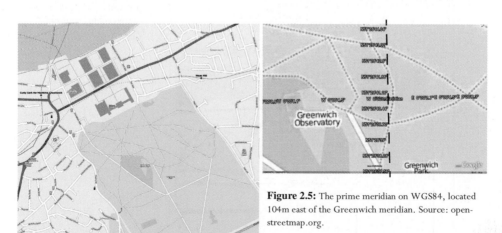

Figure 2.5: The prime meridian on WGS84, located 104m east of the Greenwich meridian. Source: openstreetmap.org.

including the entire first edition of most topographical maps at many scales. When these maps are scanned or retrieved from databases for use in GIS, obviously it becomes important to be able to change between datums.

The U.S. military has also adopted the GRS80 ellipsoid but refined the values slightly in 1984 to contribute to the world geodetic system (WGS84). It is important that when maps are to be used that the datum and ellipsoid reference information be known, as at large scales there can be major differences, especially in elevations (Figure 2.6). The datum and ellipsoid are also essential to know when using a GPS receiver, as coordinates will be different in each, sometimes by kilometers. WGS84 is the reference system being used by the Global Positioning System. It is geocentric and globally consistent within ±1 m, that is the grid and ellipsoidal horizontal differences should vary by less than that. Current geodetic realizations of the geocentric reference system called the International Terrestrial Reference System (ITRS) maintained by the International Earth Rotation Service are also geocentric, and internally consistent to within centimeters, while still being consistent with WGS84 at the one meter horizontal accuracy. While the WGS84 has found excellent levels of acceptance and use worldwide, the more precise and better maintained ITRS is the international standard.

As a final complication, the science of geodesy, which measures the earth's size, shape, and gravitational fields exactly, has mapped out all of the local variations from the ellipsoid and calls the resultant surface a geoid. Only under highly demanding circumstances would a geoid be used in a GIS. In fact, in cartography, a common reference base is the sphere. The ellipsoid becomes necessary when we deal with finer, more detailed, or "large" map scales, and differences caused by not using it can become significant at scales coarser (smaller) than about 1:100,000 but are noticeable even at scales less detailed than 1:50,000. Geoids continue to be refined, based on complex spherical harmonic models of the earth's exact shape. The EGM96 geoid, for example, was improved for the United States by the National Oceanic and Atmospheric Administration with the GEOID03 (Figure 2.7). An even more precise model, the Earth Gravitational Model EGM2008, was completed by the National Geospatial-Intelligence Agency (NGA) in 2008. This gravitational model is complete to spherical harmonic degree and order 2159.

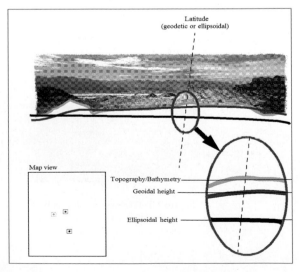

Figure 2.6: Changing datum also changes elevation and position.

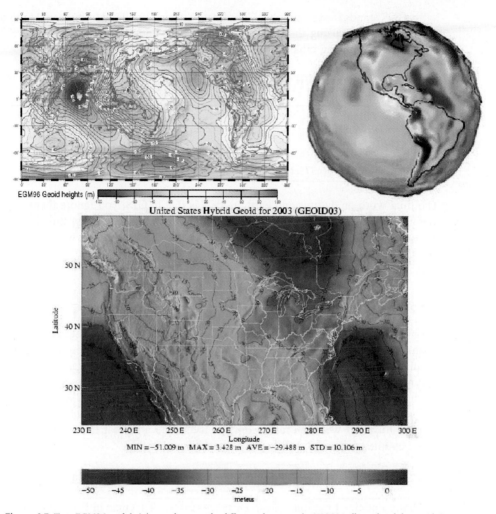

Figure 2.7: Top: EGM96, a global datum shown as the difference between the WGS84 ellipsoid and the geoid. Bottom: GEOID03, a hybrid Geoid for the US.

In GIS, often the datum parameters are required and stored for use by the system. ArcGIS, for example, uses a .prj file to store the essential numbers and values. To illustrate, California's Department of Pesticide Regulation, along with other agencies, uses Teale Albers Equal Area projection. The projection file contains the records:

```
PROJCS["Teale_Albers",GEOGCS["GCS_North_American_1927",DATUM
["D_North_American_1927",SPHEROID["Clarke_1866",6378206.4,
294.9786982]],PRIMEM["Greenwich",0],UNIT["Degree",
0.017453292519943295]],PROJECTION["Albers"],PARAMETER
["False_Easting",0],PARAMETER["False_Northing",-4000000],
PARAMETER["Central_Meridian",-120],PARAMETER
["Standard_Parallel_1",34.0],PARAME-
TER["Standard_Parallel_2",40.5],
PARAMETER["Latitude_Of_Origin",0],UNIT["Meter",1]]
```

GCS stands for Geographic Coordinate System, good old latitude and longitude. The datum is NAD27, using the Clarke 1866 ellipsoid. Some typical values for US maps are given in the following table.

Table 2.1: Ellipsoidal Parameters for Various US Datums

Ellipsoid reference	Semi-major axis a	Semi-minor axis b	Inverse flattening (1/f)
GRS80	6,378,137.0 m	6,356,752.314 140 m	298.257 222 101
WGS84	6,378,137.0 m	6,356,752.314 245 m	298.257 223 563
NAD27 (Clarke 1866)	6,378,206.4 m	6,356,583.8 m	294.978 698 200

The final word on datums and earth models is that in GIS, you ignore them at your peril. While GIS are great at integrating data across different systems, this cannot be done without information on the datum on which the data are recorded. It is not enough to notice that the data are in degrees, minutes, and seconds. We also need to know the datum used to create the data.

2.2.2 Map Scale

All maps, whether on a sheet of paper or inside a computer, are reductions in size of the earth. A map at one-to-one scale (1:1) would be virtually useless; you would barely be able to unfold it. In cartography, the term *representative fraction* is used for the amount of scaling. A representative fraction is the ratio of distances on the map to the same distances on the ground. A model airplane or train is usually at about a 1:40 scale. This means that every distance on the model is one-fortieth of its size on the real object. The world is so big that for maps we often reach some pretty small values for the representative fraction. Just a couple of examples will show why this is necessary.

Table 2.2: Lengths of the Equator at Different Map Scales

Representative Fraction	Map Distance (m)	Distance in Feet (approx.)
1:400,000,000	0.10002	0.328 (3.9 inches)
1:40,000,000	1.0002	3.28
1:10,000,000	4.0008	13.1
1:1,000,000	40.008	131
1:250,000	160.03	525
1:100,000	400.078	1,312
1:50,000	800.157	2,625
1:24,000	1,666.99	5,469 (1.036 miles)
1:10,000	4000.78	13,126 (2.486 miles)
1:1,000	40,007.8	131,259 (24.86 miles)

First, let's use the WGS84 numbers for earth's size. The two ellipsoid distances average to 6,367,444.66 meters. To calculate the circumference of a circle of this size, we multiply by two times π. This is a simplification; the circumference of an ellipse is actually quite hard to calculate. This gives a distance "around the average world" of

40,007,834.7 meters. Table 2.2 shows what this number becomes when multiplied by various representative fractions to give map distances associated with the earth's circumference. A quick look at Table 2.2 reveals a suspicious number. At 1:40,000,000 the earth's circumference maps onto a meter almost exactly. This is because the original definition of the meter was one ten-millionth of the distance from the equator to the north pole measured along the meridian passing through Paris, France. It is fairly obvious that the metric system makes these computations far easier, as we don't have to convert feet to inches and miles.

Over the range of scales shown in Table 2.2, the earth's equator would map onto about the circumference of a gumball at 1:470,000,000 (Figure 2.8), a baseball at 1:177,000,000, a basketball at 1:40,000,000, but at 1:50,000 would map onto about 10 Manhattan city blocks! At 1:1,000, a very detailed scale used in engineering and construction maps, the earth's equator would map onto twice the length of Manhattan Island. A convenient scale to hold in mind is 1:40,000,000, with the equator mapped across a 1-meter poster-size world map. Obviously, we don't use all scales in cartography. Most national mapping for GIS use is between 1:1,000,000 and 1:10,000. In the United States, key scales are 1:100,000 and 1:24,000, at which national coverages are available.

Another important factor to keep in mind is that a GIS is largely scaleless. The data can be multiplied up or reduced to any size that is appropriate. However, as we get farther and farther from the scale at which a map was made before it was captured into the GIS, problems of scale appear. As we enlarge maps, detail does not appear as if by magic. A smooth coastline, for example, remains smooth and imprecise as we enlarge it. On the other hand, if we keep reducing the scale of a map without eliminating detail, the map becomes so "dense" with data that we cannot see the forest for the trees. The proper presentation of information at a particular scale is one of the most important goals of cartographic design. Most GIS software packages and on-line mapping services show a different level of detail on a map at different scales. Some software allows control of what scales are permissible for any given map layer.

A last point to keep in mind as we finish this short discussion of map scale is that only on a globe is a scale constant. As we move the map from the curved surface of the sphere or ellipsoid to the flat surface of paper or the computer screen, we necessarily have to distort the map in some way. The part of cartography that deals with this problem of putting a round earth onto flat paper is called map projections.

Figure 2.8: Assuming a sphere, the earth maps onto a gumball at 1:470,000,000; a baseball at 1:177,000,000, and a basketball at 1:40,000,000. At 1:50,000 the earth's circumference maps on 10 city blocks.

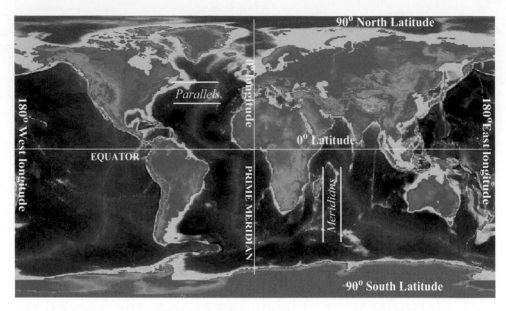

Figure 2.9: Geographic coordinates. The familiar latitude and longitude system, simply converting the angles at the earth's center to coordinates, gives the basic equirectangular projection. The map is twice as wide as high (360° east-west, 180° north-south).

2.2.3   Map Projections

Given that the earth can be approximated by a shape like the sphere or ellipsoid, how can we go about converting data in latitude and longitude into a flat map, with x and y axes? The simplest way is to ignore the fact that latitude and longitude are angles at the center of the earth, and just pretend that they are x and y values (Figure 2.9). Obviously, the map will range from 90 degrees north to 90 degrees south, and from 180 degrees east to 180 degrees west. A huge advantage of showing a map this way is that we can see the whole earth at once, something impossible to do from any point in space.

The corresponding (x,y) values are from (*-180*, *-90*) to (*+180*,*+90*). This map is now a map projection, because the earth's geographical (latitude and longitude) coordinates have been "mapped" or projected onto a flat surface. Obviously this can be done in many ways. We can "project" the sphere (or the ellipsoid) onto any of three flat surfaces and then unfold them to make the map. These can be the plane (as above), the cylinder, or the cone. Projections onto these three surfaces are called azimuthal, cylindrical, and conic, respectively. Examples of each are shown in Figure 2.10.

We can also choose how the mapping takes place with relationship to earth's surface. We can have the figure, such as the cone or the cylinder, "cut" through the earth. The resulting projection, shown in Figure 2.11, is called secant. So, for example, if a cone cuts the earth we would have a secant conic projection. The line on the map where the "cut" falls on the projection is important because it is a line along which the earth and the map match exactly, just as on a globe at the same scale, with no distortion. If this line coincides with a parallel of latitude, it is often called a standard parallel.

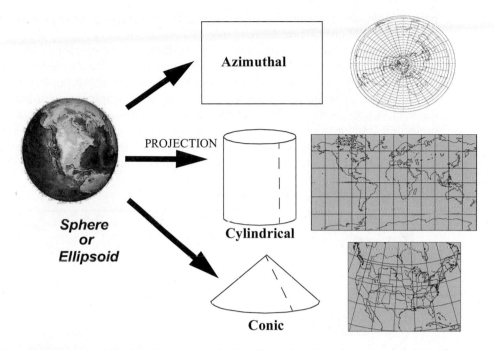

Figure 2.10: The earth can be projected in many ways, but basically onto three shapes that can be unrolled into a flat map: a flat plane, a cylinder, and a cone.

Figure 2.11 shows a secant conic projection with two standard parallels. On or near these lines the map is most accurate. Similarly, there is no hard-and-fast rule that we have to orient the figure we are using in the projection with the earth's polar or rotation axis. If we align instead to a line at 90° to this axis, we call the projection transverse, and some important projections exploit the fact that transverse projections have straight lines that run

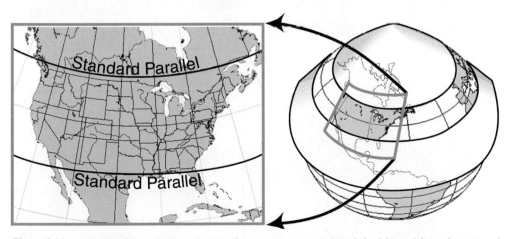

Figure 2.11: Standard parallels on a secant projection. The conic projection cuts through the globe, and the earth is projected both in and out onto it. Lines of true scale, where the cylinder and sphere touch, become standard parallels. If the touching is along one line, the projection is tangent and has one standard parallel.

pole to pole along a meridian. If we orient the projection axis at another angle, the projection is oblique (Figure 2.12).

Cartographers have devised thousands of different map projections. Fortunately, they all fall into a set of "types" that are quite easily understood. The simplest way to evaluate a projection is by how it distorts the earth's surface during the transformation from a sphere or ellipsoid to a flat map. First, some points on the earth simply map onto new points on the map, a one-to-one mapping. Others become ambiguous; for example, a cylindrical projection split along the 180th meridian maps one point onto two along the split. When multiple projections are pieced together in "gores," these distortions can be quite considerable. Such a projection is called interrupted, and an example is the Goodes

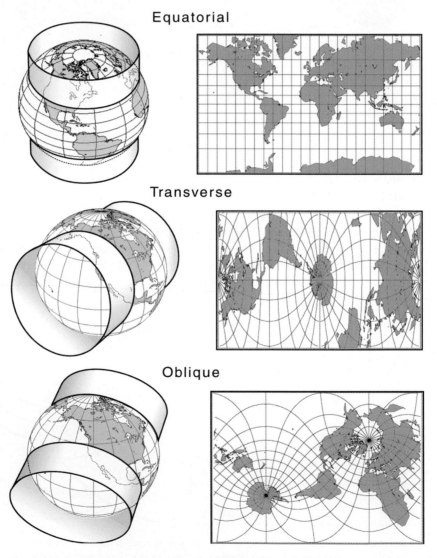

Figure 2.12: Variations on the Mercator (pseudocylindrical) projection shown as secant.

homolosine projection, shown in the bottom of figure 2.14. Lastly, the pole can become an entire line along the top or bottom of the map. On some projections, points such as the pole disappear altogether, since the projection is incapable of displaying them. The poles, for example, cannot be shown on an equatorial Mercator projection (Figure 2.13).

Some projections preserve the property of local shape, so that the outline of a small area like a state or a part of a coastline is correct. These are called conformal projections. They are easily identified, because on a conformal projection the lines of the latitude and longitude grid (called the graticule) meet at right angles, the way they do on a globe, although not all right-angle graticules mean conformal projections. Conformal projections are employed mostly for maps that must be used for measuring directions, because they preserve directions around any given point. Examples are the Lambert Conformal Conic and the Mercator projections.

At the other extreme are projections that preserve the property of area. Many GIS packages compute and use area in all sorts of analyses, and as such must have area mapped evenly across the surface. Projections that preserve area are called equal area or equivalent. On an equivalent projection, all parts of earth's surface are shown with their correct area, as on the sphere or ellipsoid. Examples are the Albers equal area and the sinusoidal projections.

A third category of projections is the set that preserves distances but only along one or a few lines between places on the map. The simple conic and the azimuthal equidistant projections are examples. These projections are useful only if distances are critical, and are infrequently used in GIS. A final category is that of the miscellaneous projections. These are often a compromise, in that they are neither conformal nor equivalent, and sometimes are interrupted or broken to minimize distortion. Similarly, projections are sometimes the average of two or more similar projections. Examples are the Goode's homolosine (made by joining a sinusoidal projection for the equatorial areas below 41° latitude with six lobes on separate Molleweide projections) and the Robinson projection (Figure 2.14).

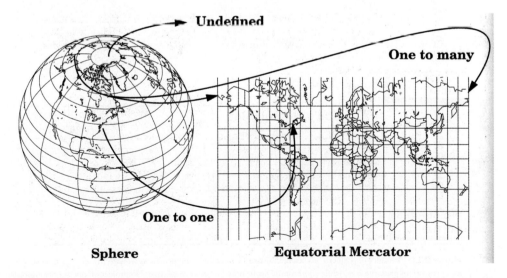

Figure 2.13: Possible transformations when a map is projected. Many points map onto a new relative location, some map onto lines, and others are impossible to map.

The most important implications of map projections for GIS are the following. First, the larger the area involved, the more important the mapping errors due to the projection become. At 1:24,000 scale, the errors are already significant, and at smaller scales like 1:1 million they are major. Second, the projection used should suit the GIS application. If directions or bearings from point to point are important, obviously a conformal projection is called for. If the analysis within the GIS consists of comparing or calculating areas or values based on areas, such as densities, then an equivalent projection is essential. Finally, to overlay or edge-match any two maps, they must be on the same map projection.

A much ignored aspect of map projections is the impact that they can have on gridded data. Many data sets, for example global topographic data, are made available by latitude and longitude increment grids, for example with an elevation every three arc seconds of latitude and longitude for the Shuttle Radar Topographic Mission data. When

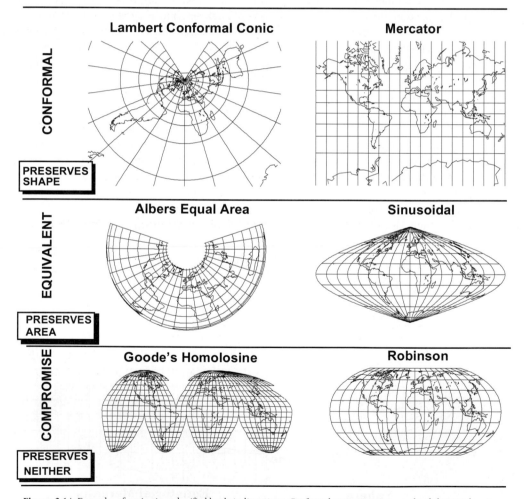

Figure 2.14: Examples of projections classified by their distortions. Conformal projections preserve local shape and equivalent projections preserve area, while compromise projections lie between the two. No projection can be both equivalent and conformal.

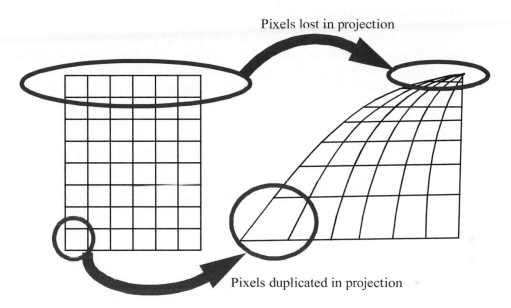

Pixels lost in projection

Pixels duplicated in projection

Figure 2.15: The problem of pixel loss and duplication when regular grids are projected.

these data are projected for use in a GIS, the array of grid points is no longer a rectangular or square grid, but a complex scatter of points with different spacings. If the GIS is used to interpolate points to a new grid on another projection, some significant distortions can arise (Steinwand et al., 1995). In some areas, cells disappear entirely, while in others the elevation points are duplicated (Figure 2.15).

Many GIS data layers are projected into different projections. Much mapping world-wide takes place on a projection chosen to minimize distortion error locally, or even nationally. Countries, regions, and even cities and counties adopt their own particular projection and create local plane coordinates. When we get maps in these local systems, they have to be converted back to latitude and longitude, or into some common system. Many GIS packages have the ability to convert geographic coordinates to several different map projections. Some allow conversion backwards, from map coordinates in a projection to latitude and longitude. Obviously, this ability is rather important to the power and capability of a GIS, because usually GIS maps come from many different sources.

Finally, certain countries, and especially certain coordinate systems, rely entirely on the ability to work in a particular map projection, with a particular ellipsoid or a specific datum. In the United States, for example, the bulk of the 1:24,000 topographic map series of the US Geological Survey uses a polyconic projection, the Clarke 1866 ellipsoid, and the NAD27 datum. The movement to the NAD83 datum and its corresponding ellipsoid the GRS80 have "moved" features by as much as 300 meters on the ground or 12.5 millimeters (0.49 inch) on a 1:24,000 map. If the GIS user makes a basic mistake in comparing or assembling maps on different projections, based on different ellipsoids, and with different datums, many complex errors can result. Layers may become misregistered at best, and completely distorted at worst. This is especially important when data are to be captured from a map into the computer, as we will see in Chapter 3.

2.3 COORDINATE SYSTEMS

When we describe where we are, we usually give the place with reference to somewhere else. Giving directions, for example, we would say, "Go down to the second traffic light, turn right, then continue until you see the diner on the left, then take the second right." When we describe the location of a house or business, we might give the street address, "695 Park Avenue," for example. A street address is also a reference to another place, simply saying, "Go to the street named Park Avenue and find the building labeled 695." Geography calls such references to locations relative location, because they give locations with respect to some other place. Later we will see that a GIS can handle some relative locations, such as US street addresses. Most GISs, however, operate by fixing locations with respect to the earth as a whole. This is called an absolute location, because it is fixed with respect to an origin, a "zero point." For latitude and longitude, we use the earth's equator and the prime meridian as the system's origin. The location of the point, actually in the ocean off West Africa, is not critical, but locations fixed using the origin are indeed important.

Converting maps into numbers requires that we choose a standard way to encode locations on the earth. Maps are drawn (whether by computer or not) on a flat surface such as paper. Locations on the paper can be given in map millimeters or inches starting at the lower left-hand corner. A computer plotter or a printer can understand these dimensions also, and usually requires that the locations be given in (x, y) format; that is, an east-west distance or easting, followed by a north-south distance or northing. This pair of numbers is called a coordinate pair or, more usually, a coordinate. Standard ways of listing coordinates are then called coordinate systems. Maps on common coordinate systems are automatically aligned with each other.

A significant problem with coordinates is that while the map dimensions are simple and the (x, y) axes are at right angles to each other, locations on earth's surface are not so simply derived. The first and foremost problem is that a flat map of all or part of earth's surface is necessarily on a map projection. Something has been distorted to make the surface flat, usually scale, shape, area, or direction. On our flat map, we would like all of the earth's curvature removed. Just how this is done depends on which of the various coordinate systems we use, how big an area we seek to map, and what projection the system uses.

We consider four of the systems in common use in the United States in more detail in this section. As we cover each of the systems, take note of what projections are used and relate them back to the categories of projections introduced in Section 2.2. As you will quickly see, none of the coordinate systems in regular use is really ideal for computer mapping. Considering how complex a shape the earth is, however, many of the systems are perfectly adequate, indeed extremely well suited, for work with GIS.

The five systems we cover are the geographic coordinates themselves; then the worldwide universal transverse Mercator (UTM) coordinate system favored in many mapping efforts; the military grid system, an alternative form of the UTM that has been adopted in many countries outside the United States and for world mapping and its variant the U.S. National Grid; and the state plane system, the basis of most surveying practice in the United States. Finally, we consider what other systems might be encountered in the GIS world and the implications of using these systems.

Figure 2.16: The International Meridian Conference in Washington, D.C., in October 1884 agreed "That the Conference proposes to the Governments here represented the adoption of the meridian passing through the centre of the transit instrument at the Observatory of Greenwich as the initial meridian for longitude."

2.3.1 Geographic Coordinates

Many GIS systems store locations as numbers using latitude and longitude or geographic coordinates. This system was standardized by the International Meridian conference, held in Washington, D.C., in 1884. At this conference, it was decided to establish the origin for longitude for the earth at the Greenwich Observatory in England (Figure 2.16). In a GIS, latitude and longitude are almost always geocoded, or captured from the map into the computer, in one of two ways. These are degrees, minutes, and seconds (DMS) and decimal degrees (DD). In both cases, latitudes go from 90 degrees south (-90) to 90 degrees north (+90) (Figure 2.9). Precision below a degree is geocoded as minutes and seconds, and decimals of seconds, in one of two formats: either in DMS as plus or minus DD.MMSS.XX, where DD are degrees, MM are minutes, and SS.XX are decimal seconds; or alternatively, in DD as DD.XXXX, or decimal degrees. Just as with time, a degree consists of 60 minutes, each with 60 seconds. Longitudes are the same, with the exception that the range is -180 to +180 degrees. In the second format, degrees are converted to radians and stored as decimal numbers with the appropriate number of significant digits.

For example, the file listed below is part of the World Data Bank, a listing of coordinates of the world's coastlines, rivers, islands, and political boundaries. The points are from the first line in the coastline of North America. The coordinates are decimal degrees, rounded to the nearest 0.000001 degree. At the equator, one degree is about 40,000 km per 360 degrees = 111.11 km.; 0.000001 degree is then 11.1 centimeters. So, these data have a resolution of about 1/10th of a meter on the ground. Their accuracy is determined by how well the line actually represents the real coastline of North America (in fact, part of Alaska) (Figure 2.17).

```
52.837778  -128.137778
52.841944  -128.137778
52.847778  -128.136944
```

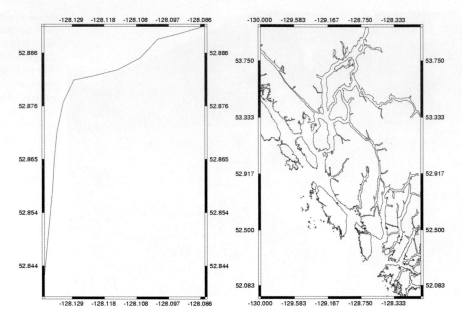

Figure 2.17: A plot of the first vector in the World Data Bank for North America. Detail at left.

```
52.853333  -128.136111
52.858889  -128.135278
52.864722  -128.134444
52.870278  -128.133611
52.876389  -128.131667
52.880556  -128.128056
52.881389  -128.121389
```

The advantage of using geographic coordinates in a GIS is that all maps can be transformed into a projection in the same way. If maps captured on a variety of projections are reprojected into geographic coordinates, there is some room for error. For example, the points in Figure 2.17 are at a resolution better than 0.1m on the ground, but were digitized at a scale of 1:2M, so cannot be considered at this level of precision when compared with other maps. If the GIS does not support transformations between projections, then working in a common coordinate system such as the UTM or state plane system is very important if the maps are expected to overlay each other. It is also important to note that even geographic coordinates still need to have their ellipsoid or datum specified, since locations change from one datum to another.

2.3.2 The Universal Transverse Mercator Coordinate System

The universal transverse Mercator (UTM) coordinate system is commonly used in GIS because it has been included since the late 1950s on most USGS topographic maps. The choice of the transverse Mercator, probably now used more than any other projection for accurate mapping, has an interesting history. The story begins with the observation that the

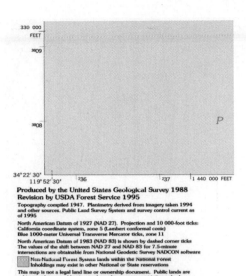

Figure 2.18: The various coordinate systems shown on the Goleta, CA, seven-and-a-half minute USGS topographic map at 1:24,000. UTM zone 11 is shown by the blue tic marks with 1-km spacing. Note that the map uses NAD27, but shows the displacement when transformed to NAD83.

Produced by the United States Geological Survey 1988
Revision by USDA Forest Service 1995

Topography compiled 1947. Planimetry derived from imagery taken 1994 and other sources. Public Land Survey System and survey control current as of 1995

North American Datum of 1927 (NAD 27). Projection and 10 000-foot ticks: California coordinate system, zone 5 (Lambert conformal conic) Blue 1000-meter Universal Transverse Mercator ticks, zone 11

North American Datum of 1983 (NAD 83) is shown by dashed corner ticks The values of the shift between NAD 27 and NAD 83 for 7.5-minute intersections are obtainable from National Geodetic Survey NADCON software

Non-National Forest System lands within the National Forest Inholdings may exist in other National or State reservations

This map is not a legal land line or ownership document. Public lands are subject to change and leasing, and may have access restrictions; check with local offices. Obtain permission before entering private lands

equatorial Mercator projection, which distorts areas so much at the poles, nevertheless produces minimal distortion laterally along the equator.

Johann Heinrich Lambert modified the Mercator projection into its transverse form in 1772, in which the "equator" instead runs north-south. The effect is to minimize distortion in a narrow strip running from pole to pole. Johann Carl Friedrich Gauss further analyzed the projection in 1822, and Louis Kruger worked out the ellipsoid formulas in 1912 and 1919 adjusting for "polar flattening." As a result, the projection is often called the Gauss conformal or the Gauss-Kruger, although the name transverse Mercator is used in the United States. Rarely, however, was the projection used at all until the major national mapping efforts of the post-World War II era.

The transverse Mercator projection, in various forms, is part of the civilian UTM system described here, the state plane system, and the military grid. It has been used for mapping most of the United States, many other countries, and even the planet Mars. The first version is the civilian UTM grid, used by the U.S. Geological Survey on its maps since 1977, and marked on many maps since the 1940s as blue tic marks along the edges of the quadrangle maps or grids over the surface (Figure 2.18). In 1977 the transverse Mercator projection replaced the polyconic for large-scale U.S. mapping. NAD27 geographic coordinates are typically converted to UTM using the Clarke 1866 ellipsoid, while NAD83 coordinates use the GRS80 ellipsoid.

The UTM capitalizes on the fact that the transverse Mercator is accurate in north-south strips by dividing the earth up into 60 pole-to-pole zones, each 6 degrees of longitude wide, running from pole to pole. The first zone starts at 180 degrees west (or east), at the international date line, and runs east, that is, from 180 degrees west to 174 degrees west. The final zone, zone 60, starts at 174 degrees east and extends east to the date line. The zones therefore increase in number from west to east. For the coterminous United States, California falls into zones 10 and 11, while Maine falls into zone 19 (Figure 2.19). Within each zone we draw a transverse Mercator projection centered on the middle of the zone oriented north-south. Thus for zone 1, with longitudes ranging from 180 degrees west to 174 degrees west, the central meridian for the transverse Mercator projection is 177

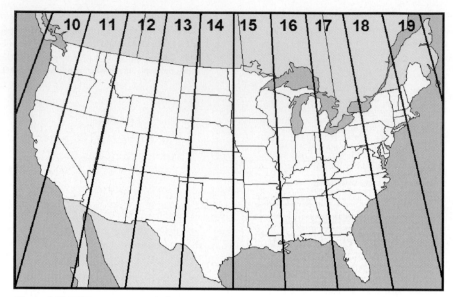

Figure 2.19: UTM zones covering the lower 48 United States.

degrees west. Because the equator meets the central meridian of the system at right angles, we use this point to orient the grid system (Figure 2.20).In reality, the central meridian is set to a map scale of slightly less than 1, making the projection for each zone secant along two lines at true scale parallel to the central meridian.

To establish a coordinate system origin for the zone, we work separately for the two hemispheres. For the southern hemisphere, the zero northing is the South Pole, and we give northings in meters north of this reference point. As the earth is about 40 million meters around, this means that northings in a zone go from zero to 10 million meters, although they actually stop short of the poles.

The numbering of northings starts again at the equator, which is either 10 million meters north in southern hemisphere coordinates or 0 meters north in northern hemisphere coordinates. Northings then increase to 10 million at the north pole. As we approach the poles, the distortions of the latitude-longitude grid drift farther and farther from the UTM grid. It is customary, therefore, not to use UTM beyond 84 degrees north and 80 degrees south. For the polar regions, the universal polar stereographic coordinate system is used.

For eastings, a false origin is established beyond the westerly limit of each zone. The actual distance is about half a degree, but the numbering is chosen so that the central meridian has an easting of 500,000 meters. This has the dual advantage of allowing overlap between zones for mapping purposes and of giving all eastings positive numbers. We can tell from our easting if we are east or west of the central meridian, and so the relationship between true north and grid north at any point is known. To give a specific example, my office at UCSB in Santa Barbara is located at 238,463mE; 3,811,950mN; Zone 11, northern hemisphere. This tells us that we are about four-tenths of the way up from the equator to the north pole, and are west of the central meridian for our zone, which is centered on 117 degrees west of Greenwich. On a map showing UCSB, UTM grid north would therefore appear to be west of true north.

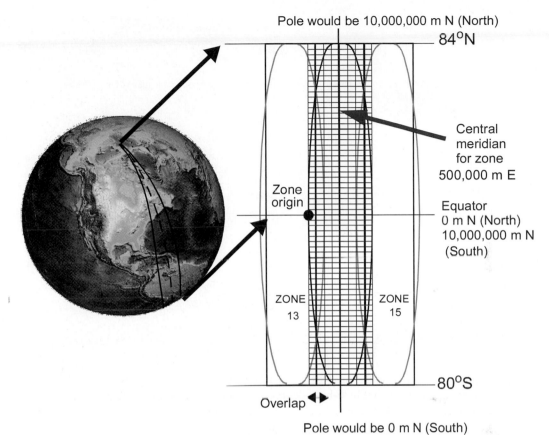

Pole would be 10,000,000 m N (North)

84°N

Central
meridian
for zone
500,000 m E

Zone
origin

Equator
0 m N (North)
10,000,000 m N
(South)

ZONE
13

ZONE
15

80°S

Overlap

Pole would be 0 m N (South)

Figure 2.20: The universal transverse Mercator coordinate system.

The variation from true scale is 1 part in 1000 at the equator. As a Mercator projection, of course, the system is conformal and preserves the shape of features such as coastlines and rivers. Another advantage is that the level of precision can be adapted to the application. For many purposes, especially at small scales, the last UTM digit can be dropped, decreasing the resolution to 10 meters. This strategy is often used at scales of 1:250,000 and smaller. Similarly, submeter resolution can be added simply by using decimals in the eastings and northings. In practice, few applications except for precision surveying and geodesy need precision of less than 1 meter, although it is often used to prevent computer rounding error and is stored in the GIS nevertheless.

2.3.3 The Military Grid Coordinate System

The second form of the UTM coordinate system is the military grid, or Military Grid Reference System (MGRS) adopted for use by the U.S. Army in 1947 and used by many other countries and organizations. The military grid uses a lettering system to reduce the number of digits needed to isolate a location. Zones are numbered as before, from 1 to 60, west to east. Within zones, however, 8-degree strips of latitude are lettered from C (80 to 72 degrees south) to X (72 to 84 degrees north: an extended-width strip). The letter

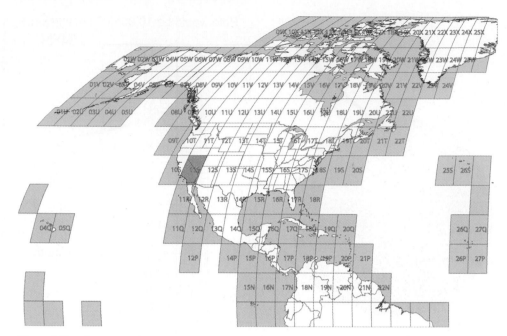

Figure 2.21: Military Grid cell designations for the US; cells are 6 longitude degrees wide by 8 latitude degrees high. Red cell is 11S.

designations A, B, Y, and Z are reserved for Universal Polar Stereographic designations on the poles. A single rectangle, 6-by-8 degrees, generally falls within about a 1000-kilometer square on the ground. These squares are referenced by numbers and letters; for example, Santa Barbara falls into grid cell 11S (Figure 2.21).

Each grid cell is then further subdivided into squares 100,000 meters on a side. Each cell is assigned two additional letter identifiers (Figure 2.22). In the east-west (x) direction, the 100,000-meter squares are lettered starting with A, up to Z, and then repeating around the world regardless of the 6 x 8 degree cells, with the exception that the letters I and O are excluded, because they could be confused with numbers. The first column, A, is 100,000 meters wide and starts at 180 degrees west. The alphabet recycles about every 18 degrees and includes about six full-width columns per UTM zone. Several partial columns are given designations nevertheless, so that overlap is possible, and some disappear as the poles are approached. The general pattern of lettering is visible for the Pacific around Hawaii in Figure 2.22. This rather curious lettering pattern is designed to make unique letter combination references within each number grid, such that repetition is avoided, especially for nearby grid cells.

In the north-south (y) direction, the letters A through V are used (again omitting I and O), starting at the equator and increasing north, and again cycling through the letters as needed. The reverse sequence, starting at V and cycling backward to A, then back to V, and so on, is used for the southern hemisphere. Thus a single 100,000-meter grid square can be isolated using a sequence such as 11S KU. Within this area, successively accurate locations can be given by more and more pairs of x and y digits. Note that the x and y components of the grid references are interwoven, i.e., the x is followed by the y without a separation. For example, 11S KU31 isolates a 10,000-meter square, 11S KU 3811 a 1000

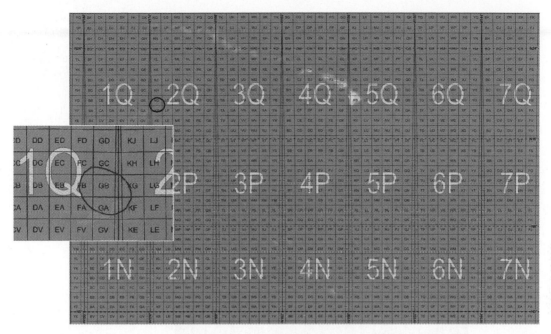

Figure 2.22: US Military Grid. After designation of 6 x 8 degree blocks (e.g., 1Q), a second repeating grid of 100,000 m squares is designated by letter pairs, e.g., 1QGB, see enlargement at left. Note that some cells get truncated.

meter square, and 11SKU3847911950 a 1-meter square. In the latter case, 38479 is the easting, and 11950 is the northing. Finally, the polar areas are handled completely separately on a different (UPS) projection.

2.3.4   The United States National Grid

There is a variant on the MGRS, used by the USGS and other agencies for mapping applications in the 50 United States and territories. This system differs only from the MGRS in that it uses the NAD 83 US datum instead of WGS84, and its extent covers only the USA. The ellipsoids, GRS80 which was used for NAD 83 compared to WGS84 are for all practical purposes identical (see Table 2.1). Very small numerical differences amount to less than 0.1 mm across all of North America, and no more than 2m worldwide. The United States National Grid (USNG) was designed to supplement other grid systems, with the needs of local government and emergency management in mind. The USNG was approved by the Federal Geographic Data Committee in 2001, with the goal of expanding the utility of street, and other, maps by providing a grid reference system that is seamless across jurisdictional boundaries; provides the foundation for a universal map index (that is it places USNG within the wider MGRS); describes point-locations on gridded paper and digital maps; allows geopositioning with the Global Positioning System, and supports Web mapping that links digital maps with locational data. It is now beginning to be used in public safety response (e.g., police, fire, rescue, National Guard), and national guard units are being trained in its use. Figure 2.23 shows an NGA-prepared training map for USNG users working on security and event planning for Washington, D.C.

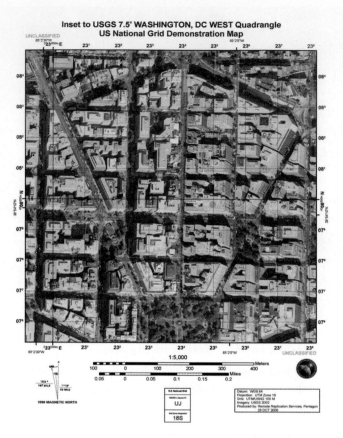

Figure 2.23: NGA-prepared inset of the USGS map for Washington, D.C. at 1:5000 showing use of the US National Grid coordinate system.

Practically, the use of the USNG is the same as the MGRS. The fact that the grid is often used over small areas means that the leading letters and numbers can be omitted, since they are almost always identical for every point. However, many US cities and important locations lie at the intersection of grid cells, both 6 x 8 degree blocks and the lettered squares 100,000 meters on a side. At the intersection points of zones, such as California's UTM zones 10 and 11 at the 120th meridian, there is compression of 100,000 cells and some unusual tiling of 10,000 meter cells, as shown in Figure 2.24.

A principal use of the USNG is in the USGS's national map viewer, an on-line web-based GIS that can browse and allow download of much Federal government data for the USA. In the viewer (Figure 2.25), the USNG coordinates are shown in white at the lower right of the main map display window. Because the USNG can be used with either GRS80 or WGS84, the datum chosen should be stated along with the coordinate. The top of the state capital building in Austin, Texas, to within 1m would therefore be:

14R PU 21164 49875 (NAD83)

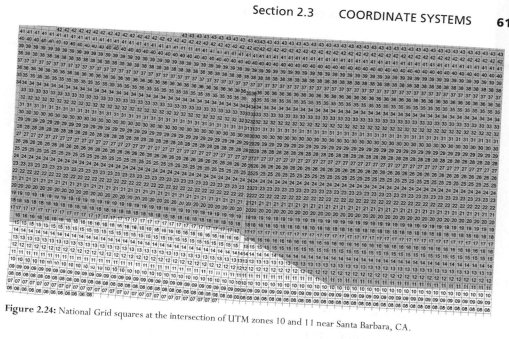

Figure 2.24: National Grid squares at the intersection of UTM zones 10 and 11 near Santa Barbara, CA.

2.3.5   The State Plane Coordinate System

Much geographic information in the United States uses a system called the state plane coordinate system (SPCS). The system is used primarily for engineering applications, especially by utility companies and local governments that need to do accurate surveying of facilities networks such as power lines and sewers, or for property designation. The SPCS is based on both the transverse Mercator and the Lambert conformal conic projections with units in meters (previously feet). The SPCS, which has been used for decades to write legal descriptions of properties and in engineering projects in many states,

Figure 2.25: National Map Viewer view of the Texas State Capitol building showing USNG coordinates (lower right) and grid.

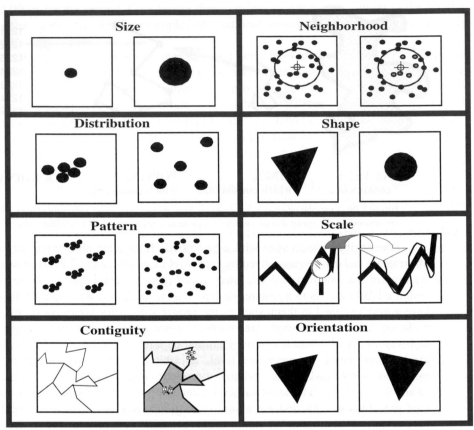

Figure 2.29: Basic spatial properties of geographic features.

what prompted the use of a GIS in the first place. For example, we can measure the areas (size) of land parcels, or the orientation of highways, or the distribution of flora and fauna in a state park. The basic properties are summarized in Figure 2.29. Even though the GIS will directly hold only the coordinates and some additional information such as contiguity, information about every one of these properties will be available by using the tools within the GIS for higher-level analysis. Part of what the GIS user does is to coax descriptions of these properties out of the data that are available in the particular GIS in use. How well you can do this will depend on your skills as an intelligent GIS user.

Our digression on the cartographic roots of GIS is now complete. As we have seen, there are numerous important considerations to bear in mind that are directly related to the geometry of the map and the geometry of the features that we will store in the GIS. We are now ready to move on and begin covering GIS concepts proper. Step one is to cover how the map is structured as sets of digits inside the computer. Step two is to examine how to get the data from the map into the computer. As we will find out, this is another basic but overridingly important step in getting started with GIS.

2.5   STUDY GUIDE

2.5.1   Quick Study

CHAPTER 2: GIS's ROOTS IN CARTOGRAPHY

O *A GIS links attributes about places to points, lines, and areas on maps* O *GIS shares the need to quantify positions with cartography and geodesy* O *The earth is about 40,000,000 m around, and about 1/300 off of a perfect sphere, flatter at the poles, a shape called an ellipsoid* O *A geometric model of an ellipsoid determines a level from which heights can be measured and on which locations can be determined* O *The initial mapping of the USA started using the Clarke 1866 ellipsoid and the NAD27 datum* O *Ellipsoids are now earth-centered, not local to a country or region* O *Much mapping today uses WGS84 globally and GRS80 (identical to NAD 83) in the USA* O *An even more accurate geodetic model is called a geoid* O *A GIS needs to know what ellipsoid and datum were used to create a map so that layers fit together perfectly* O *GIS is scaleless, but data are not* O *At 1:40M a world map is a meter across* O *A representative fraction is a ratio of distance on the map to the same distance on the ground* O *After a map is scaled, it is still on a curved surface and needs to be projected flat* O *Projections have an analog of shining a surface map onto a flat plane, which can be flat, curved into a cylinder, or into a cone* O *Projections can touch the globe at points or along lines, or can cut through (secant)* O *Secant projections have lines of minimum projection distortion; if these follow parallels they are called standard* O *The latitude and longitude grid of parallels and meridians on a map is called the graticule* O *Projections can preserve area, local shape, or direction* O *GIS applications should choose projections carefully* O *No flat map can be both equivalent and conformal* O *Coordinate systems are standard ways of applying numbers to locations in two dimensions* O *Geographic coordinates are just latitude and longitude, often in decimal degrees but including a datum* O *UTM has two versions, military and civilian; both divide the earth into 60 zones 6 degrees of longitude wide and drawn on a transverse Mercator projection* O *The USGS shows on its maps UTM eastings based on 500,000 m at the central meridian, and northings with origins at the equator and the south pole* O *UTM does not apply to the poles; the UPS is used there* O *The military grid splits the 60 UTM zones by 8 degree latitude blocks and assigns letters to blocks* O *A second letter pair is assigned to 100,000 m grids within the cells that wrap around in their lettering* O *Within the grids, numbers of meters are given depending on the desired precision, 10 digits resolve one meter* O *The USNG is the same as the MGRS, but uses NAD 83 and applies only to the US and territories* O *The US State Plane System divides the 50 states into 120 zones, with slices drawn on either a Lambert conformal conic or a transverse Mercator projection* O *Zones have false origins, and use meters east and north and the NAD83 datum* O *There are many other grids, most specific to one country* O *Geographic information can be attached to points, lines and areas, and the data items can be nominal, ordinal, interval, or ratio—the levels of scaling* O *GIS data can be feature based or field based if the data are continuous over space* O *GIS can analyze higher order spatial relations, such as size, distribution, pattern, orientation, neighborhood, contiguity, shape, and scale* O

2.5.2   Study Questions and Activities

Map and Attribute Information

1. Make a list of your friends' names and addresses. Enter the data into a spreadsheet, and shade in the columns which contain the geographical information.

Map Scale and Projections

2. Using an atlas, make a list of as many map projections as you can find. Are any of the atlas maps not annotated with their projection? Make a table listing the properties of each of the projections, plus any other information you can find out-for example, whether the projection is secant, transverse, based on an ellipsoid, conformal, and so on. In a final column, state what properties are distorted on the map; for example, "Map distorts area increasingly as one moves north and south." A good source of reference is: erg.usgs.gov/isb/pubs/MapProjections/projections.html

3. Using the worldwide web, try to find the sizes of as many ellipsoids as possible. Are any of them more or less suitable for foreign countries? Select a country, for example, Egypt or Australia, and research which ellipsoids have been used and whether any particular projection is favored for that country. Why would one projection be better than another?

4. Find the regulation sizes of either a baseball diamond or a soccer field. Draw maps of the fields at the following scales: 1:1000, 1:24,000, 1:100,000, and 1:1,000,000. What problems do you run into? What would be the effect of mapping both a winding river and an irregular patch of forest at these scales?

Coordinate Systems

5. Use the Google Maps USNG mash-up at http://www.fidnet.com/~jlmoore/usng/help_usng.html to find the USNG location of your house, school, or other another point of interest near you.

6. For a single location, such as your house or school, try to find the coordinates of two nearby positions in as many coordinate systems as possible. What might be the sources of error or confusion if you were using the systems to locate important positions during an emergency, such as an evacuation?

Geographic Information

7. Make a table of levels of measurement versus dimension. In each cell of the table, write in as many types of geographic data or features as you can think of.

8. Using the example of a lake, write out sample measurements that might describe each of the major geographic properties covered in Figure 2.29. For example, the size of the lake is its area in square meters. For which properties is it most difficult to think of representative numbers? (Hint: Can a single number describe the shape?)

2.6 REFERENCES

Bugayevskiy, L. M. and Snyder, J. P.(1995) *Map Projections—A Reference Manual*. Taylor and Francis Inc., Bristol, PA.

Campbell, J. (1993) *Map Use and Analysis.* 2nd ed. Dubuque, IA: William C. Brown.

Clarke, K. C. (1995) *Analytical and Computer Cartography.* 2nd ed. Upper Saddle River, NJ: Prentice Hall.

Department of the Army (1973) *Universal Transverse Mercator Grid*, TM 5-241-8, Headquarters, Department of the Army. Washington, DC: U.S. Government Printing Office.

Snyder, J. P. (1987) *Map Projections—A Working Manual.* U.S. Geological Survey Professional Paper 1396. Washington, DC: U.S. Government Printing Office.

Snyder, John P., and Philip M. Voxland (1989) *An Album of Map Projections.* U.S. Geological Survey Professional Paper 1453; Denver, CO.

Steinwand, D. R., Hutchinson, J. A. and Snyder, J. P. (1995) "Map Projections for Global and Continental Data Sets and an Analysis of Pixel Distortion Caused by Reprojection." *Photogrammetric Engineering and Remote Sensing*, Vol. LXI. No. 12.

Thompson, M. M. (1988) *Maps for America*, 3ed. U.S. Geological Survey; Reston, VA.

United States Defense Mapping Agency (1984) *Geodesy for the Layman*. Published online at: http://www.ngs.noaa.gov/PUBS_LIB/Geodesy4Layman/toc.htm.

2.7   KEY TERMS AND DEFINITIONS

absolute location: A location in geographic space given with respect to a known origin and standard measurement system, such as a coordinate system.

accuracy: The validity of data measured with respect to an independent source of higher reliability and precision.

attribute: A numerical entry that reflects a measurement or value for a feature. Attributes can be labels, categories, or numbers; they can be dates, standardized values, or field or other measurements. An item for which data are collected and organized. A column in a table or data file.

azimuthal: A map projection in which the globe is projected directly on a flat surface. Only one "side" of the globe can be shown at a time.

compromise: A map projection that is neither area preserving nor shape preserving. An example is the Robinson projection.

conformal: A type of map projection that preserves the local shape of features on maps. On a conformal projection, lines on the graticule meet at right angles, as they do on a globe.

conic: A type of map projection involving projecting part of the earth onto a cone-shaped surface that is then cut and unrolled to make it flat.

continuity: The geographic property of features or measurements that gives measurements at all locations in space. Topography and air pressure are examples.

coordinate pair: An easting and northing in any coordinate system, absolute or relative. Together these two values, usually termed (x, y), describe a location in two-dimensional geographic space.

coordinate system: A system with all the necessary components to locate a position in two- or three-dimensional space: that is, an origin, a type of unit distance, and axes.

cylindrical: A type of map projection involving projecting part of the earth onto a cylinder-shaped surface that is then cut and unrolled to make it flat.

database: A collection of data organized in a systematic way to provide access on demand.

datum: A base reference level for the third dimension of elevation for the earth's surface. A datum can depend on the ellipsoid, the earth model, and the definition of sea level.

dimensionality: The property of geographic features by which they are capable of being broken down into elements made up of points, lines, and areas. This corresponds to features being zero-, one-, and two-dimensional. A drill hole is a point, a stream is a line, and a forest is an area, for example.

distortion: The space distortion of a map projection, consisting of warping of direction, area, and scale across the extent of the map.

easting: The distance of a point in the units of the coordinate system east of the origin for that system.

edge matching: The GIS or digital map equivalent of matching paper maps along their edges. Features that continue over the edge must be "zipped" together and the edge dissolved. To edge-match, maps must be on the same projection, datum, ellipsoid, and scales and show features captured at the same equivalent scale.

equal area: A type of map projection that preserves the area of features on maps. On an equal-area projection, a small circle on the map would have the same area as on a globe with the same representative fraction. See also equivalent.

equatorial radius: The distance from the geometric center of the earth to the surface, usually averaged to a single value for a sphere.

equirectangular: A map projection that maps angles directly to eastings and northings. A cylindrical projection, made secant by scaling the height-to-width ratio. The nonsecant or equatorial version is called the Plate Carree. Credited to Marinus of Tyre, c. A.D. 100.

equivalent: A type of map projection that preserves the area of features on maps. On an equal area projection, a small circle on the map would have the same area as on a globe with the same representative fraction. See also equal area.

file: Data logically stored together at a location on the storage mechanism of a computer.

flattening (of an ellipsoid): The ratio of the length of half the short axis of the ellipse to half the long axis of the ellipse, subtracted from 1. The earth's flattening is about 1/300.

geocode: A location in geographic space converted into computer-readable form. This usually means making a digital record of the point's coordinates.

geodesy: The science of measuring the size and shape of the earth and its gravitational and magnetic fields.

geographic coordinates: The latitude and longitude coordinate system.

geographic property: A characteristic of a feature on earth, usually describable from a map of the feature, such as location, area, shape, distribution, orientation, adjacency, and so on.

geography: (1) A field of study based on understanding the phenomena capable of being described and analyzed with a GIS. (2) The underlying geometry and properties of the earth's features as represented in a GIS.

geoid: A complex earth model used more in geodesy than cartography or GIS that accounts for discrepancies over the earth from the reference ellipsoid and other variations due to gravity, and so on.

globe: A three-dimensional model of the earth made by reducing the representative fraction to less than 1:1.

GPS (Global Positioning System): An operational, U.S. Air Force-funded system of satellites in orbits that allow their use by a receiver to decode time signals and convert the signals from several satellites to a position on the earth's surface.

graticule: The latitude and longitude grid drawn on a map or globe. The angle at which the graticule meets is the best first indicator of what projection has been used for the map.

GRS80 (Geodetic Reference System of 1980): Adopted by the International Union of Geodesy and Geophysics in 1979 as a standard set of measurements for the earth's size and shape. The length of the semimajor axis is 6,378,137 meters. Flattening is 1/298.257.

interval: Data measured on a relative scale but with numerical values based on an arbitrary origin. Examples are elevations based on mean sea level, or coordinates.

latitude: The angle made between the equator, the earth's geometric center, and a point on or above the surface. The south pole has latitude -90 degrees, the north +90 degrees.

level of measurement: The degree of subjectivity associated with a measurement. Measurements can be nominal, ordinal, interval, or ratio.

link: The part or structure of a database that physically connects geographic information with attribute information for the same features. Such a link is a defining component of a GIS.

location: A position on the earth's surface or in geographic space definable by coordinates or some other referencing system, such as a street address or space indexing system.

longitude: The angle formed between a position on or above the earth, the earth's geometric center, and the meridian passing through the center of the observing instrument in Greenwich, England, as projected down onto the plane of the earth's equator or viewed from above the pole. Longitudes range from -180 (180 degrees West) to +180 (180 degrees East).

map millimeters: A coordinate system based on the dimensions of the map rather than those of the features represented on the earth itself, in metric units.

map projection: A depiction of the earth's three-dimensional structure on a flat map.

mean sea level: A local datum based on repeated measurements of sea level throughout all of its normal cycles, such as tides and seasonal change. The basis for elevations on a map.

meridian: A line of constant longitude. All meridians are of equal length on the globe.

metric system: A system of weights and measures accepted as an international standard as the Systeme International d'Unites (SI) in 1960. The metre (meter in the United States) is the unit of length.

military grid: A coordinate system based on the transverse Mercator projection, adopted by the U.S. Army in 1947 and used extensively for world mapping.

mosaicing: The GIS or digital map equivalent of matching multiple paper maps along their edges. Features that continue over the edge must be "zipped" together and the edge dissolved. A new geographic extent for the map usually has to be cut or clipped out of the mosaic. To permit mosaicing, maps must be on the same projection, datum, ellipsoid, and scale, and show features captured at the same equivalent scale.

NAD27 (North American Datum of 1927): The datum used in the early part of the national mapping of the United States. The Clarke 1866 ellipsoid was used, and locations and elevations were referenced to a single point at Meade's Ranch in Kansas.

NAD83 (North American Datum of 1983): The datum now used in the mapping of the United States and its territories. The ellipsoid used is GRS80, which is very close to the more recent WGS84.

nominal: A level of measurement at which only subjective information is available about a feature. For a point, for example, the name of the place.

northing: The distance of a point in the units of the coordinate system north of the origin for that system.

oblate ellipsoid: A three-dimensional shape traced out by rotating an ellipse about its shorter axis.

oblique: A map projection in which the centerline of the map is not at right angles to the earth's geographic coordinates, following neither a single parallel nor a meridian.

ordinal: A level of measurement at which only relative information is available about a feature, such as a ranking. For a highway, for example, the line is coded to show a Jeep trail, a dirt road, a paved road, a state highway, or an interstate highway, in ascending rank.

origin: A location within a coordinate system where the eastings and northings are exactly equal to zero.

parallel: A line of constant latitude. Parallels get shorter toward the poles, becoming a point at the pole itself.

polar radius: The distance between the earth's geometric center and either pole.

precision: The number of digits used to record a measurement or which a measuring device is capable of providing.

prime meridian: The line traced out by longitude zero and passing through Greenwich, England. The prime meridian forms the origin for the longitude part of the geographic coordinates and divides the eastern and western hemispheres.

ratio: A level of measurement at which numerical information is available about a feature, based on an absolute origin. For land parcels, for example, the assessed value in dollars would be an example, the value zero having real meaning.

relative location: A position described solely with reference to another location.

representative fraction: The ratio of a distance as represented on a map to the equivalent distance measured on the ground. Typical representative fractions are 1:1,000,000, 1:100,000, and 1:50,000.

scale: The geographic property of being reduced by the representative fraction. Scale is usually depicted on a map or can be calculated from features of known size.

scaleless: The characteristic of digital map data in abstract form of being usable and displayable at any scale, regardless of the scale of the map used to geocode the data.

secant: A map projection in which the surface used for the map "cuts" the globe at the map's representative fraction. Along this line there is distortion-free mapping of the geographic space. Multiple cuts are possible, for example, on a conic projection.

state plane: A coordinate system common in utility and surveying applications in the lower 48 United States and based on zones drawn state by state on transverse Mercator and Lambert conformal conic projections.

standard parallel: A parallel on a map projection that is secant and so distortion- free.

transverse: A map projection in which the axis of the map is aligned from pole to pole rather than along the equator.

USNG (United States National grid): A grid system adopted as standard by the Federal Geographic Data Committee and now used in many applications such as emergency management. Follows the MGRS closely, and can use either the NAD83 or WGS84 datums.

UTM (Universal Transverse Mercator): A standardized coordinate system based on the metric system and a division of the earth into sixty 6-degree-wide zones. Each zone is projected onto a transverse Mercator projection, and the coordinate origins are located systematically. Both civilian and military versions exist.

value: The content of an attribute for a single record within a database. Values can be text, numerical, or codes.

WGS84 (World Geodetic Reference System of 1984): A higher precision version of the GRS80 used by the U.S. Defense Mapping Agency in world mapping. A common datum and reference ellipsoid for handheld GPS receivers.

zone (of a coordinate system): The region over which the coordinates relate with respect to a single origin. Usually, some part of the earth or a state.

2.8   PEOPLE IN GIS

David Burrows. ESRI (Redlands, California) software development programmer

KC: I understand that you are Mr. Map Projections for ESRI.

DB: I am the Map Projection Development Programmer at ESRI. I write all the map projection codes, a part of the code called the projection engine that goes into all of our current products. This is the number cruncher that does all the mathematics for coordinate transformation, whether it is a geodetic latitude and longitude to map x, y or to move from one to another.

KC: What was your educational preparation for the job you found your way into?

DB: I have a BS in Geography from Penn State and also a BS in Environmental Resource Management. Once I started in

Geography at 19 years old I fell in love with the discipline. I always had a knack for mathematics, I just could do math and it didn't really scare me all that much. I took every techniques class in Geography—digital terrain modeling, satellite remote sensing, spatial analysis, computer mapping. I went on to get a Masters in Geography at UC Santa Barbara, and took a map projections course with Waldo Tobler. Even though at ESRI I did not aim to do map projections, they needed someone to take care of writing the projection engine, and they asked me if I wanted to do that. I said "I can do this. I like map projections!"

KC: What do you think is the most complex thing for students to understand about map projections?

DB: Some of the mathematics tends to scare people, especially in this country. There tends to be a math phobia, people see lots of Greek letters in the papers, lots of trigonometric functions or integrals. I see it at work too—there are a lot of people who just don't want to deal with the hard math, the level of math, to really get inside a projection algorithm takes going to another level.

KC: What sort of math classes would you recommend students take in preparation if they're serious about GIS?

DB: I'd at least go through one or two courses in calculus, understand matrix algebra, and just know trigonometry cold. In GIS, I don't think you need to go into the highest levels of theoretical math. Having a good solid background—just knowing basic math through calculus—definitely opens up a lot of opportunities, and I think it's very important.

KC: What sort of projection-related mistakes do you often see in maps that you encounter as a professional?

DB: People don't tag their data correctly. They're, like, "I have this data that's in Lambert Conformal Conic." That's great, but which datum is it on, which geographic coordinate system, what are the parameters of the projection, I mean, tell me something about your data? I think a lot of times people wind up with data and don't know what they have. It's important enough that there's a whole session at the ESRI conference on trying to figure out your unknown coordinate system.

KC: You mentioned a projection that was the most complex you've programmed.

DB: That was the Hotine Oblique Mercator, yeah. The way some of the math showed up in one of the books where I got it from presented a challenge, because, whenever math for projections is written in books, they don't take into account all of the anomalous situations. Math equations aren't written for programming, so you have make sure it actually does what it says it will do, especially forward and inverse. Another interesting projection was a complex transverse Mercator which got into using complex numbers, you know, real and imaginary parts, and using complex hyperbolic trigonometric functions. You get into some really sick stuff, with elliptic integrals.

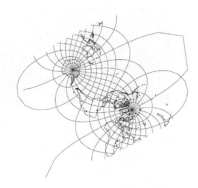

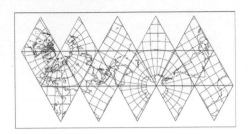

KC: Can you tell us just a couple of the things that you're doing with the latest version of the ArcGIS software?

DP: Well, we take the comments from our users, and we listen to what they say are missing. One of the things was geographic transforms, just a simple case of fixing the user interface to show users everything you can possibly show—which, for some people, might be overwhelming, but for people who really know what they're doing, they need to see it all. We're always adding new projections, new functions; we rewrote the way that images are projected now to be very quick so you can take a whole world image and throw it into these, like an interrupted Goodes Homolosine. We've increased the speed of raster projections immensely. We worked a lot on the accuracy and making sure we have a lot of really tight mathematical forwards and inverses to just nail it down so there are no anomalous situations.

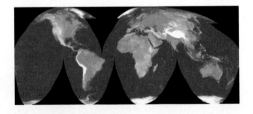

KC: What would you tell an undergraduate student about to take their first GIS class?

DP: Oh, it's a great discipline. There is so much you can do with it. It's technology that you can apply to almost any geographical problem. There's a really rich set of tools there to do so many things, but you have to understand the concepts before you can go and use the tools, so you still need a basic grounding in Geography. Some people get wrapped up in just knowing the technology itself—but for me, the fun has always been, knowing technology, to apply it towards really interesting problems. Geography is an interesting way to look at the world. It's a highly varied discipline, but, hey, it's provided me with a living, and I can say I honestly have enjoyed it all.

KC: Thank you very much.

Projection graphics from *A Gallery of Map Projections* by Paul B. Anderson (see: http://www.csiss.org/map-projections).

CHAPTER 3

Maps as Numbers

Yes raster is faster, but raster is vaster, and vector just seems more correcter.

C. Dana Tomlin

3.1 REPRESENTING MAPS AS NUMBERS

In an ideal world, a computer equipped with GIS would have a slot in the side, into which could be fed paper maps that would then appear as automatically georegistered data layers inside the software with all their attributes interpreted and stored, an analog-to-digital conversion. Unfortunately, this technical innovation is still some time away. In this and the next chapter we look at the various ways that maps can be represented using numbers. All GISs have to store digital maps somehow. As we will see, there are some critical differences in how the various types of GIS navigate on this ocean of geographic numbers. The organization of the map into digits has a major impact on how we capture, store, and use the map data in a GIS. In Chapter 4 we will see that an important first stage in working with GIS is just getting the map in the right form of numbers into the computer, and that often the means of data capture and storage determines a great deal about what we can and cannot do with the data later.

Obviously, there are many ways that the conversion of a visual or printed map to a set of digits can be done. Over the years, the designers of GIS packages have devised an amazing number of ways that maps can be converted into numbers. The differences among the ways are not trivial, not only because different types of files and codes are needed, but because the entire way that we think about the data in a GIS is affected. The link between how we imagine the features that we are working with in the GIS and the actual files of bytes and bits inside the computer is critical. To the computer, the data are stored in a physical structure that is tangible, at least to the GIS software. The physical structure is not only how computer storage or memory such as disk and RAM is used, but also how the files and directories store and access the map and attribute information.

On the physical level, the map, just like the attributes, is eventually broken down into a sequence of numbers, and these numbers are stored in the computer files. In general, two alternative ways exist of storing the numbers. In the first, each number is saved in the file encoded into binary digits or bits. A number in base 10, decimal, can be converted into base 2, binary, and the binary representation stored as on or off, 1 or 0, in the files on the computer. Eight bits in a row are termed a byte, and one byte can hold all numerical values from 0000 0000 to 1111 1111 binary or 0 to 255 decimal. Many computer programmers use shorthand to represent a byte, since in base 16 (hexadecimal) the contents of one byte can be represented as two hexadecimal digits. Hexadecimal runs out of counting digits at 9, so it fills the gaps with letters. Counting in hexadecimal then goes in the sequence 0,1,2,3,4,5,6,7,8,9,A,B,C,D,E,F. To a computer programmer, the range of numbers that can be stored in one byte goes from hexadecimal 00 to FF. Typically, as a GIS user you will never see hexadecimal values unless you crash the system or try to look at files stored in this binary format rather than the alternative.

The second way of encoding numbers into files is to treat each number the way that humans do—one decimal digit at a time. Coincidentally, this is the same way that we deal with text, punctuation, and so on, so this format is often called text or ASCII files. ASCII stands for the American Standard Code for Information Interchange, and the codes are 256 standard meanings for the values that fall into one byte. Some of the ASCII codes are numbers, some lowercase and some uppercase letters, some special characters such as "$" or ">", and some are actually keys to operations (such as the escape codes or tabs). Each fits into one byte, and we store in the file the numbers just as they come, even including indentation and spacing. These files are usually suitable for use with an editor or word processor, and they can be printed and read without a program (Figure 3.1).

It is the logical structure of the data that requires us to have a mental model of how the physical data represent a geographic feature, just as a sheet map is a flat paper symbol

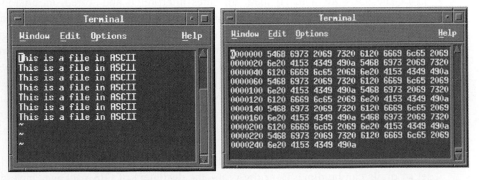

Figure 3.1: The same file listed at left as ASCII text, and at right in hexadecimal. The first column in the hexadecimal file is the line number.

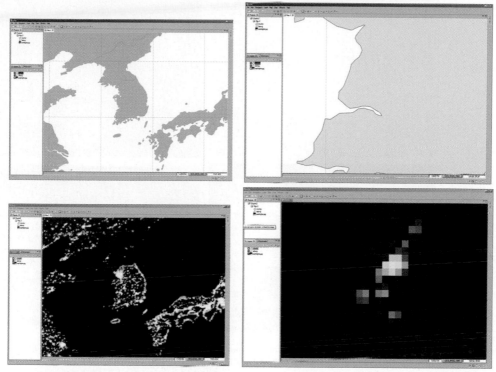

Figure 3.2: The Korean peninsula and city of Pyongyang seen at two zoom levels in both raster and vector modes. This example uses the uDig GIS software.

model of the landscape it covers. Traditionally in GIS and computer cartography, there were two basic types of data model for map data, and only one for attribute data. Map data could be structured in raster or vector format, and attributes as flat files. Take a look at the images in figure 3.2. These were generated by zooming in using the uDig open-source GIS on the Korean peninsula.

In one image and zoom, we see the vector shoreline of the world's nations. As we zoom in on the vector image, we see levels of detail not visible at the coarser scale. As we zoom in beyond the scale of the original map data (here, 1:1 million), however, we see that the coastline is a set of straight line segments connected together, like in a connect-the-dots diagram. In the other image we have a satellite image from the Defense Meteorological Satellite Program Operational Linescan System, processed to see the amount of nighttime light illumination. These data are aggregated into cells, each about 2.7 km in size on the ground. In the satellite image, we can see not only the clear division between North and South Korea, but also how "grainy" or pixelated the data become as we zoom in. In this case, the shoreline data are vector, and the satellite data are raster. Both have their advantages and disadvantages as a way of storing geographical information, as we will see in the following sections.

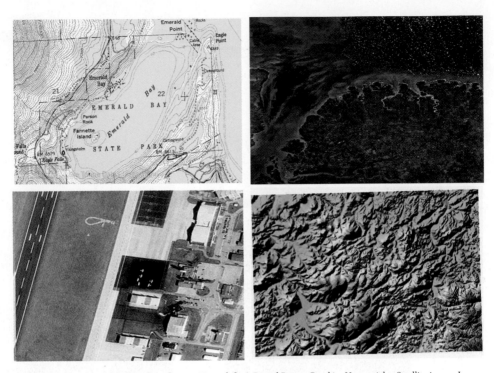

Figure 3.3: Examples of data in raster format. Upper left: A Digital Raster Graphic. Upper right: Satellite image. Lower left: Digital air photo. Lower right: A digital elevation model.

3.2 RASTER

Imagine that we could stake out the earth with tape and metal pegs so that a huge grid of square cells was formed on the ground such is often the case at archeological sites when excavating the ground. If we assigned data to the cells of this grid, then the map could be formed by representing the grid as an array of cells with their values on the map. A raster data model uses just such a grid, such as the grid formed on a map by the coordinate system, as its model or structure to hold the map data. Each grid cell in the grid is one map unit, often chosen so that each grid cell shows on the GIS map as one screen display point or pixel, or on the ground as a whole-number increment in the coordinate system. A pixel is the smallest unit displayable on a computer monitor. If you get a magnifying glass and look at a monitor or a television set, you will see that the picture is made from thousands of these tiny pixels, each made up of a triangle of three phosphor dots, one dot for red, one for green, and one for blue. When we capture a map into the raster data model, we have to assign a value to every cell in the grid. The value we assign can be the actual number from the map, such as the terrain elevation in a digital elevation model (DEM), or more usually, it is an index value standing for an attribute that is stored separately in the attribute database. Much raster data is collected by remote sensing, when an airborne instrument "scans" the ground in cells, or by digital cameras, which use pixels. A scanner can also capture an existing hardcopy map or image into a grid. Some examples of the most common GIS raster data are shown in figure 3.3.

Key elements in raster data are shown in Figure 3.4. First, the cell size determines the resolution of the data, and the cell size has both a map and a ground expression. We often talk about 30-meter Landsat data, for example, meaning that each cell in the data is 30 meters by 30 meters on the ground. On the map, we may use several pixels to display the grid cell, or on paper we may use a dot of a certain size in a given color. Second, the grid has an extent, often rectangular since a grid has columns and rows, and even if we do not wish to store data in the GIS for grid cells outside our region (such as a state), we still have to place something (usually a code for "outside") in the grid cells. Third, when we map features onto the grid there is sometimes an imperfect fit. Lines have uneven widths, points must be moved to the center or the intersection points of the grid, and areas may need to have their edges coded separately. We sometimes have to determine in advance what connections within the grid are legal. For example, taking a single cell, we can allow connections only north, south, east, and west, like the way a rook moves in chess, or we can allow diagonal connections as well. Which we choose can mean a great deal as to how the GIS works at storing and using the features. Fourth, when we deal with a grid, each grid cell can usually only be "owned" by one feature, that is, the one whose attribute it holds. In many cases, map data are not so simple. Soils, for example, are often listed by their percentage of sand, silt, and clay at every point.

Finally, when we have a grid, every cell in the grid has to be made big enough to hold the largest value of the attribute to be stored in the grid. You may have had the experience of using a spreadsheet to store people's names. Even though we can store "Jane Doe" with only eight characters, we still have to allow for the occasional very long name. Every grid cell pays the storage penalty of the extra space, and with the total number of cells being the product of the numbers of rows and columns, the amount of space needed can add up (or rather multiply up) quickly. Storage sizes for grids often increase by powers of 2 as more and more "bytes" of 8 bits are needed to store larger and larger values. The number of bits we need to store the largest value in a grid is called the bit depth. For example, we need 16 bits to store elevations, at least if we are using meters for heights.

Nevertheless, raster grids have many advantages. They are easy to understand, capable of rapid retrieval and analysis, and are easy to draw on the screen and on computer devices that display pixels. Mark Bosworth, GIS program supervisor for the Metro GIS

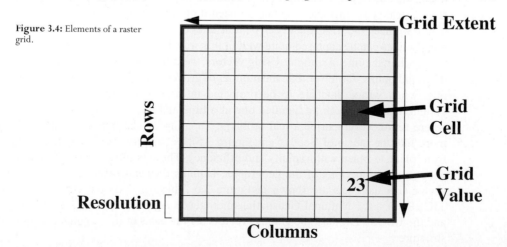

Figure 3.4: Elements of a raster grid.

Grid Extent

Grid Cell

Grid Value

Rows

Resolution

Columns

23

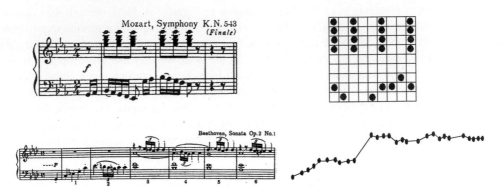

Figure 3.5: Data structures as music. *Rasters* are detailed, repetitive, highly structured, and elegant; slowly, dainty step by tiny step, they build the theme of the music into a glorious single structure, even though it may have "too many notes," like the music of Mozart. *Vectors* are bold, leaping streaks that go from place to place with rapidity and efficiency, like the music of Beethoven.

project in Portland, Oregon, has a background in music thinks of raster grids as being like the music of Mozart (Figure 3.5): detailed, repetitive, highly structured, and elegant. Slowly, dainty step by tiny step, they build the theme of the music into a glorious single structure, even though it may have "too many notes."

3.2.1   Vector

The other major type of data model for map data is the vector. The vector is composed of points, each one represented by an exact spatial coordinate. For a point or a set of points, vectors just use a list of coordinates. For a line, we use a sequence of coordinates; that is, the sequence of points in the list is the order in which they must be drawn on the map or used in calculations. Note that this gives lines a "direction" in which their points should be read. Areas in the vector model are the space enclosed by a surrounding ring of lines, either one or several of them.

Vectors are obviously very good at representing features that are shown on maps as lines, such as rivers, highways, and boundaries. Unlike the raster grid, where we have to store a grid cell's attribute whether we need it or not, we need only place points precisely where we need them. A square can be four lines connecting four points, for example. Even wiggly lines can be captured quite well in this way, by using more points for the bends and fewer when the line is straight. Using vectors, we can draw an outline map with only a few thousand points, far fewer than the number of grid cells that would be required. Better than this, when vector data are to be projected into another map projection or coordinate system, each point can be handled one at a time and with accuracy, unlike with a raster, where reprojection can be a real problem. To complete the musical analogy, vectors are more like the music of Beethoven (Figure 3.5). Vectors use bold, leaping streaks that go from place to place with rapidity and efficiency. There is little repetition, and the vectors get straight to the guts of the geographic features they represent. Some commonly used vector GIS data sets are shown in Figure 3.6. These include the U.S. Census Bureau's TIGER files, the National Geospatial-Intelligence Agency's VMAP0 world outline map, and the USGS's Digital Line graphs. Vectors are the basis of the popular arc/node GIS data structure.

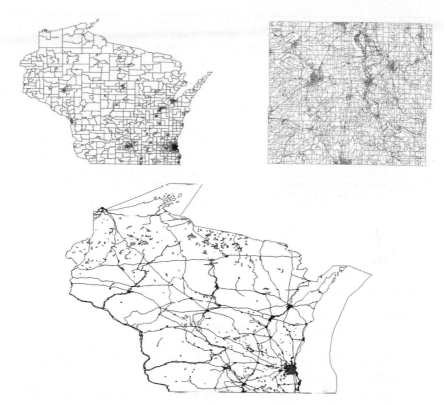

Figure 3.6: Some typical vector format data. Left top: Census Bureau's TIGER files (Wisconsin). Bottom: NGA's VMAP0. Above: USGS's DLG (Dodge Co., WI)

Vectors have the advantage of accuracy, since they can follow features very closely and so are efficient at storing features. They are very suitable for plotting devices that draw with pens or points of light, as the features can be drawn one at a time in their completeness. They can also be adjusted to store information about connectivity to other features, as we discuss under topology in Section 3.4. On the down side, the vector is not very good at representing continuous field variables such as topography, except in a way we consider later called the triangulated irregular network. Vectors are also not a good structure to use if the maps to be generated involve filling areas with shades or color.

3.3 STRUCTURING ATTRIBUTES

We have said little yet about the attribute data other than that its model is as a flat file. A flat file is how numbers are stored in tables or in a spreadsheet. The model is also a sort of grid, with rows for records and columns for attributes (Figure 3.7). A flat file doesn't really need a structure; it is simply a list of items, values for variables, written line-by-line as records or tuples into a file as text. Conceptually, it can be thought of as a table, with columns (and sometimes column headers) of attribute values and rows of records. There is no need for spatial sequencing, since the locations for objects are stored either as points in just two columns, or elsewhere. In its simplest form, a map could be stored as a flat file,

```
"City_fips","City_name","State_fips","State_name","State_city","Type","Capital","Elevation","Pop1
990","Households","Males","Females","White","Black","Ameri_es","Asian_pi","Other","Hispanic","Age
_under5","Age_5_17","Age_18_64","Age_65_up","Nevermarry","Married","Separated","Widowed","Divorce
d","Hsehld_1_m","Hsehld_1_f","Marhh_chd","Marhh_no_c","Mhh_child","Fhh_child","Hse_units","Vacant
","Owner_occ","Renter_occ","Median_val","Medianrent","Units_1det","Units_1att","Units2","Units3_9
","Units10_49","Units50_up","Mobilehome"
05280,Bellingham,53,Washington,5305280,city,N,99,52179,21189,24838,27341,48923,411,943,1453,449,1
256,2903,7101,34814,7361,16389,18950,663,3186,4576,2758,3937,3766,5237,300,1381,22114,925,10793,1
0396,89100,371,12808,368,1198,2267,3317,1229,73235050,Havre,30,Montana,3035050,city,N,2494,10201,
4027,4955,5246,9313,15,790,65,18,116,787,2017,5949,1448,1835,4401,116,714,728,497,702,1039,1076,6
8,347,4346,319,2362,1665,56000,242,2576,57,278,651,303,86,334
01990,Anacortes,53,Washington,5301990,city,N,99,11451,4669,5506,5945,10945,62,192,154,98,233,725,
1981,6276,2469,1420,5818,151,833,916,437,711,1032,1777,76,255,4992,323,3181,1488,85300,342,3724,1
21,134,380,353,0,200
47560,MountVernon,53,Washington,5347560,city,N,99,17647,6885,8459,9188,15809,78,200,245,1315,1921
,1526,3349,10322,2450,3163,7294,291,1025,1657,802,1182,1695,1772,136,583,7167,282,3914,2971,78500
,359,4138,154,248,791,1014,171,592
50360,OakHarbor,53,Washington,5350360,city,N,99,17176,5971,8532,8644,14562,757,153,1455,249,916,2
196,3582,10164,1234,1971,8481,178,538,845,421,577,2493,1580,85,388,6173,202,2379,3592,86500,411,3
315,301,177,1004,872,0,405
53380,Minot,38,NorthDakota,3853380,city,N,1580,34544,13965,16467,18077,33098,380,724,261,81,268,2
467,6276,20983,4818,7537,14938,289,2225,2215,1753,2545,3511,3735,152,1006,15040,1075,8406,5559,56
200,279,8308,400,893,1920,1901,276,1215
```

Figure 3.7: An example of a flat file with spatial content.

and, in the early days of computer mapping this was often done. Just like for the raster grids, we have to store values in the cells of the table (Figure 3.8).

As we have already noted, these values must somehow link the data in the flat file to the data in the map. For a raster grid, we could store index numbers in the grid and any number of attributes for the index numbers in the flat file. For example, on a land use map, 1 could stand for forest, 2 for farmland, and 3 for urban. For vector data, we need a little more complexity. Point data are simple; we can even put the coordinates in the flat file itself. Lines and areas, however, have variable numbers of points. We again need to number off or "identify" the lines and store an attribute for the whole line in the flat file.

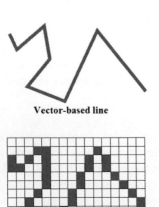

Vector-based line

Raster-based line

```
4753456   623412
4753436   623424
4753462   623478
4753432   623482
4753405   623429
4753401   623508
4753462   623555
4753398   623634
```

Flat file

```
0000000000000000
0001100000100000
1010100001010000
1100100001010000
0000100010001000
0000100010000100
0001000100000010
0010000100000001
0111001000000001
0000111000000000
0000000000000000
0000000000000000
```

Figure 3.8: Simple means of converting a graphics (line) file into a flat file.

We can do the same for an area, except that we need the line flat file as well to refer to in the polygon or area file. If we called the lines arcs, for example, we might need both a polygon attribute table file, and a file of arcs by polygon. In the arc/nodes structure, we can have a single file of polygons where one attribute set applies to one polygon, exactly what is needed for area-type data. Similarly, if the basic units are lines or points, the attributes can be stored with the features.

So far we have touched only briefly on the data models that GISs use. In the sections that follow, we first delve more closely into the way that attribute data are stored in files and how the tricks of data storage have improved over the last 40 years. Then, in the following sections we go into more detail and cover different logical and physical data models that GIS systems use to store the map data. These too have evolved over time and have improved significantly over the years. We will look at some of the formats used by GIS data providers and finish the chapter by raising some of the technical problems that have faced GIS users who need to move data between formats and between systems.

In Chapter 5, we consider in more detail the logical way that attribute data are stored in files. For now, a simple way to imagine this is to recall the flat file in Figure 3.7. A flat file is a table. Columns store attributes, and rows store records. We know in advance what sort of information is stored in each attribute, whether it is text or numbers, how large the numbers are, and so on. We can then write a sequence into the file. For each record we can write the ASCII codes for the values in each attribute (in database terms often called a field) in a consistent way. At the end of each record we could start a new line. The file then would be a sort of table or matrix with rows and columns.

It is now easy to see what some of the database operations actually do. For example, if we wished to sort the data we could renumber the lines in the file. If we wanted a particular record, we could search line by line until we found the correct one and then display it. These operations would be much faster if we could encode the numbers in binary or sort them in the file so that the most commonly referenced records were first in the file. Most database management systems (DBMSs) do exactly this, and some use very clever ways of placing records into files. If you use a database manager, or even a spreadsheet program on a personal computer (PC), you have probably noticed these files, which hold your records in one place. Increasingly, GIS software has interfaces with the common database management systems, and can use them even across Internet connections. A database that has different parts at different places across the Internet is called a distributed database. Many GIS packages also have software to manage distributed databases, often called server extensions, database engines, or data warehouses.

Another important part of structuring attributes is that the database dictionary, the list of all the attributes along with all their characteristics, must also be written into the files. This is sometimes done by using a separate file, but usually the dictionary is written into the top or header of the file, before the data begin. The advantage of this is that the types of data (e.g., text, integer or whole number, floating point number, etc.) can be known in advance and used to check the contents of the records. So the attribute database part of a GIS is fairly simple. At the most basic it consists of a file that can be logically thought of as a table. At the most complex, it might be several files in a directory. From a data management point of view, manipulating the attributes is a piece of cake. Unfortunately, the maps are a little harder to deal with.

3.4   STRUCTURING MAPS

Maps have at least two dimensions; in the earth's space they have latitude and longitude, and in the map's space they have the easting (x) and the northing (y) directions. What is stored are also scaled-down representations of features, features that can be points, lines, areas, or even volumes. Point features are very simple to deal with, and you could easily argue that you don't really even need a GIS for point features other than to draw them, since they can be contained in a list in a flat file. The x and y coordinates can be stored just as regular attributes in a standard database. Line and area features are more complicated because they can be different shapes and sizes, and therefore lengths in the records of a flat file. A stream and a road would be captured with different numbers of points, and these would not fit easily into the attribute database.

Nothing says that we have to capture points, however. In Section 3.1, we met the two basic models for map data, vector and raster. While a GIS can usually deal with data from either format, only one structure can be used for retrieval and analysis of the map when we want to do direct comparisons, such as overlays. The structure we choose can influence many GIS operations. The data structure also affects the type and amount of error involved in the process and the type of map we can use for display. It is worthwhile, then, to consider in turn each of the ways in which data can be structured in more detail.

3.4.1   Vector Data Structures

Vector data structures were the first to be used for computer cartography and GIS because they were simply derived from digitizing tablets, because they are more exact in representing complex features such as land parcels, and because they are easily drawn on pen-type output devices such as plotters. Surprisingly, few people in the early days thought of standardizing how digitizing a map was to take place, and since there were different technologies, many different formats evolved. The earliest included ASCII files of (x, y) coordinates, but these soon became very unwieldy in size, so binary files rapidly took over.

The first generation of vector files were simply lines, with arbitrary starting and ending points, which duplicated the way a cartographer would draw a map. Obviously the pen would be lifted from the paper to start a new line, but it could be lifted anywhere along a line. The file could consist of a few long lines, many short lines, or even a mix of the two. Typically, the files were written in binary or ASCII and used a flag or code coordinate to signify the end of a line. Early databases such as the CIA's World DataBank II were structured in this way.

As a computer programmer, having to follow the line from one place to another in the file was compared by early GIS programmer Nick Chrisman to following the path of a single strand of spaghetti through a pile of spaghetti on a plate. The name stuck, and to this day unstructured vector data are called cartographic spaghetti. A surprising number of systems still use this basic structure, however, and the structure survives in many data formats, such as the former Defense Mapping Agency's standard linear format. Most systems allow users to import data in this structure, but almost all now convert it to topological data after entry. An exception is ESRI's popular Shapefile format.

Just as the hierarchical system had caught on as a way of organizing attribute databases, starting in the 1960s a hierarchy for spatial data was worked out that became the arc/node model. Many first-generation systems, including POLYVRT, GIRAS, and ODYSSEY, used this system. This data structure uses the fact that each type of feature—

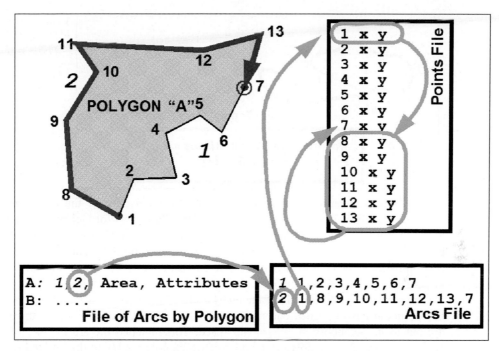

Figure 3.9: The Basic Arc/Node model. Points, lines, and polygons are stored in separate files, and linked together by cross references.

point, line, and area—consists of features with the next fewer dimensions. So area features consist of connected lines, and lines consist of connected points. The most important advantage that this buys us is that we can have separate files for areas, lines, and points. The price is that we need to keep track of links between the files in a fairly arbitrary way. For example, in Figure 3.9, a single polygon consists of two lines or arcs, each with a set of points. The points are ordered from node to node in a sequence along the arcs.

At the least, we need a file containing the attributes for the polygon, a file listing the arcs within the polygon, and, finally, a file of coordinates that are referenced by the arc file. Figure 3.9 shows that we need to store in each of these files a set of references between the files. For example, an entry in the arcs files states that to get the points for arc 2 from the points file they begin at the first coordinate point in the file, followed by the eighth, the ninth, and so on. Note that an "island," that is an isolated polygon within another polygon, needs to be stored as a polygon with only one arc, one that loops back on itself. Similarly, sometimes we have features (e.g., the state of Michigan with its upper and lower peninsulars), that require two polygons to be linked to a single attribute set, such as Michigan's population and per capita income. The arc/node structure has special means to deal with these problems, which are far more common than it first appears.

During the early days of GIS, several systems evolved different versions of this structure. Obviously, to save space the files were stored in binary. There are few ways, however, to store point, line, and area data that are as efficient. As long as the data are valid, this is a very powerful way to store data for map features. When the system breaks down, however, is when data contain errors, which is virtually always.

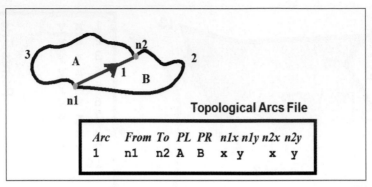

Figure 3.10: The basic arc, a line segment plus its topology.

A new generation of arc/node data structures arrived after the First International Advanced Study Symposium on Topological Data Structures for Geographic Information Systems, held in 1979. This elegant new structure used the arc as the basis for data storage and relied on reconstructing a polygon when it was needed. The way that this was accomplished turned out to have other practical benefits, as we discuss in Section 3.4. The system stored point data as before, but included in the file of arcs linked to the points file was an abbreviated "skeleton" of the arc (Figure 3.10). This consisted of just the first and last points in the arc, called end nodes, and information that related not to this particular arc but to its neighbors in geographic space. This included the arc number of the next connecting arc, and the polygon number of which polygon lay to the left and right of the arc. If the line was just a river or a road, this information was not essential. If, however, the arc was part of a network that formed enclosed areas or polygons, the polygon identifier number became the key to polygon construction.

The way that a polygon could be built was by extracting all of the arcs that a specific polygon had as a neighbor. If the polygon is the right-hand neighbor of each of the arcs, the end nodes can be tested against each other to see which sequence they should be drawn in. The use of the arc as the basic unit meant that when a map was digitized, the user only needed to trace each arc once, instead of twice if each area was traced around the edge. As exactly the same arc was used in both cases, the type of error known as a sliver was avoided completely.

One problem with the vector data structure was that it did not really deal very well with geographical surfaces such as topography or air temperature. This was corrected by a research team that devised a new data structure called the triangulated irregular network, or TIN (Figure 3.11). The TIN is really just a list of points with their coordinates; stored with the points is a file containing information about the topology of a network. The network is a set of triangles constructed by connecting the points in a network of triangles called a Delaunay triangulation. This way of drawing triangles is optimal, because changing any one triangle makes the angles within the triangle less similar to each other.

Two sorts of TIN can be built, one with a file containing information about the arcs that connect points, and one containing all the data about one triangle. The TIN became popular as a way of storing topography or land elevation data for visualization and engineering. With a TIN it is easy to draw contours, make a three-dimensional view of an area, estimate how water would run downhill over a digital landscape, or calculate how much material would have to be moved in a construction project. Many GIS programs that work with computer-aided drafting (CAD) systems or with surveying software use TIN as

Figure 3.11: TIN triangles for terrain surface covering part of the coast of New Zealand's South Island. Original data is a DEM at 90m from the SRTM. Left: Original DEM. Right: TIN. Note the increased number of triangles in the more detailed terrain areas.

their data structure, as do video games that require storing vast virtual terrains and landscapes. The TIN has proven to be both efficient in storing data and versatile in finding new uses within GIS.

3.4.2   Raster Data Structures

Raster or grid data structures have formed the basis for many GIS packages. The grid is a surprisingly versatile way of storing data. The data form an array or matrix of rows and columns. Each pixel or grid cell contains either a data value for an attribute or an index number that points to a reference in the attribute database. So a pixel containing the number 42, for example, could correspond with the number 42 or "deciduous forest" in the Anderson Level II land use/cover system, or just the 42nd record in the attribute file.

To write the numbers to a file, we can just start the file with any necessary attribute codes, perhaps the number of rows and columns and the maximum size of one value, and then write the data into the file in binary across all columns for all rows, one long stream of data with a start and an end, like an unraveled sweater. When reading the data back in, we just place the data back into a raster grid of the correct dimensions.

A major advantage of the raster system is that the data form their own map in the computer's memory. An operation such as comparing a grid cell with its neighbors can be performed by looking at the values in the next and preceding row and column of the grid cells in question. However, the raster is not very good at representing lines or points, since each becomes a set of cells in the grid. Lines can become disconnected or "fat" if they cross the grid at too shallow an angle. However, continuous surface variables that needed the TIN in vector data structures fit easily into the raster. This is particularly suited to data that come from remote sensing or scanning.

One major problem with raster data is the mixed pixel problem. Figure 3.12 provides an example. The photographs show a part of the outline of a lake in an oblique view. In the lower left cell, there is only one type of land cover, "bare ground," so the pixel clearly belongs in one class. In the cell above it, there are two types of attributes, bare ground and vegetation. Even though a wet foot is a sure indicator of a water grid cell, it is hard to assign each pixel to one or the other category. A compromise, and one often used in GIS, is to assign edge pixels, those that are not exclusively in one class or another. Finally, when multiple classes are involved, we either have to make up rules for assignment, such as assigning a mixed pixel to the class that occupies the most area within it, or we have to put up with edges and mixed pixels. Even when the boundary is absolutely clear, a vector data representation may be better than the raster.

At least three ways have been devised within GISs to deal with the problem that a grid often contains redundant or missing data. In Figure 3.13, for example, showing NOAA's MMAB Sea Ice Analysis for February 2009, there are areas with no sea ice at all, such as the land, and areas where no data on sea ice were available (shown in pink), or the ice was obscured by cloud.

Figure 3.12: The mixed pixel problem. Domination can be applied in sequence, the area covering most of the pixel assigned the pixel, or edges can be assigned a separate value. Any of the three options is acceptable, or any other, as long as the rule is applied consistently. (Water=Cyan, Vegetation=Green, Bare Ground=Black).

Figure 3.13: Example of a map with missing and redundant data. NOAA's MMAB Sea Ice Analysis for February 2009

The first of these is a compression mechanism called run-length encoding. Along each row, only changes between attributes and the numbers of pixels of that same attribute are stored. If a whole row is all one class, it is stored as the class and the number of pixels only, quite a saving in space. As the raster becomes more and more varied, however, fewer and fewer savings can be made over the original grid. Many GIS packages and many industry-standard image formats use run-length encoding.

A second way of saving space on raster grids is to use Range or R-tree encoding. In an R tree, the bounding boxes of features in the grid are stored, that is, the rectangles that contain all of the feature and are aligned with the grid. The features can then be indexed, and stored separately if necessary. A search can then check each bounding box to see whether that part of the grid needs to be searched at all, or can just be skipped. This can save a considerable amount of time when a grid is being processed. An example of a range tree applied to a land use map is shown in Figure 3.14. The R-tree is a hierarchical index based on the inclusion of minimum bounding rectangles within the map as a whole. Each rectangle level stores only those sub-rectangles (and points) that are exclusively contained within it. Each completely non-overlapping set of bounding boxes can be stored separately, with all of the overlapping rectangles within it. The polygons can be ordered, sorted, searched-for, etc. without needing to read the polygon boundary points.

Another way to save space is to use a data structure called a quad tree. A quad tree works by dividing a grid into four quadrants, saving a reference to the quadrant of the grid only if it contains data. Then the quadrant is split into four half-size quadrants, and so on until the individual pixel is reached. If we split a quadrant and all the pixels in it have the same attribute, then obviously we would not need to store anything else for that entire

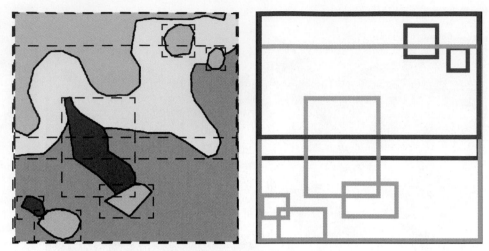

Figure 3.14: Illustration of an R-tree. The map polygons on the left have their minimum bounding rectangles computed (dashed lines). These are then grouped so that each box is stored with those others that are wholly enclosed, with the overlaps being assigned to the smallest or closest box. Colors on right are one possible R-tree ordering. Note that the method works for points, lines, or polygons.

quadrant. Each attribute becomes a list of quadrants needed to get to the area, like a coordinate reference (Figure 3.15). Quad trees have been used more in image processing than in GIS; they are used in packages such as Oracle Spatial. A variant on a quad tree is the image pyramid. In this structure, the image is successively divided into smaller and smaller sub-grids, just as in a quad tree, except that the average of the values is stored in place of all the individual grid values. When a map is zoomed out, for example, it is not necessary to display all pixels and an average is fine. When the user zooms, the higher level data can replace the averages. In the image pyramid, the image is preprocessed to generate all the levels at once, and the result stored for use when necessary.

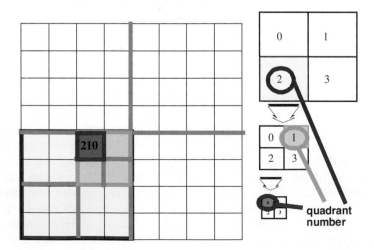

Figure 3.15: The quad-tree structure. Reference to code 210.

3.5 WHY TOPOLOGY MATTERS

When topological data structures became widespread in GIS, some significant benefits resulted, enough that today the vector arc/node data structure with topology probably is the most widespread for GIS data. Typically, a GIS maintains the arc as the basic unit, storing with it the polygon left and right, the forward and reverse arc linkages, and the arc end nodes for testing. This means that each line is stored only once and that the only duplication is the endpoints. The disadvantage is that whenever areas or polygons are to be used, some recomputing is necessary. Most programs save the result, however, such as the computed polygon areas, so that recalculation is unnecessary.

Topology allowed GIS for the first time to do automatic error detection. If a set of polygons is fully connected, and there are no gaps at nodes or breaks in the lines defining the areas, the set of areas is called topologically clean. When maps are first digitized, or converted from scanned maps, however, this is rarely the case. The topology can be used to check the polygons. Polygon interiors are usually identified by digitizing a point inside a polygon, a label point, and by keeping track of the arcs as they are entered. A polygon gets the label from the label point when the point is found to be inside the polygon. A GIS will have the ability to build the topology from the unconnected arcs. First, each endpoint is examined to see if it is "close" to another. If it is, the points are "snapped" together; that is, their (x, y) coordinates are averaged and each is replaced with exactly the same values (Figure 3.16).

In addition, arcs that connect the same nodes are tested to see if they are duplicates, and the user is asked which to delete. Any small areas, probably the result of errors called slivers, caused by double digitizing, are also eliminated. So, an error automatically

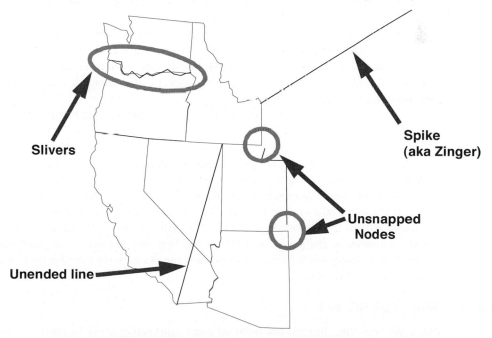

Figure 3.16: Examples of slivers (unmatched nodes along two lines), unsnapped nodes (endpoints of two lines that should be the same point), spikes (erroneous coordinates), and unended lines.

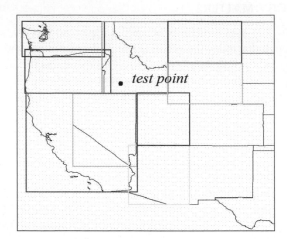

test point

Figure 3.17: Bounding rectangles: rectangles that contain a polygon completely. If a test point lies wholly outside the bounding rectangle, it must be outside the enclosed polygon.

detected can become an error automatically eliminated. Obviously, the separation between nodes required before they are treated separately and the size that a polygon must reach before it is retained are critical values for the map. Sometimes called fuzzy tolerances, these values should be handled with caution, because they allow the map's features to move around. Short lines, small polygons, or precisely measured point locations have critical significance and should not be deleted by automatic testing for topological completeness. For example, a map of Europe should include Andorra, Monaco, and Liechtenstein.

The primary advantage of having a topologically consistent map is that when two or more maps must be overlain, much of the initial preparation work has been done. What still has to be established are where new points must be added along lines to become nodes, and how to deal with any small or sliver polygons that are created (Figure 3.16). The latter can be a real problem. Many borders between regions, states, counties, and so on match along lines such as rivers, which are generalized differently at different map scales. Although the line should be the same, in fact it is not. Some packages allow the extraction of a line from one map to be "frozen" for use on another. This seemingly small difference can be very significant, especially if areas or densities are being calculated.

The final advantage to topology is that many operations of retrieval and analysis can be conducted without having to continuously deal with the (*x*, *y*) data. In some cases, pretesting can be done. For example, say that a point is to be tested to see if it falls inside an area. If the point falls outside the bounding rectangle of all the endpoints, the point is most likely outside the region (Figure 3.17). So useful are these bounding rectangles, the highest and lowest x and y values along an arc, that they are often computed once and saved with the topological information in the arcs file. As we have seen, for raster data they can be the basis of R-trees.

3.6 FORMATS FOR GIS DATA

With a decades-long history and with so many alternative ways to structure map and attribute data, it is hardly surprising that most GISs use radically different approaches to handling their content. The data structures used are often invisible as far as the GIS user is concerned. We might not even need to understand exactly what is happening when two maps are overlain. However, if we are to be objective, scientific GIS users, at the very least we must have a full understanding of the errors and transformations involved. Regardless of how a GIS structures its maps as numbers, it must be able to import data from other GIS packages and from the most common data sources, as well as scanned and digitized data, and to convert the result into its own internal format. In some cases this is an open process. Most GIS companies have published and documented their internal or exchange data formats, including Intergraph and Autodesk. Others protect their internal data as a trade secret, in the hope of being able to sell data and data converters as well as their GIS. Increasingly, GIS software is moving toward open standards for data, allowing easy data integration. This is an important trend since most data are now acquired via the World Wide Web.

The most common data formats for GIS data have been used by so many GIS operations and for so much existing data that a GIS user ignores them at their peril. Some are so common that utility programs and even operating systems read, process, and display these formats automatically. These formats include some that have arisen because they are a common data format, such as TIGER and DLG. Others are industry-standard formats, proprietary formats that have been used so much that they are documented and published, although their use may have restrictions.

In the GIS world, a small subset of these formats has become commonplace, and we cover them here for completeness. We then finish the chapter by discussing some of the issues of data exchange between GIS systems and take a look at the issue of data interoperability.

3.6.1 Vector Data Formats

A general distinction between industry and commonly used standards for GIS data is that between formats that preserve and use the actual ground coordinates of the data and those that use an alternative page coordinate description of the map. The latter are the coordinates used when a map is being drafted for display in a computer mapping program or in the data display module of a GIS (Figure 3.18).

The Hewlett-Packard Graphics Language (HPGL) is a page description language designed for use with plotters and printers. The format is simple and the files are plain ASCII text. Each line of the file contains one move command, so a line segment connects two successive lines or points. The format works with a minimum of header information, so that files can be written or edited easily. However, the header can be manipulated to change the scaling, size, colors, and so on. The HPGL is an unstructured format and does not store or use topology.

Another industry-standard format is the PostScript page definition language, developed by the Adobe Corporation for use in its desktop and professional publishing products and now in widespread use. So common is this format that most laser-quality printers use it as the printer device control format. PostScript, at least in its vector mode, is a page description language. This means that its coordinates are given with respect to a printed page, say an 8-1/2-by-11-inch sheet.

PostScript uses ASCII files but has particularly complex headers controlling a very large number of functions, such as fonts, patterns, and scaling. PostScript is really a sort of programming language, and to be viewed it must be interpreted. There are many commercial and shareware packages for viewing PostScript files, and many word processors and graphics packages will both read and write the format, although many more write it than read it. In GIS, PostScript is usually used to export or print a finished map rather than data as such.

```
IN;IP0 0 8636 11176;
SC-4317 4317 -5586 5586;
SP1;
SC-4249 4249 -5498 5498;

SP1;
PU-2743 847;PD -2743 3132 608 3132
608 847 -2743 847;
```
HPGL

```
%%BeginSetup
11.4737 setmiterlimit
1.00 setflat
/$fst 128 def

%%EndSetup
@sv
/$ctm matrix currentmatrix def
@sv
%%Note: Object
108.58 456.98 349.85 621.50 @E
0 J 0 j [] 0 d 0 R 0 @G
0.00 0.00 0.00 1.00 K
0 0.22 0.22 0.00 @w
 0 O 0 @g
0.00 0.00 0.00 0.00 k
%%RECT 241.272 -164.520 0.000
108.58 621.50 m
349.85 621.50 L
349.85 456.98 L
108.58 456.98 L
108.58 621.50 L
@c
B
@rs
@rs
%%Trailer
end
```
PostScript

```
POLYLINE
  8
7
  6
CONTINUOUS
 66
  1
  0
VERTEX
  8
7
  10
-2.742
  20
3.132
  0
VERTEX
  8
7
  10
0.608
  20
3.132
  0
VERTEX
  8
7
  10
0.608
  20
0.848
  0
VERTEX
  8
7
  10
-2.742
  20
0.848
  0
VERTEX
  8
7
  10
-2.742
  20
3.132
  0
SEQEND
  0
ENDSEC
  0
EOF
```
AutoCAD DXF

Figure 3.18: Some alternative industry-standard vector formats. Headers have been removed. Graphic is the same four-point rectangle in each case.

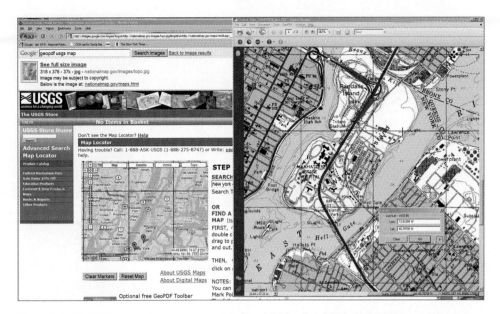

Figure 3.19: The USGS topographic map e-store, with a Google Maps search window for GeoPDF format files. When displayed with the TerraGo GeoPDF desktop, many basic GIS functions can be used on the map. Note that coordinates can be displayed for points.

Another common Adobe document-based format is the Portable Document Format, or PDF. Many print documents, tickets, programs, and itineraries delivered over the Internet use this format, which includes graphical objects. A proprietary enhancement to this format is the GeoPDF format, created by Adobe's partner company TerraGo Technologies. This format has been adopted for distribution of maps under the National Geospatial-Intelligence Agency's eChart program via the US Army Corps of Engineers. The principal difference is that the base reference information on the digital map can use a cartesian coordinate system, which can be transferred to the GIS at the receiving end. This means that the maps are simultaneously paper maps, but also can function as GIS data sources (Figure 3.19). The TerraGo GeoPDF Desktop even embeds some limited GIS capability into the data viewer.

The popular CAD package AutoCAD by Autodesk has made commonplace the AutoCAD digital exchange format (DXF) for drawing data. The AutoCAD Map GIS software also makes use of these formats. While Autocad uses its own internal format for data storage, it uses DXF for transfer of the files between computers and between packages. Again, these are simple ASCII files (although there is a binary mode), but in the DXF case there is a very large and mandatory file header containing significant amounts of metadata and file default information.

Although DXF does not support topology, it does allow the user to maintain information in separate layers, a familiar GIS concept. There is considerable support for details of drawings, line widths and styles, colors, and text, for example. DXF is importable by almost all GIS packages other than those that use raster formats. Some GIS packages work directly with the Autocad or other CAD software and can manage these files internally.

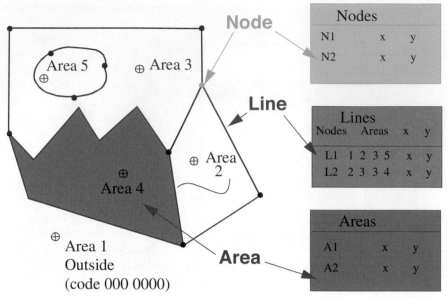

Figure 3.20: Sample digital line graph coding format.

Two formats have become widespread largely because so many important data have been made available using them: the DLG and the TIGER formats. The digital line graph (DLG) format of the US Geological Survey's (USGS) National Mapping Division. DLG files are available at two scales, the scales of the map series from which they were captured (Figure 3.20). These scales are 1:100,000, for which almost all of the country has some data available, and 1:24,000, with only a small portion of the country but in extreme detail.

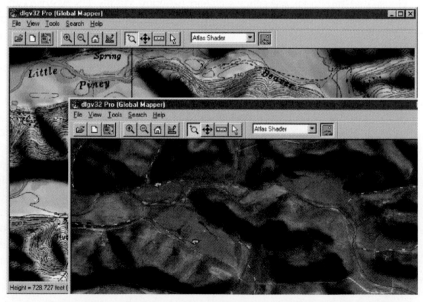

Figure 3.21: Sample DLG obtained from the USGS and displayed with the dlgv32 DLG viewer software.

The formats of the data are documented formally, and the files are ASCII. They use the ground coordinates in UTM, truncated to the nearest 10 meters to reflect their locational precision and to save space. Features are handled in separate files—for example hydrology, hypsography (contours and topographic features), transportation, and political boundaries. Many GIS packages will import these files, but often some extra data manipulation is necessary, such as requiring records to have some fixed line length in bytes (Figure 3.21).

The TIGER formats are from the enumeration maps of the U.S. Census Bureau, as used for the decennial census (Figure 3.22). They are vector files and contain topology; in fact, their predecessor, the GBF/DIME files, were highly instrumental in popularizing topological data structures. They consist of an arc/node type arrangement, with separate files for points, lines, and areas linked together by cross references. The TIGER terminology calls points zero cells, lines one cells, and areas two cells. The cross indexing means that some features can be encoded as landmarks, and these include rivers, roads, permanent buildings, and so forth, which allow GIS layers to be tied together.

The TIGER files exist for the entire United States, including Puerto Rico, the Virgin Islands, and Guam. They are block level maps of every village, town, and city, and include geocoded block faces with address ranges of street numbers (Figure 3.23). This means that the address matching function is possible, and a large proportion of GIS use depends highly upon this capability, including map applications on the Internet such as Google Maps. The data are also obviously referenced to the U.S. Census, so that many population, ethnicity, housing, economic, and other data can also be used with TIGER. These data are already in spreadsheet tables, and can be added as attributes to the features fairly simply, allowing maps of the census data to be made.

Furthermore, the whole country is available at minimal cost and over the Internet. Most GIS vendors, and some independent data suppliers, offer updated and enhanced TIGER files as their own products. Although topologically correct, TIGER has been criticized as not being particularly geographically correct. More recent GIS functions have made the addition of higher levels of geographic accuracy possible, and many data

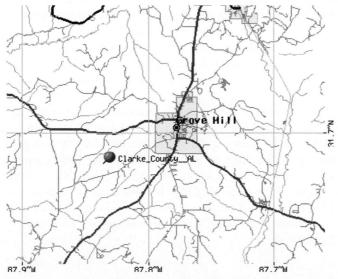

Figure 3.22: 1999 census TIGER files plotted for part of Clarke Co., Alabama.

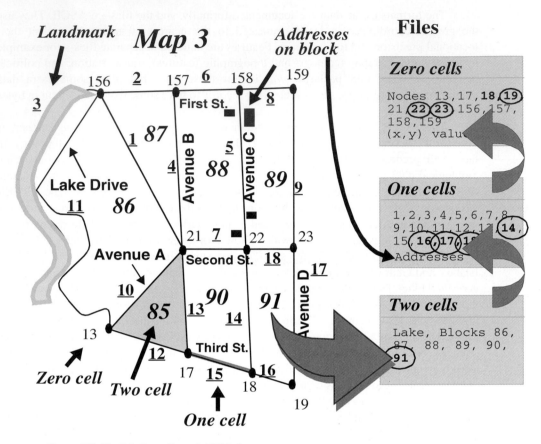

Figure 3.23: The U.S. Census Bureau's TIGER data structure.

suppliers have enhanced the TIGER files. Some of the 2010 census additions are discussed in the *People in GIS* section at the end of this chapter. Lastly, a suite of vector formats has been created due to the popularity of software called geobrowsers that serve GIS-like data through client software such as Worldwind, ArcExplorer, GoogleEarth, and Microsoft's Bing Maps. The vector format used by Google, the Keyhole Markup Language (KML), allows users to easily store points, lines, and areas, and to register raster images into the Google applications interface.

A typical KML file is shown in Figure 3.24. Another more standardized format is an extension of the Extensible Markup Language (XML) called GML, for the Geography Markup Language. This allows place coordinates to be embedded in web pages, so that mapping applications can sort web data by location and interface with mapping programs like Google Maps, and MapQuest. Yet another standard format is SVG, for scalable vector graphics, which is structured the same as KML, but works in all web browsers except Microsoft's Internet Explorer. This XML language allows the definition of vectors in web pages that redraw at increased resolution when the user zooms, avoiding many of the disadvantages of other formats.

```
<?xml version="1.0" encoding="UTF-8"?>
<kml xmlns="http://www.opengis.net/kml/2.2">
<Placemark>
  <name>New York City</name>
  <description>New York City</description>
  <Point>
   <coordinates>-74.006393,40.714172,0</coordinates>
  </Point>
</Placemark>
</kml>
```

Figure 3.24: An example of the Keyhole Markup Language (source: Wikipedia).

3.6.2 Raster Data Formats

Raster data formats have been much more widely used than vector, especially since the arrival of networking, because many of them are the same formats that are used to store digital images and pictures for use on the Web. Image formats are particularly simple to create, and as a result there are many. Some of the formats have been optimized for passing the images through networks, and it is these formats that are now most common. Raster formats are mostly similar in how the files are arranged. Few of the formats use ASCII; most use binary. Usually the file structure is a header with a fixed length and a keyword or "magic number" to identify the format. Included in the header is at least the length of one record in bits (called the image depth), the number of rows, and the number of columns in the file.

Optionally, the file contains a color table. The color table allows the data file to consist of indices, say the numbers 1, 2, 3, 4, and so forth. Each index corresponds to a value, usually three values in each of three bytes as intensities between 0 and 255 in the red, green, and blue (RGB), respectively. For example, if an image file were to consist of only four colors, red, green, blue, and white, there is no need to store values in each pixel like (255,255,255) for white's RGB value. Instead, the color table could assign white to 0, red to 1, green to 2, and blue to 3. Now no pixel need be more than two bits in size (2 bits can store decimal 0 through 3), as opposed to 24 bits for the full colors. This saves a significant amount of space, since the record size is multiplied up for all rows times all columns.

The final part of the image file is the data, usually all columns for each row. Some formats are padded at the end so that the total number of bytes is a multiple of a key factor, such as 512 bytes. For many GIS files, we do not want color but want to store single-band files, using all of the bits for data. Elevations, for example, which may be in the thousands of feet or meters, need more than one byte per pixel. Alternatively, if the pixel data are simple attributes such as land cover type with only a few categories, then the same value can be assigned to each of the red, green, and blue values if the format requires it.

A surprising number of utility programs exist to convert between raster formats. Among them are Image Alchemy and xv. Many packages also read and write a huge number of formats. Some will convert from raster to vector and vice versa. In some cases, this capability is included within the GIS package.

Some common raster formats are the Tagged Interchange format (TIF), which can use run length and other image compression schemes and has a number of different forms that are publicly available; the Graphics Interchange Format (GIF), popularized by the online network services CompuServe, but now an open format, which uses a quite sophisticated compression scheme on the data part of the image; and the Joint Photographic Experts Group (JPEG) format, which uses a variable resolution compression system offering both partial and full resolution recovery depending on the space available, as also does the common Internet image format PNG, or Portable Network Graphics. Of these, only TIF supports geodata, through the popular GeoTIFF formats, that include a second WRL file that contains coordinates for the TIF file. The USGS's scanned equivalents of the 1:24,000 series maps for the US are available in this format. Many GIS packages can import GeoTIFF files with ease, and recover the georegistration from the WRL data.

Finally, the PostScript industry standard includes a specification for images called encapsulated PostScript. This is a very simple image format indeed, which literally encodes the hexadecimal values to be placed into the image inside an image macro within a regular PostScript file. Many PostScript devices and programs will not handle this format, but those with higher amounts of memory and more advanced capabilities can. This is far better for storing images than maps for GIS use, and it is rarely used to store data; rather, it is used to drive printers and plotters to generate GIS maps.

One raster format has gained widespread acceptance and is often read directly by GIS packages or stand-alone utilities that come with the software. This is the Digital

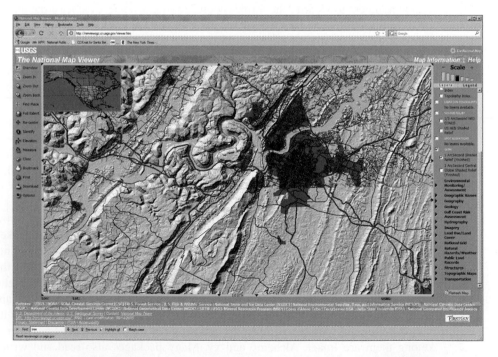

Figure 3.25: National Map Viewer scene of Chattanooga, TN. Shaded relief is from the 30m Shuttle Radar Topographic Mapping data (SRTM). Right hand menu shows the DEM format data layers available for the USA. SRTM data cover much of the world at the coarser 3-arc second or 90m resolution.

Elevation Model format of the USGS. This format is one in which two types of data are distributed, the 30-meter elevation data from the 1:24,000, 7.5-minute quadrangle maps, and the 1:250,000, 3-arc second digital terrain data originally supplied by the Defense Mapping Agency but now distributed by the USGS. These are documented formats and are somewhat complex because of the map projections involved. All DEM data for the USA were mosaiced into a seamless 30m coverage called the National Elevation Database (NED). The National Map Viewer makes available DEM data from the NED, and from other sources, such as high resolution LiDAR data for some states (Figure 3.25). Other DEM files, such as the Shuttle Radar Topographic Mapping Mission data, are available in this format, at 30m for the US and 90m for much of the world land surface.

3.7   EXCHANGING DATA

Exchanging data can be thought of in two ways. First, as we have seen in this chapter, the vector and the raster formats often store similar GIS data in very different ways. The GIS software adopts one of two strategies for dealing with the two types of data. Some systems use only one format exclusively and provide utilities or import options to bring in and convert the data to the format to be used. Raster-based GIS programs especially use this approach. Alternatively, the system can support the native format of each type of data, and it can require the GIS operator to explicitly change formats when operations requiring communality of formats are executed. In both cases, a computer program, part of the GIS, either performs a raster-to-vector or a vector-to-raster conversion. While a full discussion of how these operations work is beyond the scope of this book, it should be clear that going from vector to raster, filling in grid cells as lines cross them or as polygons include them, is relatively simple. The opposite is quite complex (Clarke, 1995).

Raster format data are often output from scanners, when the GIS requires vector data. In many cases, a special suite of software or even a special-purpose computer is used. When done by the GIS, especially for a large data set and for fine resolutions, the process can be very time-consuming. The program must try to follow each line from pixel to pixel, figure out where the end nodes are, and generate a vector equivalent of the line. Often the lines are jagged from the raster effect and must be smoothed. If the lines are too fat, they sometimes must be thinned first, and this can generate false connections and loops in the lines (Figure 3.26).

The second way to envision data exchange is to consider the issue of transferring data not between formats, but between entirely different computer systems potentially using different GIS packages. This situation is quite ordinary. Different local authorities

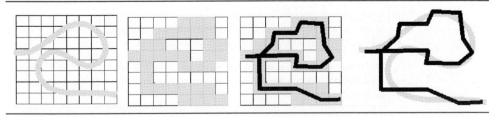

Figure 3.26: Errors caused by converting between vector and raster data formats. The original cyan line is superimposed on the vectorization of the raster equivalent. Major topological errors can result.

may have developed their GIS operations around different software. Different projects may have delivered GIS data in a huge variety of formats. There is also a real need for data exchange. At state or local boundaries, for example, data should be able to be matched for continuity across the borders, just as the data should match from map sheet to map sheet.

The history of GIS has ensured that this commonality and sharing rarely have taken place. Even state and city GIS efforts have often had contradictory or even competing GIS data, and more than a few projects have found it easier to start digitizing and data assembly all over again rather than convert GIS data from an exchange-unfriendly data set. Added to this has been the marketing philosophy of many GIS vendors, which has left data formats as proprietary in nature, undocumented and therefore unable to be used as import/export modules for other packages. In the past, data exchange could only be characterized as haphazard and chaotic.

Not exchanging or reusing data between projects, especially within a single organization, is a good example of duplication and waste. There is a real advantage in starting from a single, standard data set for all development or enhancement, especially when GIS operations will eventually bring the data back together for analysis and display. Some industry standards have been quite useful for data exchange, as we saw in Section 3.5. However, two aspects remain a problem. First, none of the industry standards exchange topology with the data, transferring instead only the graphic information. Second, with many different formats, each package has to include a large number of format translators.

A parallel exists between GIS data formats and spoken languages (Figure 3.27). We can get by in isolation by knowing English alone, or perhaps we learn a little French. If we wish to speak to someone who speaks only Russian, however, we need someone who speaks either English and Russian or French and Russian. In the latter case, as I speak my words in French, they arrive at the destination having moved twice between languages. Anyone who has played the children's game "telephone" knows the outcome of this process. From a GIS context, we finish with data of unknown accuracy, source, projection,

---

We can get by in isolation by knowing English alone, or perhaps we learn a little French. If we wish to speak to someone who speaks only Russian, however, we need someone who speaks either English and Russian or French and Russian

Мы можем обойтись в отрыве зная английский в одиночку, или, возможно, мы узнаем немного французский. Если мы хотим говорить товарищу которые говорят только в России, однако, мы нуждаемся которые кто-либо говорит на английском и русском или французском и русском языках.

Nous pouvons le faire sans connaître l'anglais, ou peut-être nous en apprendre un peu français. Si nous voulons parler camarade qui parlent seulement en Russie, cependant, nous avons besoin de quelqu'un qui parle l'anglais et le Russe ou le français et les langues Russe.

We can do it without knowing English, or maybe we learn a little french. If we want to talk comrade who speak only in Russia, however, we need someone who speaks English and Russian or french and Russian languages.

Figure 3.27: An example of multiple translations. Two sentences from the previous paragraph filtered through Google Translate. English to Russian, Russian to French, then French to English.

and with unspecified or imperfectly matched attributes. On one county map, perhaps, all streams and rivers are shown, on the other only the major ones. One might think that the climate was wetter, but in fact the difference is one of interpretation and lack of standardization.

The GIS industry in the United States began a standardization effort in the mid-1980s, which led to a federal standard approved in 1992. This standard, the Federal Information Processing Standard 173, called the Spatial Data Transfer Standard (SDTS), had to be quite broad because of the huge degree of complexity involved. Not only did the standard have to produce a bibliography, a terminology, and a complete list of geographic and map features, it also had to address the problems of data accuracy and the broader metadata issues for data description. The terminology created sets of terms for features and data structures that have become commonplace.

Best of all, the standard included a mechanism for file exchange. As a result, different implementations of the standard, called profiles, have been developed, for vector, raster, and point data. Data sets have been made available in the vector profile of the standard for DLG and for TIGER data. These have been termed DLG-SDTS and TIGER-SDTS. At the same time, many GIS vendors have incorporated input and output utilities that read and write the data in SDTS format and to SDTS specifications. Back to the spoken language analogy, I have convinced the Russian to learn English, and with luck most of the rest of the world too! The cost was many hours spent debating exactly what was meant by each and every word that was to be used in discussion in everyone's language. Still managed by the Federal Geographic Data Committee, the standards have been extended to include many aspects of geospatial data, including metadata.

The United States civilian mapping agencies have not been alone in seeking standardization of GIS information for data exchange. Within the members of NATO, an exchange standard called DIGEST was developed, with a vector data profile called the Vector Product Format (VPF). This format is best known as the format in which the Digital Chart of the World was released on CD-ROM. Similar efforts have standardized data exchange in Germany, Australia, South Africa, the European Union, and for worldwide nautical chart data by the International Hydrographic Organization (DX-90). As data become used for global instead of local and national projects, the ability to exchange data internationally will increase in importance. International peacekeeping and disaster relief efforts, for example, involve cooperation and therefore exchange of GIS data among many quite different countries and organizations (Moellering and Hogan, 1997; Moellering, 2005).

As the diverse world of proprietary data formats proliferated with the growth of GIS, government and industry sought a solution to the problem of "interoperability," the ability of data to move with no modification among GIS systems and applications. One result was the Open Geospatial Consortium, Inc. (OGC), an international industry consortium of several hundred companies, government agencies and universities participating in a consensus process to develop publicly available interface specifications. OpenGIS Specifications provide interoperable solutions for the Internet and location-based services. The overall standards are created by working groups, and past efforts have included web services, emergency mapping, and geographical objects. Side benefits are a large number of open source and freely distributed tools and plug-ins that have greatly helped in making GIS data more ubiqitous, and minimized the problem of "lost in translation."

We will return to this issue in Chapter 10, when we discuss the future of GIS. Obviously, the open exchange of data can help those developing and using GIS considerably. The SDTS took many years to develop and met with considerable resistance. Nevertheless, the advantages for a GIS future when data can be imported and exported at will promises a more effective use of GIS, and allows GIS users to concentrate on the science of data analysis and the effectiveness of common-sense information use, rather than the politics of data acquisition.

3.8   STUDY GUIDE

3.8.1   Quick Study

CHAPTER 3: MAPS AS NUMBERS

O *All GIS software has to convert maps and geographical data into numbers inside the computer* O *How we change a map into numbers impacts how useful it is in the GIS, and what analyses can be done later* O *Data are stored in a physical structure of bits, bytes, ASCII codes, files, and directories* O *Map data also need a logical structure—that is how the numbers represent the features they depict* O *GIS historically has used raster, vector, and flat file data structures* O *Raster maps are grids, with arrays of rows and columns and a value for each cell* O *Cell values can be actual values or indexes pointing to the actual value* O *Grids have resolution, extent, rows, columns, and null values* O *Each cell can only have one "owner" attribute* O *Rasters are easy to understand, easy to retrieve from storage, and fast to display, but the files are big* O *Vectors are points, represented by coordinates, linked together in different ways to depict lines, areas, and surfaces* O *Some vectors have direction* O *Vectors are efficient, do not duplicate features, and accurate* O *Examples are the TIGER files, VMAP0, and DLGs* O *Vectors can store topology, and form a network of triangles to represent a surface* O *Attributes can logically be thought of as flat files, tables with rows and columns, in which cells hold values of attributes* O *Flat files cannot hold maps well, because vector lines can have any number of points* O *Flat files and their data dictionaries can be managed in database management systems* O *Points can simply be lists containing x and y in a flat file* O *Vector best represents land parcels, roads, streams, and boundaries* O *Vectors without topology are called cartographic spaghetti* O *Each features consists of features at a lesser dimension—surfaces, areas, lines, points* O *The arc/node model has a points file, a file of arcs by polygon, and a polygon file* O *Holes in polygons and islands are special cases in the vector model* O *All map data contains errors, but storing topology helps find them, such as duplicate lines, slivers, and unsnapped nodes* O *A unit for map topology is a line without its points, but with information about connecting arcs and bounding polygons* O *TINs work well for CAD, terrain, 3D, and video games* O *Raster is vaster, but corresponds well to computer memory* O *Points and lines don't mix well with rasters* O *The raster model suffers from the mixed pixel problem, for polygon interiors and edges* O *Rasters can be handled more efficiently by run-length encoding, by using R-trees, quad trees, and image pyramids* O *Topological structures can allow automated error correction and cleaning, within fuzzy tolerances* O *Topology allows many analyses without having to read all of the points in features* O *GISs use different data structures, and so store data in different formats* O *GISs should be able to read and write at least the most common data formats* O *Formats can be open, common, industry-standard, or proprietary* O *Vector formats often describe position on a page rather than on a map, and include PostScript, PDF, and HPGL* O *Examples of geographic vector formats are GeoPDF, DXF, GML, DLG, TIGER* O *Geobrowsers and WebGIS use KML, XML, GML, and SVG* O *Raster files include a header, a color table, and data* O *Examples of raster formats are TIFF, JPEG, PNG, Encapsulated Postscript, and DEM* O *Geographic raster formats include GeoTIFF* O *Geodata should be easily movable from one GIS function to another and even from one computing environment to another ◯ Often this means converting from raster-to-vector and vice versa ◯ All data are changed by structure conversions, like speech in the multiple translator problem ◯ Many standards efforts in the US and elsewhere have led to better GIS interoperability ◯ Very important for interoperability has been the OpenGIS specifications ◯ When data moves without error and structural change due to transfer, GIS analysts can concentrate on analysis rather than deal with bothersome data issues ◯*

3.8.2 Study Questions

Representing Maps as Numbers

1. Make a map or diagram that shows the levels of abstraction in data storage. Mark "physical" and "logical" representations. Label bit, byte, file, directory, database, data format, data structure, data model, and GIS.

2. Brainstorm the reasons why different GIS packages might have different data structures.

3. Make a list of attributes that might go along with entries in the Yellow Pages section of your phone book. How many fields are necessary? Which are spatial attributes?

Structuring Maps

4. Using one of the online sources of articles on GIS applications, or the examples in Chapter 10, make a list of how many applications use vector versus raster in the analysis.

5. Explain Dana Tomlin's statement about raster and vector at the beginning of the chapter to a child.

Vector Data Structures

6. Write personal definitions of the following: cartographic spaghetti, point file, arc, polygon, topology, forward link, polygon left, TIN.

Raster Data Structures

7. Draw figures illustrating the following: pixel, resolution, grid extent, fat line, mixed pixel, polygon boundary, array.

8. Why might the attribute in a particular pixel be wrong as soon as it is geocoded?

Why Topology Matters

9. Write your own definition of topology. Make a simple diagram of one or two polygons, connected by arcs. Label the polygons A, B, C, etc. Label the arcs 1, 2, 3, etc. Now create the arcs file as a table. Make the first arc 1, second 2, and so on. Add extra columns to the table for forward and reverse links, and polygon left and right. How do you deal with the "outside"? How might you deal with a "hole" inside a polygon?

Formats for GIS Data

10. List three characteristics of each of the following GIS data formats: TIGER, DLG, DEM, TIF, GIF, JPEG, KML, DXF, PostScript.

Exchanging Data

11. Make a list of the advantages and disadvantages of sharing GIS data. What obstacles to data sharing exist at the level of one company, a municipality, a state, or between nations?

12. Visit the website for the OpenGIS Consortium, and outline what data structures their standards have dealt with.

3.9   REFERENCES

3.9.1   Chapter References

Burrough, P. A. and R. A. McDonnell (1998) *Principles of Geographical Information Systems*. Oxford: Oxford University Press.

Clarke, K. C. (1995) *Analytical and Computer Cartography* 2 ed., Englewood Cliffs, NJ: Prentice Hall.

Dutton, G., ed. (1979) *Harvard Papers on Geographic Information Systems. First International Advanced Study Symposium on Topological Data Structures for Geographic Information Systems*. Reading, MA: Addison-Wesley.

Moellering, H (ed.) (2005) *World Spatial Metadata Standards: Scientific and Technical Characteristics, and Full Descriptions with Crosstable.* International Cartographic Association, Elsevier.

Moellering, H. and Hogan, R. (1997) *Spatial Database Transfer Standards 2: Characteristics for Assessing Standards and Full Descriptions of the National and International Standards in the World.* International Cartographic Association, Elsevier.

Peucker, T. K. and N. Chrisman (1975) "Cartographic Data Structures." *American Cartographer*, vol. 2, no. 1, pp. 55–69.

Peucker, T. K., R. J. Fowler, J. J. Little, and D. M. Mark.(1976) *Digital Representation of Three-dimensional Surfaces by Triangulated Irregular Networks (TIN)*. Technical Report No. 10, U.S. Office of Naval Research, Geography Programs.

Tomlin, D. (1990) *Geographic Information Systems and Cartographic Modelling*, Englewood Cliffs, NJ: Prentice Hall.

Samet, H. (1990) *Design and Analysis of Spatial Data Structures*, Reading, MA: Addison-Wesley,

3.10 KEY TERMS AND DEFINITIONS

address range: The range from the highest to the lowest street number on one side of a street, on one block.

arc: A line that begins and ends at a topologically significant location, represented as a set of sequential points.

arc-node: Early name for the vector GIS data structure.

area: A two-dimensional (area) feature represented by a line that closes on itself to form a boundary.

array: A physical data structure for grids. Arrays are part of most computer programming languages, and can be used for storing and manipulating raster data.

ASCII: The American Standard Code for Information Interchange. A standard that maps commonly used characters such as the alphabet onto one-byte-long sequences of bits.

attribute: An attribute is a characteristic of a feature that contains a measurement or value for the feature. Attributes can be labels, categories, or numbers. Attributes can be dates, standardized values, or field or other measurements. Item for which data are collected and organized. A column in a table or data file.

Autocad: A leading CAD program by Autodesk, often interfaced with GIS packages and used for digitizing, especially floor plans and engineering graphics.

block face: One side of a street on one block, which is between two street intersections.

bounding rectangle: The rectangular region defined by the maximum extent of a map feature in the x and y directions. All parts of the feature must lie within or on the edge of the bounding rectangle.

byte: Eight consecutive bits.

bit: The smallest storable unit within a computer's memory with only an on and an off state, codable with one binary digit.

CAD: Computer aided design. Computer software used in producing technical and design-type drawings.

cartographic spaghetti: A loose data structure for vector data, with only order as an identifying property to the features.

color table: Part of the header record in a digital image file that stores specifications of colors based on simple index values, which are then stored in the data part of the image file.

computer memory: A holding location within a computer into which a sequence of nonrandom bytes can be placed then retrieved intact.

data analysis: The process of using organized data to test scientific hypotheses.

database: Any collection of data accessible by computer.

data dictionary: The part of a database containing information about the files, records, and attributes rather than just the data.

data exchange: The exchange of data between similar GIS packages or among groups with a common interest.

data format: A specification of a physical data structure for a feature or record.

data model: The logical means of organization of data for use in an information system.

data retrieval: The ability of a database management system to get back records that were previously stored.

data structure: The logical and physical means by which a map feature or an attribute is digitally encoded.

data transfer: The exchange of data between non-communicating computer systems and different GIS software packages.

DBMS: Database management system. Part of a GIS, the set of tools that allow the manipulation and use of files containing attribute data.

decennial census: The effort required by the U.S. Constitution that every 10 years all people be counted and their residences located.

decimal: The counting system when people have 10 fingers.

Delaunay triangulation: An optimal partitioning of the space around a set of irregular points into non-overlapping triangles and their edges.

DEM: Digital elevation model. A raster format gridded array of elevations.

DIGEST: The NATO transfer standard for spatial data.

DIME: Dual Independent Map Encoding. The data model used for the Census Bureau's Geographic Base Files, the predecessor of TIGER.

DLG: A vector format used by the USGS. for encoding lines on large-scale digital maps.

digital elevation model: A data format for digital topography, containing an array of terrain elevation measurements.

double digitized: The same feature captured by digitizing twice.

DXF: Autocad's digital file exchange format, a vector mode industry-standard format for graphic file exchange.

editor: A computer program for the viewing and modification of files.

elevation: The vertical height above a datum, in units such as meters or feet.

enumeration map: Map designed to show one census enumerator the geographic extent and address ranges within one's district.

encapsulated PostScript: A version of the PostScript language that allows digital images to be included and stored for later display.

end node: The last point in an arc that connects to another arc.

export. The capability of a GIS to write data out into an external file and into a non-native format for use outside the GIS, or in another GIS.

fat line: Raster representation of a line that is more than one pixel wide.

feature: A single entity that composes part of a landscape.

field: The contents of one attribute for one record, as written in a file.

file: A collection of bytes stored on a computer's storage device.

file header: The first part of a file that contains metadata rather than data.

FIPS 173: The Federal Information Processing Standard maintained by the USGS and the National Institute of Standards and Technology that specified a standard organization and mechanism for the transfer of GIS data between dissimilar computer systems. FIPS 173 specifies terminology, features types, and accuracy specifications, as well as a formal file transfer method.

forward/reverse left: Moving along an arc, the identifier for the arc connected in the direction/opposite direction of the arc to the immediate left.

forward/reverse right: Moving along an arc, the identifier for the arc connected in the direction/opposite direction of the arc to the immediate right.

fully connected: A set of arcs in which forward and reverse linkages have identically matching begin and end nodes.

GBF: Geographic Base File. A database of DIME records.

geographical surface: The spatial distribution traced out by a continuously measurable geographical phenomenon, as depicted on a map.

GIF: An industry standard raster graphic or image format.

grid cell: A single cell in a rectangular grid.

grid extent: The ground or map extent of the area corresponding to a grid.

hexadecimal: The counting system that would be used if people had 16 fingers.

hierarchical: System based on sets of fully enclosed subsets and many layers.

HPGL: Hewlett Packard Graphics Language. A device-specific but industry-standard language for defining vector graphics in page coordinates.

industry standard format: A commonly accepted way of organizing data, usually advanced by a private organization.

image depth: The numbers of bits stored for each pixel in a digital image.

import: The capability of a GIS to extract data from an external file and in a non-native format for use within the GIS.

internal format: A GIS data format used by the software to store the data within the program, and in a manner unsuitable for use by other means.

label point: A point digitized within a polygon and assigned its label or identifier for use in topological reconstruction of the polygon.

landmark: TIGER term for a geographic feature not a part of the census features.

layer: A set of digital map features collectively (points, lines, and areas) with a common theme in coregistration with other layers. A feature of GIS and most CAD packages.

line: A one-dimensional (length) map feature represented by a string of connected coordinates.

logical structure: The conceptual design used to encrypt data into a physical structure.

magic number: Any number that has a specific value for a specialized need.

matrix: A table of numbers with a given number of rows and columns.

metadata: Data about data, usually for search and reference purposes.

mixed pixel: A pixel containing multiple attributes for a single ground extent of a grid cell. Common along the edges of features or where features are ill defined.

missing data: Elements where no data are available for a feature or a record.

node: The end of an arc. At first, any significant point in a map data structure. Later, only those points with topological significance, such as the ends of lines.

page coordinates: The set of coordinate reference values used to place the map elements on the map, and within the map's own geometry rather than the geometry of the ground that the map represents. Often page coordinates are in inches or millimeters from the lower left corner of a standard size sheet of paper, such as A4 or 8 1/2-by-11 inches.

physical structure: The mechanical mapping of a section of computer memory onto a set of files or storage devices.

pixel: The smallest unit of resolution on a display, often used to display one grid cell at the highest display resolution.

point: A zero-dimensional map feature, such as a single elevation mark as specified by at least two coordinates.

polygon: A many-sided area feature consisting of a ring and an interior. An example is a lake on a map.

polygon interior: The space contained by a ring, considered part of a polygon.

polygon left: Moving along an arc, the identifier for the polygon adjacent to the left.

polygon right: Moving along an arc, the identifier for the polygon adjacent to the right.

PostScript: Adobe Corp.'s page definition language. An interpreted language for page layout designed for printers but also an industry standard for vector graphics.

quad tree: A way of compressing raster data based on eliminating redundancy for attributes within quadrants of a grid.

RAM: The part of a computer's memory designed for rapid access and computation.

raster: A data structure for maps based on grid cells.

ring: A line that closes upon itself to define an area.

run-length encoding: A way of compressing raster data based on eliminating redundancy for attributes along rows of a grid.

R-tree: A spatial data structure that organizes features by the overlapping nature of their bounding boxes.

SLF: An early Defense Mapping Agency data format.

sliver: Very small and narrow polygon caused by data capture or overlay error that does not exist on the map.

snap: Forcing two or more points within a given radius of each other to be the same point, often by averaging their coordinate.

Spatial Data Transfer Standard: The formal standard specifying the organization and mechanism for the transfer of GIS data between dissimilar computer systems. Adopted as FIPS 173 in 1992, SDTS specifies terminology, features types, and accuracy specifications, as well as a formal file transfer method for any generic geographic data. Subsets for the standard for specific types of data, vector, and raster, for example, are called profiles.

spreadsheet: A computer program that allows the user to enter numbers and text into a table with rows and columns, and then maintain and manipulate those numbers using the table structure.

table: Any kind of organization by placement of records into rows and columns.

TIGER: A map data format based on zero-, one-, and two-cells, used by the U.S. Census Bureau in the street-level mapping of the United States.

TIF: An industry-standard raster graphic or image format.

TIN: A vector topological data structure designed to store the attributes of volumes, usually geographic surfaces.

tolerance: The distance within which features are assumed to be erroneously located different versions of the same thing.

topologically clean: The status of a digital vector map when all arcs that should be connected are connected at nodes with identical coordinates, and the polygons formed by connected arcs have no duplicate, disconnected, or missing arcs.

topology: The property that describes adjacency and connectivity of features. A topological data structure encodes topology with the geocoded features.

USGS: The United States Geological Survey, part of the Department of the Interior and a major provider of digital map data for the United States.

vector: A map data structure using the point or node and the connecting segment as the basic building block for representing geographic features.

volume: A three-dimensional (volume) feature represented by a set of areas enclosing part of a surface, in GIS usually the top only.

VPF: Vector product format, a data transfer standard within DIGEST for vector data.

zero/one/two cell: TIGER terminology for point, line, and area, respectively.

3.11 PEOPLE IN GIS

Tim Trainor, Assistant Division Chief for Geographic Areas and Cartographic Products in the Geography Division at the US Census Bureau:

KC: What is your job and how does it involve GIS?

TT: I'm responsible for all the geographic areas that the Census Bureau manages. That includes legal, statistical, and administrative geographic areas, and I'm also responsible for all the cartographic products that we generate to both collect the data—we send field enumerators out to collect data—as well as publish the results of the Census, two very different things. GIS we use for many different facets. We're the developers and keepers of the TIGER database, so we are using the technology more to do our work on a day to day basis. That's something we haven't done much in the past, but now we're starting to really exploit that. We've been using GIS technology to do mapping, and we've pushed the envelope from a software development perspective to improve our automated processes for cartographic design and production.

KC: Does the Census Bureau employ students who have taken classes in GIS or Geography?

TT: The Census Bureau may be the single greatest employer of geographers in the world, I've been told at the International Geographical Union meeting. I don't know if that's true or not, but we have about 300 people in our division and about half of them are geographers or cartographers. We look for people who have degrees or advanced degrees in Geography, with a specialization in GIS, since that's the technology that we try to exploit for working. The answer is we don't

just hire geographers; we need geographers with technical skills as well.

KC: Is the Census Bureau planning anything to extend TIGER for the 2010 Census?

TT: Yes, we're doing a lot in that area. We are realigning TIGER to a much greater precision and accuracy, realigning all the street center lines to a 7.6 meter accuracy level. Most of the data we're getting is actually much better than that. The reason for that is that we're planning to use GPS technology to capture each housing unit in the nation through an address canvassing operation to take place in 2009. In order to put a housing unit in the correct census block, we need to be sure that the streets are aligned to the right geography. At the same time, we have taken our traditional in-house database structure and are moving it to a commercial environment, an Oracle database. We've developed a new data model for TIGER, and that means, from our perspective, that all the applications development, all that software development, needs to be redone. We are in the middle of redoing that work.

KC: What educational background do you have in GIS?

TT: My undergraduate degree is in History, and I did graduate work at Glasgow University in Scotland in Cartography in the Department of Topographic Science, so I have a little bit of surveying, a little bit of photogrammetry, and mostly cartography. I learned how to manage mapping projects. Knowing the whole mapping process, all the pieces that go into it and how people contribute to design and production of good products, no matter what the technique—I think that was what I learned the most from.

KC: What would you say to a freshman or sophomore in college taking their first GIS class?

TT: What I would say is that there is very little that you encounter in life that doesn't have something to do with Geography. Everything has a sense of place—or everything is related to a sense of place—and you should explore that perspective. You should think about it, go to the root cause of what it is you see and observe and experience, and, if you do that, you'll really fall in love with Geography.

KC: Thank you very much.

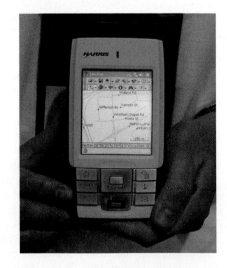

C H A P T E R   4

Getting the Map into the Computer

"He's leading us to disaster!" cried the Head of the Air Force. He was shaking with fear. In the seat behind him sat the Head of the Army who was even more terrified. "You don't mean to tell me we've gone right out of the atlas?" he cried, leaning forward to look. "That's exactly what I am telling you!" cried the Air Force man. "Look for yourself. Here's the very last map in the whole flaming atlas! We went off that over an hour ago!" He turned the page. As in all atlases, there were two completely blank pages at the very end. "So now we must be somewhere here," he said, putting a finger on one of the blank pages. "Where's here?" cried the Head of the Army. The young pilot was still grinning broadly. He said to them, "That's why they always put two blank pages at the back of the atlas. They're for new countries. You're meant to fill them in yourself."

Roald Dahl, *The BFG, p.161–163 (1982)*

4.1   ANALOG-TO-DIGITAL MAPS

Most people think of maps as drawings on paper. Maps hang on walls, lie in map drawers, and fill the pages of books, atlases, street guides, newspapers, and magazines. Maps roll off the nation's printing presses in the millions each year, and they fill the spaces in every car's glove compartment, neatly folded or not! The traditional paper maps of our everyday world can be called *real maps*, because they are touchable. We can hold them in our hands, fold them up, and carry them around. The computer, in contrast, has forced us to reconsider this simple definition of a map. In the digital era, and especially within GISs, maps can be both real and *virtual*.

A *virtual map* is a map waiting to be drawn. It is an arrangement of information inside the computer in such a way that we can use the GIS to generate the map however and whenever we need it. We may have access to map information about roads, rivers, and forests, for example, but may decide that only the forests and rivers need be shown on any map that the GIS produces. Every real map is simply a rendering of the virtual map on a display medium, which gives the map the form that it takes. In many cases, the medium we use is paper, but increasingly we view the map on a computer screen.

Unless new field data have been collected, using maps within a GIS means that somehow they have already been turned from real into virtual maps. Another way to say this is that a paper map has gone through a conversion, from a paper or analog form into a digital or number form. We start with paper, or sometimes film, Mylar, or some other medium, and we end up with a set of numbers inside files in the computer. This conversion process is called *geocoding*, which we can define as the conversion of spatial information into computer-readable form. Some GIS vendors would be pleased to help you acquire the data you need, but at an immense price. Studies have shown that finding the right maps, and converting these maps from real to virtual form by geocoding, takes up anywhere between 60% and 90% of both the time and money spent on a typical GIS project. Fortunately, this is a once-only cost. As soon as we have the map in digital form, we can use it in a GIS over and over again for different uses and projects unless it needs an update. As time goes by, fewer and fewer maps will exist solely on paper and not digitally.

Digital map data for use in GIS really falls into three categories. Either the data already exist and all we have to do is find or buy them, or they don't exist and we have to geocode paper maps or maps on some other medium. A third case is that the maps don't even exist, perhaps because the earth's surface has changed, and here we often turn to remote sensing, aerial photography, or field data collection by surveyors or the global positioning system (GPS) to get our first map of a new location. Also, sometimes the maps we need already exist, but whoever geocoded them is not interested in sharing the data with you, even for a price! Even when we can get the maps we need in digital form, they may not suit our particular type of GIS, or may be out of date or not show the features we want. The bottom line is that sooner or later, and usually sooner, we end up geocoding at least some of our own maps.

Before we cover the ways that maps can be converted into numbers—scanning and digitizing—we will take a look at how we might go about finding digital map data that already exist. If we are successful, with a little effort, some conversion programs, and knowledge of GIS data formats, we can reuse one of the many maps already available to us. Many of these maps can be read straight into a GIS, sometimes without any need to research the way the files and numbers are structured. In this chapter we take a guided tour of the various flavors of data, their formats, and the way the information in the maps has been structured during geocoding.

These days, very few GIS projects have to start with no data at all. The vast amount of data that is collected and made available by the various branches of government is an excellent base on which to start building. The trick is knowing where to look, what to do when you find what you want, and how to get the data into your GIS.

4.2   FINDING EXISTING MAP DATA

The search for paper maps is often started in a map library or on a map library website. The libraries most likely to carry maps and to support cartographic research are the research libraries in the largest cities or those attached to major universities. Map librarians make use of computer networks to share information and conduct searches. They are increasingly making census and other digital maps available both in libraries and via computer networks.

Another place to look for map information is in books. A starting point is *GIS Data Sources* by Drew Decker (2000). *Maps for America*, by M. M. Thompson (1987) of the USGS, is a good survey of existing published maps for the United States. Another information source, especially internationally, is the *Inventory of World Topographic Mapping* (Bohme, 1993). The appendices in John Campbell's book *Map Use and Analysis* (Campbell, 2001) show how to use map series and their indices, and many other information sources are listed, especially Chapter 21, "U.S. and Canadian Map Producers and Information Sources," duplicated on the book website at: `http://auth.mhhe.com/earthsci/geography/campbell4e/links4/appalink4.mhtml`. A great deal of information not only about map data online, but also open source software for display, is contained in *Mapping Hacks: Tips & Tools for Electronic Cartography* (Erle, Gibson and Walsh, 2005).

In many cases, state and local governments keep collections of paper maps. A local planning or building permit office can often find maps of your property or of parks and business properties. Make sure to call ahead. How good the service of providing maps to the public is depends a great deal on the office and its policies and services. Some larger agencies have their own map division. A state highway authority, park service, or industrial development organization may have its own maps available, sometimes free or at little cost.

Commercial companies sell cartographic data and some will conduct map data searches. Imagery from most commercial vendors can be searched and browsed using an online database. Many commercial services offer not only packaged existing data for your use, but will digitize or scan data and even write the data in GIS format for you at a cost. Two companies offering such services are TeleAtlas (www.teleatlas.com) and Yahoo, which references locations with its own WOEID (Where On Earth Identification) georeferencing system.

Obviously, each company has its own strengths and types of map for sale. Commercial companies are not, however, for the novice. They are primarily used by large corporations, governments, the real estate industry, and so on. For a first cut, the usually free public data are the best starting point, and in many cases enough, even many times more than you will ever need to work with your GIS.

Digital map data by public agencies have been dominated by data from the federal government. In the United States, digital map data created at the federal level for its own use are the property of the American people, with the obvious exception of sensitive data of use in national security—although recently even spy satellite data have been made available. The Freedom of Information Act guarantees everyone the right to get copies of digital map data used by the federal government, subject to a distribution or copying cost that may not exceed a reasonable marginal cost of providing the data.

Not all data has to be extracted from the government using the act, however. Government agencies have made it their mission to make map data as freely accessible as possible to any interested party. Computer networks have made this not only accessible to almost any computer user but have also made it more flexible.

4.2.1 Finding Data on the Web

The best way to begin a data search is to use the World-Wide Web (WWW). Most computers are equipped with web browsers, such as Opera, Firefox, Safari, Chrome, or Internet Explorer, that allow you to search the web using keywords or in other ways. The WWW is an interlinked set of computers and servers, or data repositories on the Internet. A new generation of geobrowsers is making Internet search increasingly geographic. Each major agency has a WWW server, or *gateway*, through which data can be searched and downloaded. Simply enormous amounts of data are available through this mechanism.

While many U.S. government agencies create and distribute digital maps, data from three agencies, each one with its own different types, have been most used in GISs. The agencies are the U. S. Geological Survey (USGS), part of the Department of the Interior; the U. S. Bureau of the Census; and the National Oceanic and Atmospheric Administration (NOAA), both part of the Department of Commerce. Data they supply cover the land and its features, the population, and the weather, atmosphere, and oceans across the United States. There are many other agencies that supply on-line spatial data, including NASA, EPA, FEMA, and so forth.

Each of the three main agencies is worth covering here in detail. Finding information in any of them has been made much easier by several public information service and computer network services, especially over the Internet. In many cases there are data clearinghouses that locate data across agencies.

4.2.2 U.S. Geological Survey

Digital cartographic data from the USGS are distributed publicly through an on-line depository with its own search system that links to the Seamless Data Server called the National Map Viewer (`http://nmviewogc.cr.usgs.gov/viewer.htm`). The USGS digital data fall into six categories: *digital line graphs* (DLGs), *digital elevation models* (DEMs), *land-use and land-cover digital data, digital cartographic text* (Geographic Names Information System, GNIS), *digital orthophotoquads* (DOQ), and *digital raster graphics* (DRG). The National Map viewer portal is shown in Figure 4.1. The USGS continues to improve coverage of the United States and distributes the map data products via a seamless server that is linked to the National Map viewer. The user first specifies an area of interest by browsing the map on-line, then sets up a query where data sets are identified. The server "clips" these data from the master data files, and transfers them via the file transfer protocol directly to a user in many common formats, including those supported by the majority of GIS packages. These data models and structures were covered in chapter 3, and most GIS packages support them directly.For example,a search for data for Santa Cruz Island, California, produced 38 data sets in seven categories (boundaries, hydrography, imagery, land use/land cover, structures and transportation) supplied by USGS, NOAA, the U.S. Forest Service, and the U.S. Fish and Wildlife Service (Figure 4.2). Clicking on the Landsat data, a query was sent automatically to the National Map Seamless Data server, which delivered a Landsat image from 2002, with two images joined together in GeoTIFF

Figure 4.1: The National Map Viewer. See: http://nmviewogc.cr.usgs.gov/viewer.htm.

format and on NAD83. These data were imported straight into both ArcView 3.2 and uDig (Figure 4.3) by uncompressing the transfer file and simply reading it from the software. Many GIS packages support direct importing of data in the most commonly used

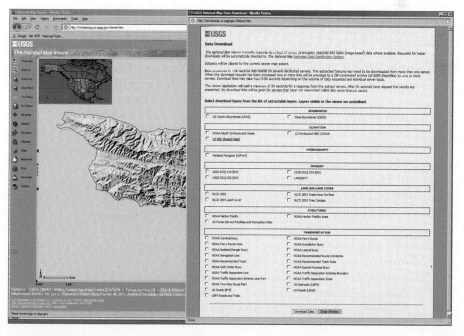

Figure 4.2: Searching for and downloading data from the national map via the USGS seamless data server.

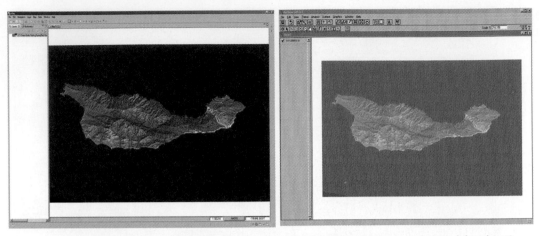

Figure 4.3: Landsat data for Santa Cruz Island, California, as downloaded in Figure 4.2, introduced into the uDig (left) and ArcView 3.2 (right) GIS software packages.

government data formats, including use of metadata to provide coordinates and projection information.

The USGS also distributes data on land cover derived from classifications of NOAA's AVHRR (*advanced very high resolution radiometer*) measurements. These data are distributed by the EROS Data Center on the Internet, with a ground resolution of 1 kilometer. Biweekly composites show a vegetation index for North America and the world at this resolution. There are also Landsat images and the Multi-resolution Land Characterization database, a land use coverage consistent nationwide. Data sets of topical interest are often added to the National Map viewer, for example hurricane tracks, as are data sets acquired from collaborating agencies and local governments.

4.2.3   National Oceanic and Atmospheric Administration (NOAA)

The NOAA concentrates on marine and aeronautical navigation systems that electronically integrate digital charts, global positioning system-based locations, and real-time environmental information. Examples are the daily weather map, satellite and radar images, and maps used by pilots and air traffic control. The NOAA charts must be carried aboard all large ships in U.S. waters. NOAA includes agencies such as the National Geodetic Survey, which maintains accurate GPS control for the country, the National Weather Service, and the Satellite and Information Service, which operates and distributes data for several important mapping satellites.

The National Geophysical Data Center, part of NOAA, has released numerous digital map data sets, both global and national, most recently including detailed bathymetry of the ocean and land-surface topography at one minute of latitude and longitude resolution, as well as geodetic and magnetic data for the earth's surface (Figure 4.4). NOAA maintains several map data portals that support interactive query for data using ESRI's Internet Map Server. Data are supplied in various formats, again well supported by GIS software. For example, the world topographic dataset shown in Figure 4.5 displayed in QuantumGIS was downloaded, unzipped, and opened directly by the software.

NOAA also hosts the National Geodetic Survey, responsible for the nation's geodetic control data and coordinator of datums, grid systems and more. A particularly

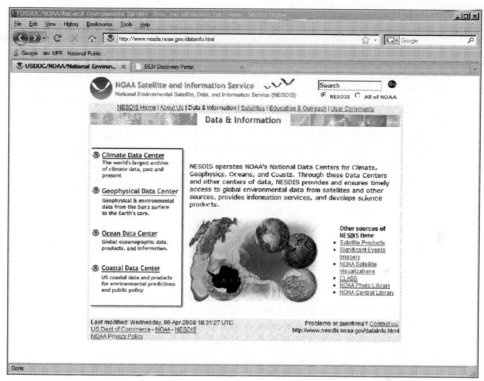

Figure 4.4: Geospatial data portal for NOAA.

useful set of tools is on-line and available at `http://www.ngs.noaa.gov/TOOLS/`. These tools support conversion of geospatial data among the many map projections, coordinate systems and datums in common use with GIS.

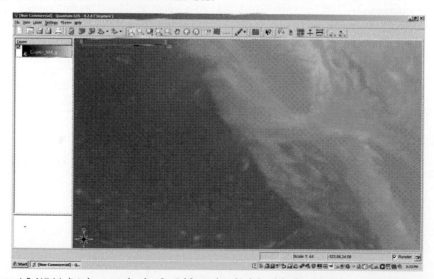

Figure 4.5: NOAA digital topographic data for California downloaded and displayed in Quantum GIS software.

4.2.4   U.S. Bureau of the Census

The mapping of the U.S. Census Bureau is to support the decennial census by generating street-level address maps for use by the thousands of census enumerators. For the 1990 census, the Census Bureau developed a system called TIGER (*topologically integrated geographic encoding and referencing*). As we saw in Chapter 3, the TIGER system uses the block face or street segment as a geographic building block and recognizes cartographic objects of different dimensions, points (nodes), lines (segments), and areas (blocks, census tracts, or enumeration districts). In TIGER terminology, points are *zero cells*, lines are *one cells*, and areas are *two cells* (Figure 4.6). Objects can have geometry (G), topology (T), or both (GT). The figure shows a generalized block that consists of three GT-polygons (GT stands for geometry and topology). The block contains a point landmark (Parkside School) inside GT-polygon 2 and an area landmark (Friendship Park) that is coextensive with GT-polygon 3 (Census Bureau, 2000). Geometry-only objects are usually landmarks, the features that fill in the details on the TIGER maps. TIGER files by state in the ESRI shapefile format as updated to 2006 can be downloaded by state from the TIGER website at `www.census.gov/geo/www/tiger/tiger2006se/tgr2006se.html`.

A large-scale cooperative effort prepared these maps for the 1990 census, and the files were updated for the 2000 census. Map digitizing was initially performed in collaboration with the U.S. Geological Survey but is being completely revised for the 2010 census. The maps are distributed along with the census data over the Internet. Virtually every GIS allows TIGER files to be imported directly into the system, although not all GISs handle the attribute data as well. The TIGER was the first comprehensive GIS database at street level for the entire United States. An important ability of TIGER is to do *address matching*: the search for street addresses through the attribute files to match a block or census tract in the TIGER graphic files; that is, finding its geographic location on the map solely from a street address listing. Address matching takes a street number and name, plus a city name, and uses these to determine where the address falls geographically. For example, odd and even numbers are on opposite street sides, and house numbers increase by 100 for each city block, so a guess at where house number 7262 fell on the block would be 62/100ths of the way along the block 7200 on the even street number side. This is how many on-line and GPS-based address matching systems locate street addresses

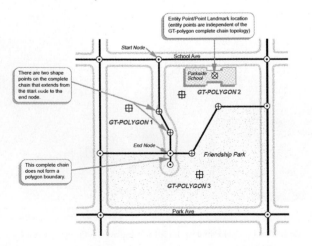

Figure 4.6: Basic units in the TIGER files, zero cells, one cells, and two cells.

from user information, by using the TIGER files. The files are also a good base map of feature information against which other data can be plotted (Figure 4.7). The TIGER files are being improved and updated for the 2010 census to improve accuracy and to include use of field GPS in data collection (See *People in GIS,* Chapter 3). The TIGER data are already the foundation of most GIS and mobile map applications in the U.S. and will increasingly play a critical role in the nation's economy and government.

4.2.5   Other Federal Data

The U.S. government, under the Federal Geographic Data Committee's leadership, has created a national spatial data infrastructure (NSDI), a suite of data warehouses and catalogs, all indexed together. Portals to the NSDI include the GeoSpatial OneStop at www.geodata.gov. This means that finding data across agencies is easier. There are also a large number of non-government data sources that also have clearinghouses and data-sharing depositories that is, places where GIS users can archive and distribute the data they have already collected and processed. Naturally, supplying good metadata—data about the data—is essential for the clearinghouses to survive. There are many examples of readily available geobrowsers that serve out Federal geographic data and facilitate data discovery, that is, finding the data layers that fit your needs. Some of these examples are USGS's Natural Hazards Support System (nhss.cr.usgs.gov) (Figure 4.8), NASA's Worldwind (worldwind.arc.nasa.gov), Google Earth and Google Maps, and the Alexandria Digital Library Globetrotter (clients.alexandria.ucsb.edu/globetrotter) (Figure 4.9) as portals that serve mostly US government data. As you will quickly discover, not only is there an extraordinary amount of US government data out there, it is fairly easy to find,

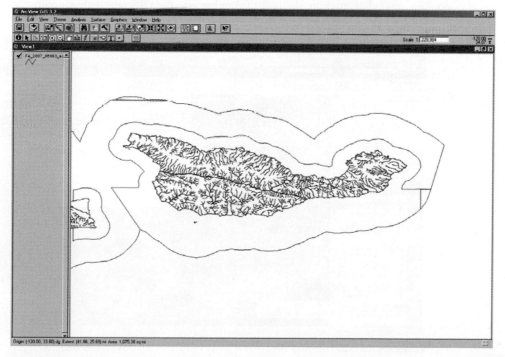

Figure 4.7: Santa Cruz Island, California. TIGER stream features.

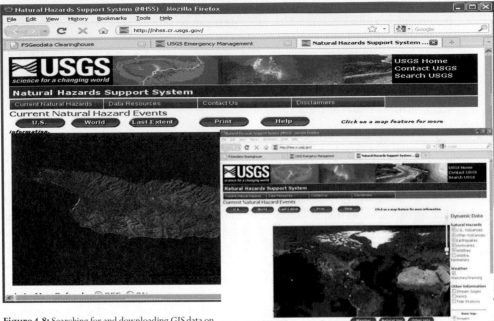

Figure 4.8: Searching for and downloading GIS data on Natural Hazards for Santa Cruz Island from the USGS's Natural Hazards Support System.

download, and use in a GIS. As publicly created data, it is usually documented and contains metadata, making the task of moving into the GIS relatively smooth.

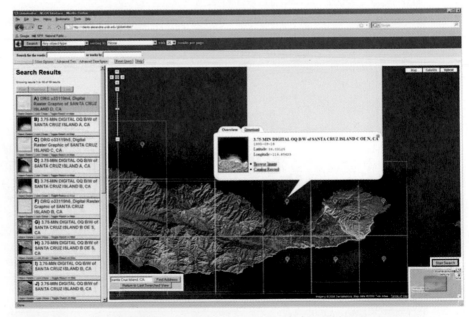

Figure 4.9: Search for Santa Cruz Island maps in the Alexandria Digital Library Globetrotter (clients.alexandria.ucsb.edu/globetrotter).

4.2.6   Creating New Data

Wonderful as it is to find an existing digital map, the myriads of different data formats is usually the least of a GIS analyst's problems. Digital maps, like their analog sources, are specific to a given map scale. Boundary lines, coastlines, and so on all reflect the degree of generalization applied to the lines when the map was originally digitized. In addition, maps were usually digitized with different levels of precision, from source maps that were out of date, or that have become out of date since they were digitized, or sometimes even have errors or problems with their accuracy. Two different maps of the same area rarely agree over every detail, yet the computer is unable to resolve the differences in the same way that the human mind can reason about the reliability of information, its timeliness, and so on.

In summary, like it or not, sooner or later if you are involved with GIS you will find yourself digitizing a map. Although this is a tedious, time-consuming, and potentially frustrating exercise, the learning process involved will greatly increase your awareness of the limitations of digital maps for GIS use. It is far better to persevere and learn, than to make a million errors and misjudgments for the lack of a little hands-on experience. Time, then, to get a little digital (or at least virtual) "mud" on our boots!

4.3   DIGITIZING AND SCANNING

Historically, many different means have been used to geocode. At first, some very early GIS packages required maps to be encoded and entered by hand. The hours of monotonous work required for this task made errors common and their correction difficult. Since special-purpose digitizing hardware became available, and especially since the cost of this hardware fell substantially, virtually all geocoding has been performed by computer.

Two technologies have evolved to get maps into the computer. Digitizing mimics the way maps were drafted by hand and involves tracing the map over using a cursor while it is taped down onto a sensitized digitizing tablet. The second method involves having the computer "sense" the map by scanning it. Both approaches work and have their advantages and disadvantages. Most important, the method of geocoding and the scale of the initial map stamp their form onto the data in such a way that almost all subsequent GIS operations are affected somehow.

4.3.1   Digitizing

Geocoding by tracing over a map with a cursor is sometimes called *semi-automated digitizing*. This is because in addition to using a mechanical device, it involves a human operator. Digitizing means the use of a digitizer or digitizing tablet (Figure 4.10). This technology developed as computer mapping and computer-aided design grew and placed new demands on computer hardware. While still sometimes used, the heads-up method using scanning has largely now replaced this process. This discussion is included here, however, because the majority of legacy data sets available for GIS were captured in this way. The inherent errors in the process are now a permanent part of the existing data.

The digitizing tablet is a digital and electronic equivalent of the drafting table. The major components are a flat surface, to which a map is usually taped, and a stylus or cursor, with the capability of signaling to a computer that a point has been selected. The mechanism to capture the location of the point can differ.

Figure 4.10: Semi-automated map digitizing with tablet and cursor.

To register the coordinate system, the coordinate locations of three or more points are digitized, usually the upper right easting and northing, the lower left easting and northing, and at least one other corner. From these points, with their map coordinates and their raw digitizer coordinates, all the parameters can be computed for converting the data into the map's coordinates. Many map entry and digitizing software packages require four of these control points for computing the map geometry, and it is advisable to repeatedly digitize control points and to average the coordinates to achieve higher accuracy.

Points are usually entered one at a time, with a pause after each to enter attributes such as labels or elevations. Lines are entered as strings of points and must be terminated with an end-of-chain signal to determine which point forms the node at the end of the chain. Areas such as lakes or states are usually digitized as lines. Sometimes an automatic closure for the last point (snapping) can be performed. Finally, the points should be checked and edited. The digitizing software or GIS may contain editing features, such as delete and add a line or move and snap a point. After editing the data are ready for direct integration into the GIS. Usually, a separate module of the GIS is used for digitizing and editing, and the map can now be passed on for use. Often map errors in GIS are attributable to this former digital capture process and its limitations.

4.3.2 Scanning

The second digitizing process is *automated digitizing,* or more usually, just *scanning.* The scanner you may have seen at a computer store or in an advertisement, or perhaps the one you use for scanning documents, is a *flatbed scanner.* The *drum scanner* is most commonly used for maps. This type of scanner receives an entire sheet map, usually clamped to a rotating drum, and scans the map with very fine increments of distance, measuring the amount of light reflected by the map when it is illuminated, with either a spot light source or a laser (Figure 4.11). The finer the resolution, the higher the cost and the larger the data sets. A major difference with this type of digitizing is that lines, features, text, and so on are scanned at their actual width and must be preprocessed for the computer to recognize specific cartographic objects. Some plotters can double as scanners, and vice versa. For scanning, maps should be clean and free of folds and marks. Usually, the scanned maps are not the paper products but the film negatives, Mylar separations, or the scribed materials that were used in the map production. An alternative scanner is the *automatic line follower,* a scanner that is manually moved to a line and then left to follow

Figure 4.11: Large format drum scanner in use. Source: USGS.

the line automatically. Automatic line followers are used primarily for continuous lines, such as contours. These and other scanners are very useful in CADD (computer-aided drafting and design) systems, where input from engineering drawings and sketches is common.

Simple desktop scanners are becoming important geocoding devices as their resolutions improve and their prices fall. The process of scanning usually begins with preparing the section of map, which obviously needs to be as clean and with as solid and crisp lines as possible. Next, the map is placed on the desktop scanner. The software is told which window to scan, the scan is previewed, and the scan is then saved to the resultant scan file. The process can be very quick; nevertheless, care and attention can save considerable work later on. Desktop or low-resolution scanning is rarely adequate for GIS purposes but can be used to put a rough sketch into a graphic editing system for reworking. In this way, a field sketch can be used as the primary source of information for developing the final map for the GIS.

It is important to have a clear concept of *scale* and *resolution* when scanning. In Figure 4.12, the same map, part of the Little Pine Mountain, California, 7.5-minute USGS quadrangle map, was scanned at four different resolutions. The square scanned on the map was 100 millimeters on a side. At 1:24,000, this distance is 24,000 x 100 = 2,400,000 millimeters, or 2400 meters.

Although Figure 4.12 shows a scanned 100mm-by-100mm segment of the map, the pixel density of the scan is given per inch. The scan at 200 dots per inch (DPI) translates to 7.87 pixels per millimeter, making the scanned map square about 787 by 787 pixels. The

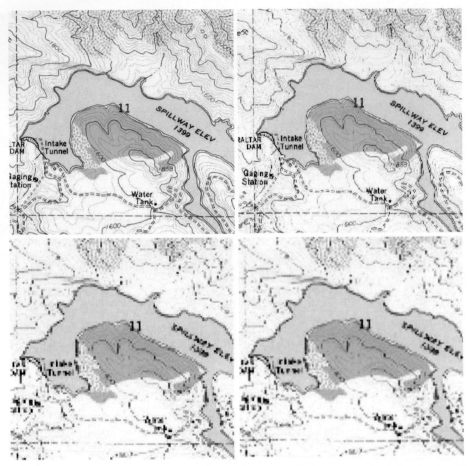

Figure 4.12: Drum scanner images of a section of the Little Pine Mountain, California, topographic quadrangle at 1:24,000. Top left 200 dpi, top right 100 dpi, lower left 50 dpi, lower right at 25 dpi.

same area was also scanned at 100 DPI, or 3.937 pixels per millimeter, for an image that is 394 by 394 pixels. On the two scans at their ground equivalent, one pixel is about 3 meters on the first but 6 meters on the second. It is not this print density but the equivalent scale that is important for the map's accuracy. The width of a very thin line on the map, such as a small stream, is about 0.2 millimeter.

At 1:24,000, this means that the stream would be 4.8 meters wide if it were painted on the ground, more than the pixel size on the 200 DPI scan but less than a single pixel of 6 meters on the 100 DPI scan. Most of the line would be skipped, and only occasionally would the pixel and the line coincide. This can be shown clearly in the insets in Figure 4.12. Losing features in this way is called *dropout*. Dropout can virtually eliminate a feature on the map, or at best make it seem like background "noise." This could be critical if projection graticules, grid tic marks, or detailed features that are essential for later processing are lost. Another fact to note about scanning is that pencil lines, coffee stains, paper discoloration, and, in particular, wrinkles and folds all show up. Also, if the map to be scanned has printing on both sides, you may get a double image on the scan, i.e. of both sides together. This can lead to problems, as we will see in Section 4.6.

4.4   FIELD AND IMAGE DATA

4.4.1   Field Data Collection

An increasing amount of data for GIS projects comes from a combination of data collected in the field, global positioning system data, and imagery. Field data are collected using standard surveying methods. In these methods locations are first established in the field as *control points*. Next, additional locations are collected to trace out features or covering the terrain, for example, the edges of features are traced out by sequential measurements using instruments designed to measure angles and distances. The highest accuracy standard survey instruments, called *total stations,* are digital recorders as well as measurement instruments. They use laser ranging to prism reflectors to calculate distance. Less expensive instruments such as theodolites, engineers' transits, and levels often use a technique for measuring distances called *stadia*, which involves reading the numbers on a calibrated pole through the lens of the instrument.

Data are recorded in notebooks, and usually the data are then entered into a computer program to turn the bearings, angles and distances into eastings, northings, and elevations. The type of software used is called COGO, for "coordinate geometry," and many COGO packages either write data directly into GIS format or are capable of writing files that can transfer directly. Field surveys are common in land surveying, ecology, archeology, geology, and geography. Many less detailed surveys usually use rough field sketches, grids, transects, point sampling, and hand-held GPS. In many cases, GIS software runs directly on the field collection device, for example the ESRI ArcPad software.

Figure 4.13: Total station in use at an archeological survey in Bantay Srei, Cambodia.

4.4.2   Heads-Up Digitizing

The benefits of scanning for GIS data collection are many. A standard procedure for map capture and update exploits these advantages by manual digitizing over scanned maps. This can be done in several ways. A student might find it easiest to bring the image into a visual editing program, such as Adobe Illustrator or Coreldraw, or into a CAD program such as AutoCad. If a map with crisp lines or a tracing has been scanned, there are automated software packages and GIS modules such as ArcScan which can extract continuous lines and replace them with vectors. Things like printed text and overlapping line features (e.g., a road crossing another over a bridge) often make this difficult. Often an old vector map is overlain with an updated image that reflects a few new features to be added, such as new neighborhoods and roads built at the city edge as it expands. In some cases, features are extracted from an image to create a new map. In the latter case, finding features in the image of known location is very important so that the map can be georegistered into the same geometry as the existing GIS data.

Figure 4.14 shows the stages by which vectors can be extracted from an image and then imported into the GIS. With this being a bit of a subjective process, it is important to visually check features and the fit of the new image with any existing data. When data must match perfectly, it may be best to edit the existing coverage rather than to create new vectors that may be in error. Lastly, since the scanned map has a finite resolution, it is best to work on capturing vectors without changing the map scale, resolution, or coordinates. Then, when the vectors are ready, they can be transformed using the control points. Warping the raster scan to fit the map is a good way of revealing any underlying errors in the plane geometry as the map goes from image coordinates to those used in the GIS.

4.4.3   GPS Data Collection

The first stage in the surveying process, that of setting the control, is usually done by locating a USGS control point (a bench mark) or by using GPS. Using two GPS receivers together in *differential mode* it is possible to locate control points to sub-meter accuracy. GPS data can also be post-processed to higher accuracy using software, or corrected in real-time using the Wide Area Augmentation System (WAAS). Accuracy of less than a meter is relatively simple to acquire. These points are then used as the basis for continued extension of the survey network going outward and between these points. A GPS unit being used for control surveying at extremely high precision by the USGS after hurricane Katrina in 2005 is shown in Figure 4.15.

The GPS is a system of 24 orbiting satellites in medium earth orbits (about 20,000 km), each transmitting a time signal. At any given time, at least four of the satellites are above the local horizon at every location on earth 24 hours a day. When a GPS receiver is activated, the nearest satellites are located and the signals are received from each visible satellite. By decoding the time differences between the signals from each satellite, combined with data from the satellite itself about its orbit (called *ephemeris data*) it is possible to solve the three unknowns of latitude, longitude, and elevation. Many receivers can do direct conversion into any of several coordinate systems and datums, and most can download the data directly to a computer. Some GPS equipment can download directly in common GIS formats.

Prior to May 2000, the GPS signal was accurate in its coarse acquisition (C/A) code mode to only about 75 to 100 meters because the signal was deliberately degraded under

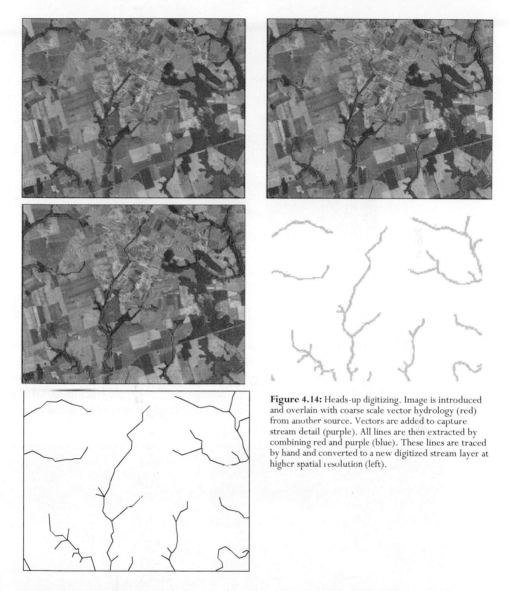

Figure 4.14: Heads-up digitizing. Image is introduced and overlain with coarse scale vector hydrology (red) from another source. Vectors are added to capture stream detail (purple). All lines are then extracted by combining red and purple (blue). These lines are traced by hand and converted to a new digitized stream layer at higher spatial resolution (left).

a system called *selective availability*. The use of selective availability has now been determined to be no longer of national security interest and has consequently been turned off, so that accuracies of 10 to 25 meters are normally possible.

By using two units, one at a known location and one "rover" unit, the degradation can be measured and eliminated, usually by processing the data from the two units on a computer after collecting the data. This is called *differential-mode use* of the GPS. It is possible by using a radio receiver or a cellular telephone link to receive real-time differential corrections in the field. The corrections are broadcast as an aid to navigation in the United States and elsewhere as the wide area augmentation system (WAAS), and are

Figure 4.15: GPS control being established by continuously reading data at a fixed point. Averaging a large number of readings gives accurate coordinates. These data can then be used to correct other locations using differential correction. Photo; USGS

also available from private services. Many GPS receivers can process these signals, greatly improving accuracy.

Most hand-held GPS receivers are capable of downloading their data to computer software, either for post-processing for accuracy enhancement or for direct integration into GIS. In some cases, GPS receivers have elaborate map displays integrated into the portable units. In others, GPS units work in conjunction with software (including GIS) to display GPS locations directly onto maps. Several GPS vendors now offer software for portable

Figure 4.16: GPS personal tracking device (Trackstick) in use mapping the UCSB campus. Photo of Gargi Chaudhuri by Julie Dillemuth.

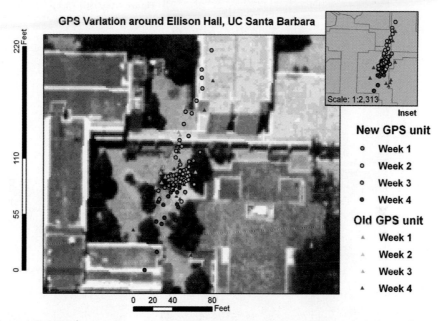

Figure 4.17: GPS data collected at two locations on the UCSB campus, with two receivers and over four weeks. Note the scatter of errors, and the surrounding buildings and vegetation. Map and data collection by Adeline Dougherty.

digital assistants, and even cellular telephones, which can come equipped with their own GPS receivers inside. Companies manufacture GPS chips connected to data loggers, allowing tracking of animals, humans, and cars, for example the Trackstick (Figure 4.16).

GPS is a highly effective way of collecting data for GIS. There are alternative systems under development, the Chinese COMPASS and European Galileo, for example. A broader term is Global Navigation Satellite System. A problem with GPS is that it is prone to error when the sky is obscured by mountains or vegetation. The signals also vary according to the geometry of the satellite at any given time. A measure of this error is the position dilution of precision, or PDOP. Many hand-held receivers warn the user when errors are too high for accurate results. In Figure 4.17, data from two GPS receivers taken at exactly the same place over various times are shown. The impact of vegetation cover, PDOP and the reflection of signals from tall buildings, called *multipath error*, is evident.

4.4.4   Image and Remote Sensing Data

Imagery data are very common input layers to a GIS. They are most frequently air photos such as the USGS's digital orthophoto maps or satellite images. The National Airphoto Program makes photography at a variety of scales available in the United States via the National Map, and private vendors also sell images. Digital orthophoto quarter quadrangles (DOQQ) are at an equivalent scale of 1:12,000 and have a 1-meter ground resolution, with some areas, such as cities, at resolutions up to 0.16m. The current program includes national coverage and frequent update after that. An example of a digital orthophoto is

Figure 4.18: USGS Digital Orthophoto Quarter Quadrangle covering the ellipse and White House, viewed in ESRI's ArcView. Note that the resolution has been degraded over the White House (zoom, lower image). Reference point in the photo below is the zero milestone marker.

shown in Figure 4.18. This DOQQ covers the zero milestone marker in the ellipse in Washington D.C., just south of the White House. Note the different resolutions of the data, with the coarser imagery covering only the White House and its immediate city block. We will return to this in the final chapter of the book. These data, and other commercial images, are commonly used in online map search tools such as Google Earth.

The Landsat program has been generating imagery of many locations in the world since 1972. Three scanning instruments, the multispectral scanner, the thematic mapper, and the enhanced thematic mapper plus, image areas on the ground at 79, 30, and 15 meters, respectively. The images are geometrically corrected into the space oblique Mercator projection and are available for use in GIS projects via the National Map

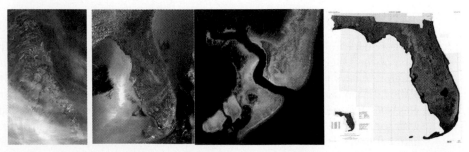

Figure 4.19: A selection of some of the many types of satellite imagery available for use with GIS, all of Florida. Left to right: ASTER, MODIS, Ikonos, Landsat composite. Sources: USGS, NASA.

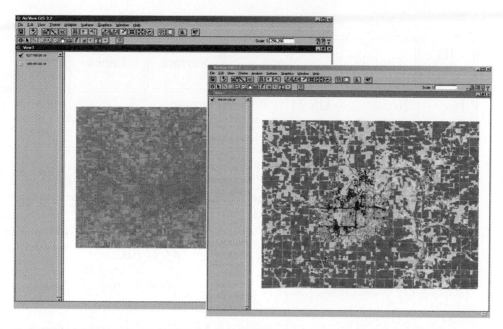

Figure 4.20: Landsat data brought into GIS and classified to show land use and land cover, an example from the National Land Cover Database (Sioux Falls, South Dakota). Source: USGS.

Seamless database. Global coverage can be quite discontinuous due to gaps in the program. Other satellites also generate imagery, including the French SPOT satellite series and the formerly Canadian RADARSAT. Other commercial suppliers include Digital Globe's Quickbird, and Ikonos. In addition, GIS projects at small scales also often use the NOAA polar orbiting satellites carrying the AVHRR (advanced very high resolution radiometer), NASA's MODIS, and NOAA's geostationary GOES weather satellite, the one seen every evening on the television weather report (Figure 4.19). The imagery is often used in GIS to extract map features, such as roads, buildings, and lakes. It is often automatically processed to show vegetation, land cover, and other map layers. Figure 4.20 shows a Landsat 7 image, and its derivative product, classified land use and land cover from the National Land Cover database.

4.5   DATA ENTRY

Geocoding is the part of GIS data input that results in getting a map into the computer. It is not the entire story, however, for as yet we have not dealt with getting the attributes into the GIS. An attribute is a value, usually a number, containing information about the features stored in the GIS. If the feature we are geocoding is a road, for example, then capturing the route of the road from a map as it winds from intersection to intersection is pure geocoding. We also have to tell the computer what this long and winding line is: a road, and anything else that the GIS needs to know about it. Relevant attributes for a road might be its state route number, the year it was built, what the surface is made of, how many traffic lanes are on the road, if the road is one-way or two-way, how many bridges it goes over, how many cars travel along the road per hour, and so on. These values are the

Features on Adams, NY, Map

ID #	Feature	Name	Surface	Lanes	Traffic per hour
1	Road	US 11	tarmac	3	113
2	Road	I 81	concrete	4	432
3	Road	Lisk Bridge Road	tarmac	2	12

value, is the number or text associated with a record for an attribute

attribute has a name and a value for each record

record, all attributes for one feature

Figure 4.21: An attribute table organized as a flat file.

road's *attributes*. They are the very meat and potatoes of GIS analysis. Somehow, we have to get them into the computer, too.

The simplest way to think of attributes is in a flat file. A flat file is really like a table of numbers. The columns of the table are the attributes, and the rows of the table are the features themselves. Each line in the table is a record, but the name used depends on who you talk to. A computer scientist would call a row a *tuple*, a statistician would call it a *case* or an *observation*. A programmer might call it an *instance of a geographic object*. They are all pretty much the same. *Record* sounds simpler.

Take a look at the flat file in Figure 4.21. The records and attributes relate to the example we discussed above, a road. The attribute table then consists of several parts. First, it has attributes with their names. Setting up the attributes means deciding what values are going to be associated with each of our features. At the time of setup, it is easy to anticipate something we may want to collect in the future and to leave a column in the table for it. Second, there are records. A record usually has a value in every one of the columns. Software programs such as spreadsheets and some databases allow you to click into a cell in the table and put in a value. Nevertheless, setting up the table has to be a little more formal than that. Each attribute has more than simply a name associated with it. For example, if we try to put "US11" into the attribute column "Surface," something is obviously wrong. Each attribute should have several characteristics, all of which usually have to be known in advance. The following is a list of what has to be considered.

1. What is the *type* of the value? For example, values could be text, number, decimal value or units such as meters, vehicles per hour, and so on.

2. What is the legitimate *range* of the values? For example, percentages should be between 0 and 100. Are negative numbers allowed? For text values, what spellings

or range of choices (known as *categories*) are allowed? For text, how many characters long is the longest string?

3. What happens when there is a missing cell in the table? For example, a record could be missing an attribute such as the traffic count in Figure 4.21 because nobody was available to make the count. Often a *missing value flag*, such as the value –999 or NULL, is used in these cases. We obviously would not want these to count if we summed or averaged the rows or columns.

4. Are duplicates allowed? What if we had two road entries for Interstate 81 on Figure 4.18, one for the northbound and one for the southbound lanes? The traffic counts, road surfaces and so on may be different and worthy of their own record. In this case, the values entered in the attribute column under "Name" would be identical.

5. Which attribute is the *key*? The key is the link between the two databases. So in the example of Figure 4.21, the attribute "ID #" should match the tag that was placed on the road when it was digitized from the map. Otherwise, all our attributes would be "lost in space."

Many of these questions must be answered when we set up the database to begin with. The tool within the database manager that allows this attribute setup is called the *data definition module*. It often has its own menus, language, and so on, and may need a programmer rather than a typical GIS user to set it up. In some cases, just as with the digital map data, the attribute data will have been found from an existing source, such as the Census Bureau's data files that link to the TIGER files. In this case, the links will already be made between the attributes and the features on the map. If there are new data, however, or if we make our own database for our own purposes, we have to make the links and check them ourselves.

```
Attribute_labels = "ID #", "Feature",
"Name" , "Surface" , "Lanes", "Traffic" , "per hour"
"1",
"Road",
"US 11",
"tarmac",
"3",
"113"

"2",
"Road",
"I 81",
"concrete",
"4",
"432"

"3",
"Road",
"Lisk Bridge Road",
"tarmac",
"2",
"12",
"4"
```

Figure 4.22: An unraveled or ASCII text form version of the flat file in Figure 4.21.

A complete listing of all of the above information is called a *data dictionary*. Having the data dictionary in advance allows the part of the GIS that handles data entry, or the spreadsheet or database program we choose to use, to check each value as we enter it. Sometimes we enter the numbers and values one by one into a special part of the database manager called the *data-entry module*. Often, we import into our GIS data manager all the records in a preexisting setup. Some of the more common databases and spreadsheets support specific formats for data exchange to allow this. The simplest form is to write a file with each of the attributes and their labels written as text, one per line, sometimes separated by commas and quote marks so that blanks and other symbols can be included. For example, the data in Figure 4.21 could be "unraveled" into the file in Figure 4.22.

For new GIS data, the process of entering attributes eventually comes down to someone (usually the lowest-paid person) entering the attribute values one by one into a database manager's data-entry module. The data usually come from a data form of some kind, onto which they had been recorded painstakingly by the person collecting the data.

Some data-entry systems are better than others. At the very least, the system should check the type and range of the value for each attribute at the time of entry. At best, it is helpful if the software allows things like copying a record but then changing it to reflect a new value, deletion, or changing of values that are wrongly entered at the time of entry, and if the software brings errors to your attention with beeps and messages so that correction can take place immediately. No software package should allow data to be lost if the computer crashes, the file fills up, or the user presses the wrong button.

Most GIS packages allow the use of almost any spreadsheet, such as Microsoft Excel, or database systems such as MySQL, PostgreSQL, or Access. Some require that you use the database entry system that comes with the GIS and no other. Each is slightly different, although all share the items discussed in this section.

4.6   EDITING AND VALIDATION

Many early geocoding systems had only limited editing capabilities. They allowed data entry, but error detection was by after-the-fact processing, and correction was by deletion of records or even whole data sets and reentry. Anything we can do in the geocoding process that reduces errors, or that makes errors easily detectable, we should indeed do. As an absolute minimum, data for lines and areas can be processed automatically for consistency, and any unconnected lines or unclosed polygons can be detected and signaled to the user. The connection between lines, known bordering of areas, and inclusion of points in areas is called map *topology*. Topology really comes into its own during the map validation stage.

The easiest way to avoid errors in geocoding is to ensure that errors are detected as soon as possible and then to make their correction easy. Video display during digitizing and audio feedback for error messages is essential. GIS software should spell out exactly what will happen in the case of an error. A common geocoding error is to overflow a disk-size quota while digitizing. It helps also to be able to recognize errors when they appear and to be able to understand their origin.

Some easy-to-detect errors are *slivers*, *spikes*, *inversions*, *lines that are not ended*, and *unsnapped nodes*, which we discussed in Chapter 3. Scaling and inversion errors are when the map appears squashed, like the titles at the beginning of a wide-screen movie shown on TV, or flipped. These are usually due to an incorrect digitizer setup procedure;

that is, they are systematic errors caused by incorrectly entering the control points for establishing the map geometry. *Spikes* are random hardware or software errors in which a zero or extremely large data value erroneously replaces the real value in one of the coordinates. Spikes are also sometimes known as *zingers*. Errors in topology, missing or duplicate lines, and unsnapped nodes are operator errors.

Plotting the data becomes a useful aid because unplottable data often have bad geocodes. Similarly, attempting to fill polygons with color often detects gaps and slivers not visible in busy polygon networks. The best check for positional accuracy is a check against an independent source map of higher accuracy. The equivalent of a plot for the attribute data is a *data listing* or *report*. Most data management systems have the ability to generate a report, listing the attributes as a table, or formatting them neatly for printing and checking. You should go line by line, checking the attributes and their values. However, even when the attributes and the map are validated by checking, it is still likely that errors exist in the links. One New York City database had more than 20 spellings for a single street name, for example.

The GIS often allows check plots to be generated that simply plot the label or identification number of the key within a polygon or next to a line. These maps and the tedious process of checking them should never be skipped. Moving straight on to making elegant graphics or doing a GIS-based analysis with erroneous data can be anything from embarrassing to dangerous, or even life-threatening. A data set that is correctly geocoded both positionally and with attributes is not necessarily logically consistent. Logical consistency can be checked most easily for topological data. Topologically, data can be checked to see that all chains intersect at nodes, that chains cycle correctly in a ring around polygons, and that inner rings are fully enclosed within their surrounding polygons. Otherwise, attributes can be checked to ensure that they fall within the correct range and that no feature has become too small to be represented accurately.

Everyone would like to say that the data in his or her GIS are accurate and correct. Obviously, this means several things. Accuracy of position means that the locations shown on the map are in their correct locations with respect to the real world. Of course, there may be a difference between the map that was geocoded and the "best possible" map. Positional error is sometimes tested or measured, and this is best done against another map of higher accuracy or against accurate field measurements such as GPS fixes. Another aspect of data is the accuracy of the attribute. A map may be perfect as far as appearance is concerned, but the roads and rivers could both be mislabeled as power lines. This type of error can be treated as a misclassification. Testing can also be conducted and can even be automated, as GIS data are already in a database management system.

A final issue is that of scale and precision. A map used for geocoding has a particular scale, such as 1:24,000. If this is the case, while the GIS allows us to compare data from another scale, say 1:250,000, it may not be appropriate to do so, as attributes, generalization of the features, and other properties of the map may be different at the two scales. Also, all data in the GIS have a degree of precision associated with them. If a highly detailed line is geocoded only to the nearest 10 meters on the ground, comparison with more detailed data becomes a problem. Generally, we should apply the same concerns and considerations of limitations to digital maps as we do to paper maps. Unfortunately, many people treat digital maps as absolutely correct instead of the digital alternative form of the analog maps to which they owe their humble origins.

The intelligent GIS user should know and understand the amount and distribution of error in a GIS database. Many of the sources of error are due to the method and process of geocoding. Some of the errors multiply as we move through the stages of data management, storage, retrieval, GIS use, and analysis. An understanding of error is essential to working effectively with GIS.

4.7   STUDY GUIDE

4.7.1   Summary

CHAPTER 4: GETTING THE MAP INTO THE COMPUTER

Analog maps are real, and displayed on a medium such as paper, while virtual maps consist of organized digital numbers O Geocoding is the conversion of spatial information into computer-readable form O Getting the map into the computer and dealing with input data is often the majority of the time and cost of GIS projects O Maps can exist in digital form and be both available or unavailable, or they can exist just on paper, or not at all—anyway sooner or later a GIS user will need to digitize a map O It makes most sense to start a GIS project using freely available data O The U.S. Federal Government makes immense amounts of digital map data available to anyone over the Internet O Finding out that data exist and are available can be by consulting books, libraries, and the Internet O Most agencies supplying map data use web portals and on-line search systems that allow data to be downloaded and used with GIS software O The USGS supplies DLGs, DRGs, DEMs, DOQQs, GNIS and LULC data via the National Map and its viewer and seamless distribution system O NOAA supplies data on nautical charts, GPS, geodesy, and real time data of use in navigation, such as weather maps O The United States Bureau of the Census distributes digital TIGER street maps that match the census attribute information O TIGER files are the basis for most geocoding by address-matching, i.e., finding a position using a house number, street, and city name O The U.S. Federal Government runs the NDSI, with data portals that can search across agencies for data O Geobrowsers are the latest way to browse and search for geographic information on the Internet O   Creating new data means semi-automated digitizing, scanning, or field data capture O Digitizing uses a digitizing tablet, the GIS equivalent of a drafting table O Inherent errors in the source map and its scale become embedded into the GIS data that digitizing creates O Scanning creates a raster map of the input map or image O If scans are done at too low a resolution, features drop out and cannot be recovered O Heads-up digitizing uses a scanned image or map to update or create a new map using manual vector editing O Field data can come from survey instruments, field notes, GPS, or field mapping O GPS is a very accurate way of collecting the location of points and lines in the field, and the data can be downloaded to GIS at high levels of accuracy O Many imagery and remote sensing programs supply data for use in GIS, which can be used in heads-up capture, or automatically processed to show land use, vegetation, man-made features, and more O Attribute data can be placed into GIS via DBMSs data entry modules, or entered into files using tools such as spreadsheets O   Data dictionaries allow automated checking and data validation, reducing error O A GIS should make errors easy to spot and easy to fix O GIS-based analysis with erroneous data can be embarrassing, dangerous, or even life-threatening O Accuracy should be assessed against independent sources of higher authority O

4.7.2   Study Questions

Analog-to-Digital Maps

1. List examples of maps that you would only find in analog form, and note some of the problems you would face in getting them into the computer. For example, ancient historical maps, 1920s road maps, and globes.

Finding Existing Map Data

2. Using the examples in the chapter, find and download on-line public data for your area of interest. How successful were you at finding data? How easy was it to get the data onto your computer? How easy was it to get the data into your GIS? Compare finding and acquiring data for a U.S. county with a foreign country.

Digitizing and Scanning

3. Make a table of various GIS applications that require non-standard data. What data capture methods would best suit these applications? For example, field data from an archeological dig, utility data on sewage pipe leaks, and data from the Christmas bird count.

Data Entry

4. Design a simple survey form, and have ten friends fill it out. Then design a database on paper to accept the information from the forms. What problems do you encounter? How might they be overcome or at least their impact minimized?

Editing and Validation

5. What software tools might be used in data editing? Name some of the common errors in geocoding that can be corrected by editing. Why might the value of an attribute in a record be invalid? What part of the database manager allows data editing and validation?

4.8   REFERENCES

4.8.1   Books

Bohme, R. (1993) *Inventory of World Topographic Mapping*. New York: International Cartographic Association/Elsevier Applied Science Publishers.

Campbell, J. (2001) *Map Use and Analysis*, 4nd ed. New York: McGraw Hill.

Clarke, K. C. (1995) *Analytical and Computer Cartography*, 2nd ed. Upper Saddle River, NJ: Prentice Hall.

Decker, D. (2000) *GIS Data Sources*. New York: Wiley.

Erle, S., Gibson, R.,. and Walsh, J. (2005) *Mapping Hacks: Tips & Tools for Electronic Cartography*. Sebastopol, CA: O'Reilly.

Thompson, M. M. (1987) *Maps for America*. 3rd ed. U. S. Geological Survey, Washington, DC: U.S. Government Printing Office.

United States Census Bureau (2000) *TIGER/Line File Technical Documentation*. On-line at: http://www.census.gov/geo/www/tiger/tiger2k/tiger2k.pdf

4.8.2   Internet Addresses

U. S. Geological Survey `http://www.usgs.gov`

National Map Viewer `http://nmviewogc.cr.usgs.gov/viewer.htm`

U.S. Census Bureau `http://www.census.gov/tiger/tiger.html`

NOAA `http://www.noaa.gov`

John Campbell's information sources `http://auth.mhhe.com/earthsci/geography/campbell4e/links4/appalink4.mhtml`

4.9 KEY TERMS AND DEFINITIONS

address matching: Using a street address such as *123 Main Street* in conjunction with a digital map to place a street address onto the map in a known location. Address matching a mailing list, for example, would convert the mailing list to a map and allow the mapping of characteristics of the places on the list.

analog: A representation where a feature or object is represented in another tangible medium. For example, a section of the earth can be represented in analog by a paper map, or atoms can be represented by Ping-Pong balls.

attribute: A characteristic of a feature that contains a measurement or value for the feature. Attributes can be labels, categories, or numbers; they can be dates, standardized values, or field or other measurements. An item for which data are collected and organized. A column in a table or data file.

data dictionary: A catalog of all the attributes for a data set, along with all the constraints placed on the attribute values during the data definition phase. Can include the range and type of values, category lists, legal and missing values, and the legal width of the field.

data entry: The process of entering numbers into a computer, usually attribute data. Although most data are entered by hand, or acquired through networks, from CD-ROMs, and so on, field data can come from a GPS receiver, from data loggers, and even by typing at the keyboard.

data-entry module: The part of a database manager that allows the user to enter or edit records in a database. The module will normally both allow entry and modification of values, and enforce the constraints placed on the data by the data definition.

digitizing: Also called semi-automated digitizing. The process in which geocoding takes place manually; a map is placed on a flat tablet, and a person traces out the map features using a cursor. The locations of features on the map are sent back to the computer every time the operator of the digitizing tablet presses a button.

digitizing tablet: A device for geocoding by semiautomated digitizing. A digitizing tablet looks like a drafting table but is sensitized so that as a map is traced with a cursor on the tablet, the locations are identified, converted to numbers, and sent to the computer.

drop-out: The loss of data due to scanning at coarser resolution than the map features to be captured. Features smaller than half the size of a pixel can disappear entirely.

drum scanner: A map input device in which the map is attached to a drum that is rotated under a scanner while illuminated by a light beam or laser. Reflected light from the map is then measured by the scanner and recorded as numbers.

editing: The modification and updating of both map and attribute data, generally using a software capability of the GIS.

flat file: A simple model for the organization of numbers. The numbers are organized as a table, with values for variables as entries, records as rows, and attributes as columns.

flatbed scanner: A map input device in which the map is placed on a glass surface, and the scanner moves over the map, converting the map into numbers.

FTP (File Transfer Protocol): A standardized way to move files between computers. It is a packet switching technique, so that errors in transmission are detected and corrected. FTP allows files, even large ones, to be moved between computers on the Internet or another compatible network.

gateway: A single entry point to all the servers and other computers associated with one project or organization. For example, the U.S. Geological Survey, though spread across the country and throughout dozens of computers, has a single entry point or gateway into these information sources.

geocoding: The conversion of analog maps into computer-readable form. The two usual methods of geocoding are scanning and digitizing.

Internet: A network of computer networks. Any computer connected to the Internet can share any of the computers accessible through the network. The Internet shares a common mechanism for communication, called a protocol. Searches for data, tools for browsing, and so forth ease the tasks of "surfing" the Internet.

medium: A map medium is the material chosen on which to produce a map; for example, paper, film, Mylar, CD-ROM, a computer screen, a TV image, and so on.

network: Two or more computers connected together so that they can exchange messages, files, or other means of communication. A network is part hardware, usually cables and communication devices such as modems, and part software.

NOAA (National Oceanic and Atmospheric Administration): An arm of the Department of Commerce that is a provider of digital and other maps for navigation, weather prediction, and physical features of the United States.

real map: A map that has been designed and plotted onto a permanent medium such as paper or film. It has a tangible form and is a result of all of the design and compilation decisions made in constructing the map, such as choosing the scale, setting the legend, choosing the colors, and so on.

report: A listing of all the values of attributes for all records in a database. A report is often printed as a table for verification against source material, and for validation by examination.

scanning: A form of geocoding in which maps are placed on a surface and scanned by a light beam. Reflected light from every small dot or pixel on the surface is recorded and saved as a grid of digits. Scanners can work in black and white, in gray tones, or in color.

server: A computer connected to a network whose primary function is to act as a library of information that other users can share.

stream mode: A method of geocoding in semi-automated digitizing, in which a continuous stream of points follows a press of the cursor button. This mode is often used for digitizing long features such as streams and coastlines. It can generate data very quickly, so excessive or deviate points are often weeded out immediately by automated line generalization within the GIS.

TIGER: A map data format based on zero, one, and two cells; used by the U.S. Census Bureau in street-level mapping of the United States.

topology: The numerical description of the relationships between geographic features, as encoded by adjacency, linkage, inclusion, or proximity. Thus a point can be inside a region, a line can connect to others, and a region can have neighbors. The numbers describing topology can be stored as attributes in the GIS and used for validation and other stages of description and analysis.

U.S. Census Bureau: An agency of the Department of Commerce that provides maps in support of the decennial (every 10 years) census of the United States.

USGS (United States Geological Survey): A part of the Department of the Interior and a major provider of digital map data for the United States.

validation: A process by which entries placed in records in an attribute data file, and the map data captured during digitizing or scanning, are checked to ensure that their values fall within the bounds expected of them and that their distribution makes sense.

virtual map: A map that has yet to be realized as a tangible map; it exists as a set of possible maps. For example, the same digital base map and set of numbers can be entire series of possible virtual maps, yet only one may be chosen to be rendered as a real map on a permanent medium.

4.10 PEOPLE IN GIS

Alan Millais: GIS student and future Air Force navigator

KC: How did you first get interested in GIS?

AM: I've been around maps as long as I can remember. Both of my parents have huge map collections, and when GIS came out, of course, they were some of the first to start using it. In high school, my Mom would drive me to her job and show me GIS maps—wow, this was pretty cool. I'd already decided to do Geography when I was in high school.

KC: So you actually met GIS when you were still in high school?

AM: Yeah. My Mom took me to the ESRI Conference my junior year in high school.

KC: Now that you've taken the whole course sequence in GIS at UCSB, what did you gain from it?

AM: More of a kind of an inner workings and understanding of it. The first lab in the first class, you think, "Oh gosh, this is going to be long," but, by the end, you're into theory and the abstract side of GIS, more like the science of it all. By the end, not only did I have a more practical understanding of how to use GIS and which buttons to push, but also how it came about, what need it arose from, and what niche it filled. On the graduation evaluation survey, I listed the advanced GIS projects class as the course that I learned the most from out of all my college years, because the project we did—on Santa Cruz Island—was probably the most comprehensive project I'd done in college in terms of research and the final output. We're also showing our project at the ESRI Conference this year, taking it to the Map Gallery.

KC: What was the purpose of the Santa Cruz Island project?

AM: We worked a lot with hand-drawn 1800s era maps that we scanned, and then a lot of newer satellite photos. We georeferenced data and scanned a lot of maps. It seemed to me about as complicated as a subject could get with GIS, because we were using every data source you could possibly imagine to come up with a project. That definitely showed the versatility of it, because a lot of the time you kind of get lost in your own little section of the project as opposed to viewing the whole GIS and the questions being asked.

KC: Tell me a little bit about your summer job, what you do, and how you got the job.

AM: I work in the GIS office of the City of Ventura, I'm not doing anything GIS directly. They still use ArcView 3, though they've heavily modified it. But I was building 3-D models of the downtown redevelopment zone with the hope of putting it into a GIS. We had picture facades and correct roof lines and correct building heights and all that sort of thing, with the hope of putting it into a 3-D dataset to see how it would look for the corresponding area with the hills and the background. But the project never really got off the ground. I was able to finish it all in Google Sketchup, but we never found a GIS that was capable of handling that much data.

KM: What do you intend to do now that you have graduated, and will it involve GIS?

AM: I'm going into the Air Force as a navigator. I will be using GPS—it is GIS work but not in a traditional sense of me sitting at a desk. I joined the Air Force with the intention of doing the pilot thing and then maybe getting out and working for the NGA. I don't know how it's going to change over the course, because I have a relatively long service commitment and then once I get out I'll have left behind GIS—I'll have a lot of catching up to do. I still do plan on getting my Masters in Geography though.

KC: What advice would you give to a new student just taking their first class in GIS?

AM: I would say have patience. No matter what level you're at either it's too easy or it's too hard. If you know computers, the start menu lab is kind of frustrating, but if you don't it's useful. Once you get learning the program in depth it takes a lot of patience, because it has so many capabilities that you can get very easily lost in GIS. Also, I'd say spend as much time as possible in the lab, because the lectures are useful but, unless you're actually on the computer doing GIS all the time, you miss something.

KC: Thanks, and good luck in your new career.

CHAPTER 5

What Is Where?

If you don't know where you're going, you'll end up somewhere else.

Yogi Berra

5.1 BASIC DATABASE MANAGEMENT

A GIS can answer the two questions "what?" and "where?" More importantly, a GIS can answer the question: What is where? *Where* relates to the map behind all GIS activities, and the means of using numbers to refer to precise locations. The *what* relates to the features, their size, geographical properties, and, above all else, their attributes. Getting information about features by locating them on the map is what the toolbox definition of GIS in Chapter 1 meant by *retrieval*.

These are not trivial questions. Other forms of data organization often fall apart when dealing with "where." The telephone book, for example, a list organized alphabetically by last name, gives only relative locations (street addresses) and a house number. An entirely new directory is necessary for each new district, and to retrieve the telephone number of a friend in another town, perhaps just across the river, becomes a major problem because you require a different telephone directory than the one covering your own neighborhood.

Geographic search, for example finding all the phone numbers of people on a single city block, is not available easily to the user of a telephone directory. The secret to *data*

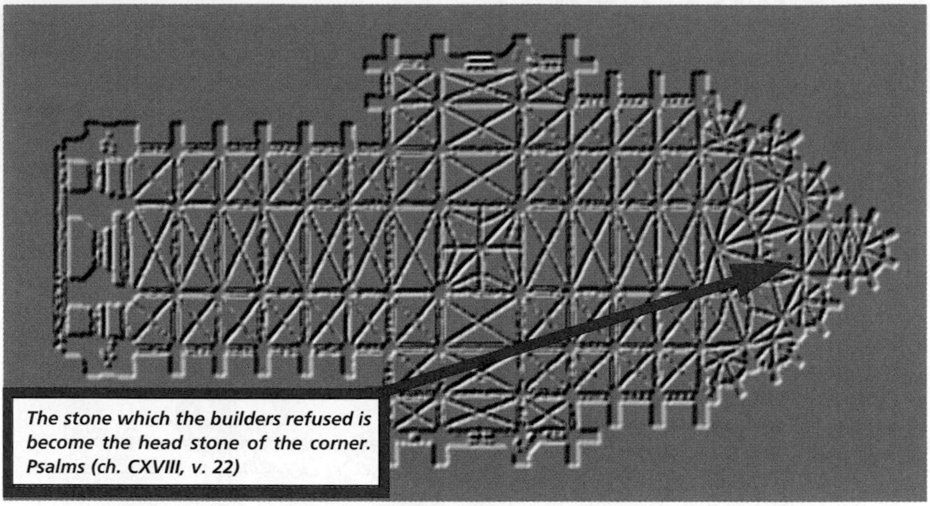

The stone which the builders refused is become the head stone of the corner. Psalms (ch. CXVIII, v. 22)

Figure 5.1: A medieval memorization aid, associating text to be remembered with locations. A form of geographical data model.

retrieval, the ability to gain access to a record and its attributes on demand, is in data organization. In Chapter 3 we examined the various ways that the graphic part of a GIS can be structured for storage inside the computer. The structure that any given GIS uses, and how the map is encoded into that structure, fully determine how easy it is to find a record and extract its values for use. Once again, the attribute and the map data have different means of access. At the most simplistic level, the GIS is a computer program that accesses data stored in files. Obviously, with a great deal of data, the ability to access the files quickly is important if the user is to receive interactive control of the GIS.

At the logical level, access requires a *data model*, a theoretical construct that becomes the key for unlocking the data's door. In medieval times, when much information had to be memorized, monks would train their powers by spending hours committing to memory the parts of a specific cathedral, capturing a mental picture of the interior and exterior so that as they later learned text, such as the contents of the Bible, they could mentally "place" a chapter, a verse, a line, or even a word in a location that they had already memorized (Figure 5.1). Reciting that verse, then, became the memorization in reverse. The monk would mentally walk to the right place in the cathedral, and the words would be stored there. The cathedral, not in substance, but as a visual memory, became the data model. Without such a data model, data cannot be searched or extracted and therefore become worthless.

We can define a data model, then, as a logical construct for the storage and retrieval of information. It is the computer's way of memorizing all the GIS data that we need to use. This is different from the data structures we examined in Chapter 3, because these deal primarily with how the data are physically stored in files on the computer system. As we have seen, this means that a GIS must have at least two data models, and that the two must have a bridge or link between them to tie the attributes and the geography together. These are the *map data model* and the *attribute data model*. In Chapter 3 we illustrated some of the storage and organization aspects of map data models. In this chapter, we deal with the attribute models and then look closely at how the models help in locating and then extracting data from both map and attribute databases.

Database management system's (DBMS) heritage is from within computer science, but the user community is as broad as that of GIS, literally millions of firms, accountants, colleges and universities, banks, and so forth that need to keep and organize records by computer. The earliest database management systems date from the efforts of the early 1970s, when large mainframe computers were used, data-entry was by key punch and punched cards, and the technology was called *automatic data processing*.

Database management went through its own revolutions due to the technological trends we have already discussed from the GIS viewpoint: the microcomputer, the workstation, the network, low-cost bulk storage, interactive and graphical user interfaces, and so on. Database management, however, has also been significantly influenced by the intellectual breakthroughs that led to radical changes in the way that attribute data can be stored in files. The latest revolution, the object-oriented database system, is sometimes used for GIS, and is covered in Chapter 11.

The parts of a DBMS have remained fairly consistent over time, regardless of how the attribute data are actually placed into files. The *data definition language* is that part of the DBMS that allows the user to set up a new database, to specify how many attributes there will be, what the types and lengths or numerical ranges of each attribute will be, and how much editing the user is allowed to do. This establishes the *data dictionary*, a catalog of all of the attributes with their legal values and ranges. Every DBMS has the ability to

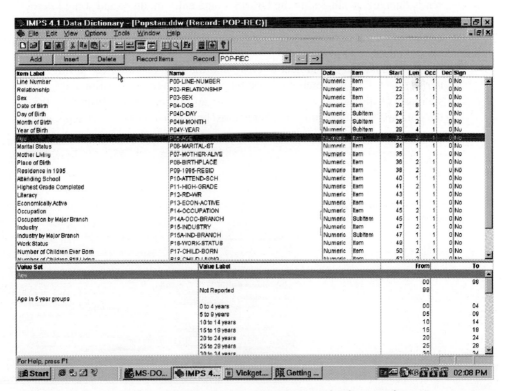

Figure 5.2: The Census Bureau's Integrated Microcomputer Processing System data dictionary editor.

examine the data dictionary, and the data dictionary itself is a critical piece of metadata (data about the data) that is often required when the database has to move between systems. For example, the U.S. Census Bureau maintains and makes available a database manager called the Integrated Microcomputer Processing System (IMPS). In this system, it is possible to design the data fields to be collected in a survey, to give them labels, types, values, etc. A screen-shot of the data dictionary module is in Figure 5.2. Once the user has built the data dictionary, it creates tables that can be used for data collection and that will import into GIS. Another important function is the maintenance of a database catalog. Most GIS packages have a way of opening a new set of related maps and tables, often called a *Catalog*. The GIS uses the catalog to track the data, the files, and the updates to the files, and often has tools for creating, editing, and deleting catalogs or projects. This is important, because in a professional GIS environment, often several people need to work on one GIS project at the same time, and all want to see the latest data configuration.

The most basic management function is *data entry*, and because most entry of attribute data is fairly monotonous and may be by transcription from paper records, the DBMS's data-entry system should be able to enforce the ranges and limits entered into the data dictionary by the data definition language. For example, if an attribute is to contain a simple percentage in the range 0–100, and the data-entry person types a value of "110", the DBMS should refuse to accept the value and alert the person. In the Internet era, much data entry is done in forms over the web. Every time you fill out a book order, make a plane reservation, or update your Myspace profile (Figure 5.3), you use an on-line form that sends data to a database via the DBMS's data entry module (and some Internet protocols). The DBMS enforces the values in the data dictionary at the point of entry. For example, if a state name is required, you can only choose from a scrolling list of the 50 states, and so cannot make a misspelling.

All data entry is subject to error, and the first step after (or during) entry should be *verification*. This is sometimes done by producing a report or printed copy of the data in a standard form that can be checked against an original, or confirmed by the user. Even if the data are error free, most databases must be *updated* to reflect change, when you change

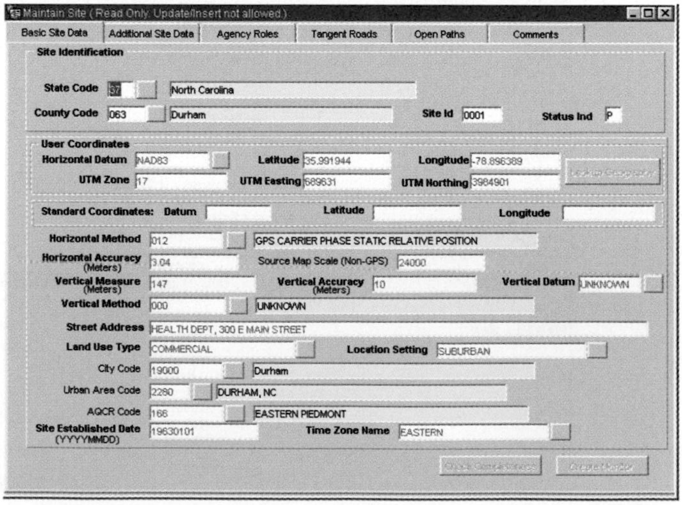

Figure 5.3: On-line forms-based data entry system for geospatial data as part of the Technology Transfer Network Air Quality System, maintained by the Environmental Protection Agency. See: www.epa.gov/ttn/airs/airsaqs/

address or phone number for example. Deletion, insertion, and modification of records, or sometimes changes to the data dictionary itself, must be made frequently. This is done by using the data maintenance part of the DBMS. Care must be taken when updates are made, because changes create a new updated version of the entire database. Sometimes modifications are done in batches and a new version of the database is "released," reflecting a whole suite of changes, perhaps to reflect the calendar year. Increasingly, computer programs handle updates automatically, for example changing your phone number across multiple files and records.

   With the preceding tasks complete, the DBMS can then be used to perform its more advanced functions. These are the *sorting, reordering, subsetting,* and *searching* functions. For example, a database of student records could be sorted by grade-point average to find all students below a "C" average. A database of students containing zip codes in their address could be reorganized by zip code to make mailing easier for the mailroom staff, an example of reordering. Subsetting involves using the *query language*, the part of the DBMS that allows the user to interact with the data to perform these tasks, to create a new data set that meets certain search criteria, such as all students who have more than 100 credits toward their degree. Finally, a *search* for a specific record is often needed. At a public terminal, for example, students may be allowed to type in their student id numbers to allow them to examine the grades they got in a given semester. All of these functions are part of regular DBMS and are performed differently in each different system. Nevertheless, all of these functions are common to both DBMSs and GISs. Examples of these queries are in Figure 5.4. The first-generation DBMSs used a hierarchical structure for their file organization. For example, a university might contain a division or school, the

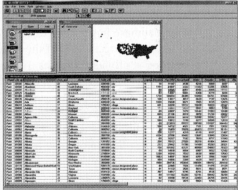

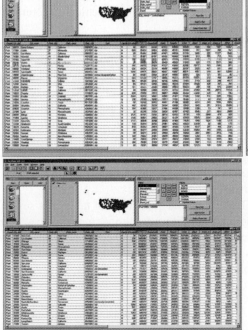

Figure 5.4: Database operations on tables in ArcView 3.2. Above left: *Search* for city name = "Santa Barbara"; above: U.S. cities *sorted* alphabetically by name; left: *Find* cities over 1M population. In each case the query result is visible in the table and on the map.

school might contain a department for a single discipline such as Geography, and a department may have a group of students who are majors in that department. All students who are declared majors must be assigned to a department, each department belongs to one division, and the entire university consists of a group of divisions. The file structure of a hierarchical system can be organized in the same way. A top-level directory could contain a list of divisions and a set of divisional directories. Changing down to the next level, the divisional level, reveals a list of departments and departmental directories. Down in a departmental directory lie the files containing the student records, one student per record, with attributes such as the student's name and address, year of expected graduation, and grades in various classes. The first generation of database managers used exactly this type of organization of files and records as its data model. We are familiar with this as the folders and files mechanism for organizing data on a PC.

Life is not as simple as the hierarchical model would like it to be, however. In many cases, relationships between records overlap. This is even more so for geographic data, as we will see below. For now, consider a simple hierarchy for the political administration of a state. The order could go state-county-city-district. For much of the country, this would work. In New York City, however, the city of New York consists of the five "boroughs," each of them a county in its own right as far as the state of New York is concerned. So the simple hierarchy model already fails, because the hierarchy is literally stood on its head in one case. The data model would require that the county file contain records for cities, but the reality would be the opposite, that the city should contain the county.

Another complex case would be multiple membership. Consider a single house that falls into a fire district, a police district, a school district, a voting district, and a census tract. Five databases, each structured using a different hierarchy, would have five completely independent ways of getting to the attributes for the single house. Although each database might store different kinds of information, at some stage it might be useful to assemble together all of the data to be used. Under a hierarchical structure, this cannot be done at all in some cases.

The revolution in DBMS that untangled this database logjam was the use of relational database management systems. Beginning with a set of theoretical break-throughs in the 1970s, relational DBMSs swept the field and replaced almost all existing systems during the 1980s. They remain the dominant form of DBMS today. The relational model is rather simple, and from the user's standpoint is an extension of the flat file model. The major difference is that a database can consist of several flat files, and each can contain different attributes associated with a record. Using the example of the house above, the house can now be a single record in several databases, none of which requires a hierarchy. If the hierarchy still makes sense, we can keep separate files by district or use codes such as zip codes to reveal districts.

For example, take a look at Tables 5.1 and 5.2. The first table lists several types of pizza available from a popular seller in California, although they have pizza parlors nationwide and overseas. This table looks like a menu, and could be used, for example, to choose by price, or to select pizzas with or without a particular ingredient. There are three attributes, the pizza name, the toppings, and the price.

Table 1: Pizzas

Pizza	Topping	Price
BBQ Chicken	Barbecue sauce, Gouda and Mozzarella cheeses, BBQ Chicken, red onions, and cilantro.	12.99
Chicken Buffalo	Grilled chicken marinated in Buffalo sauce with Mozzarella cheese, carrots, celery, and Gorgonzola cheese.	12.49
Gorgonzola with Pear	Caramelized pears, Gorgonzola, Fontina and Mozzarella cheeses, caramelized sweet onions, and chopped hazelnuts.	12.49
Mango Curry Chicken	Tandoori chicken, mango, onions, red peppers, and Mozzarella cheese in curry sauce.	11.99
Carne Cilantro	Grilled steak, chilies, onions, cilantro pesto, Monterey Jack, and Mozzarella cheeses, tomato salsa, and cilantro.	12.99
Jamaican Chicken	Grilled Jamaican Jerk spiced chicken breast with Caribbean sauce, Mozzarella cheese, applewood smoked bacon, mild onions, red and yellow peppers, and green onions.	12.49
California Special	Applewood smoked bacon, grilled chicken and Mozzarella cheese, Roma tomatoes, chilled chopped lettuce tossed in mayonnaise, and fresh sliced avocado.	12.79
Chipotle Pizza	Grilled chipotle chicken, mild chilies, chipotle sauce, Mozzarella and Enchilado cheeses, roasted corn & black bean salsa, cilantro, and lime cream sauce.	11.99

Table 5.2 is a table of locations where this pizza menu is available. This table also has attributes, but they are related to where the pizza parlor is, and what some of its services are. This table might be used for choosing a location nearby, with special facilities, etc. One way to make a single table to use in our GIS as a flat file would be to join the tables together. In the joined table, we would need a row for every pizza type at every one of the locations. With 8 types of pizza and 9 locations, we would need a flat file with 8 x 9 = 72 rows, and 8 columns. But this table would carry quite a bit of repetitive information. It is better to have two tables; not only is the result smaller, but the tables are easier to organize, and could be used to maintain different types of information and serve different needs. Should we make a pizza order, we would have to select a location, one or more particular pizzas, and we might need yet another type of table, with data on delivery location, name, phone number, number of pizzas ordered, etc.

The strength of this system is that we can cross-reference the multiple files as necessary, and even change them around. Perhaps the types of pizza, the actual toppings available, and the price would be different by location (as might be expected). No problem, since each pizza parlor could maintain its own menu and price lists without reference to others. In fact the aggregate file, with all 72 rows and 8 columns, would be very inflexible; should one location run out of one of the pizzas, they would have to update the one file for every location.

Table 2: Pizza Parlors

Location	Street Address	Zip Code	Kids menu	Private Rooms
Anaheim	123 W. Longley Ave.	92802	Yes	No
Arcadia	909 S. Goodchild Ave.	91007	Yes	No
Bakersfield	31 Rhind Hwy.	93311	Yes	Yes
Beverly Hills	212 S. McGuire Dr.	90212	Yes	No
Brea	1458 Geo Mall	92821	Yes	No
Brentwood	1555 San Fernando Blvd.	90049	Yes	No
Burbank	202 N. San Valerio Road	91502	Yes	No
Canoga Park	931 Liberty Canyon Blvd.	91306	Yes	No
Cerritos	444 Los Osos Mall	90703	Yes	No

The price of this flexibility is that should we have a particular pizza, say a Buffalo Chicken Pizza from Brea with extra cheese, we would need a walk-between. This is done by having a unique identifier that can be found in every table. For example, a particular pizza could have a time and date stamp with a code for the location. This would uniquely identify each pizza, and allow cross-referencing of the tables.

Critical to each part of a relational database is the special attribute that serves as a marker rather than a regular attribute. We could assign to every record a unique identifier, for example, the pizza time and date stamp or a transaction number (for example, order number 546), to make it different from all others. This "key" attribute can then serve as a link between the flat files. Because the key is unique it allows us to extract various attributes and records from one database and others from another as required. We can store the data in a group of linked files, each of which can be used, have data entered. These files can be edited, updated and searched separately without affecting the others.

The relational database manager contains a new set of data management commands that allows the keys and links to be exploited. These typically include such actions as *relate*, to select from two flat files that have a common key attribute, and *join*, to take the relate operation output and merge them into a single database. So the relational data model, while permitting records and attributes to be separated into different files for storage and maintenance, also allows the user to assemble any combination of attributes and records, as long as they are linked by a key attribute. A *join* can create many unneeded sub-records for a single feature, such as multiple records with different dates, so care should be exercised when joining databases.

5.2 SEARCHES BY ATTRIBUTE

Most GIS systems include as part of the package a fairly basic relational database manager, or simply build on the existing capabilities of a database system. Searches by attribute then are controlled by the capability of the database manager. All DBMSs include functions for basic data display; that is, show all attributes in a database, show all records with their attributes, and show all existing databases. Most also allow records to be output in a standard form, with a particular page layout and style, called a *report generator.* If we need a paper copy of the database, perhaps for checking and verification, then the report generator is used.

As far as actually doing retrieval is concerned, the DBMS must support functions that fall into the category of *query.* As we have seen, a DBMS should allow sufficient data query that any record can be isolated and any subset required for mapping found easily. We may also sometimes wish to reorder or renumber an attribute. A *find* is the most basic attribute search. Find is usually intended to get a single record. We might find record "Santa Barbara", for example. Finding or locating attributes can be done by *search* or by *browse.* Browse searches record by record, displaying each, until the user finds the one needed. Sorting can sort alphabetically for a field, or numerically for a number. Note that a sort may or may not deal with missing values, and where it places them may be significant.

A *restrict* operation allows the user to retrieve a subset of the total number of records by placing a restriction on the attributes' values. For example, we could restrict a search to all records with a date more recent than 12/31/1999, or to cities with a population of more than 100,000 people. A *select* operation allows us to choose what attributes will be taken out from another database to form a new database with fewer "selected" attributes. We usually do this to *join* these records and attributes onto another database in the relational system. As we will see in Chapter 6, a *compute* operation allows us to compute a value for an attribute, to assign a value, or to do mathematical operations between attributes—divide one by another, for example. We can also usually *renumber* an attribute, that is, change the values to our specifications. We might want to find all percentages in an attribute and change them to a zero if they fall below 50% or a one if they are greater, so that we can do a binary combination with another layer.

For example, using a command-line syntax, in a database of state populations and areas called states, we could use compute to create a new attribute called population density.

```
compute in states population_density=population/area
<50 records in result>
```

This creates a new attribute, which we can recode into high (3), medium (2), and low (1)

```
restrict in states where population_density > 1000
<20 records selected in result>

recode population_density = 3
<20 values recoded in result>

join result with states replace
<20 records changed in state>

restrict in states where population_density > 100
<12 records in result>
```

```
recode population_density = 2
<12 values changed in result>
join result with states replace
<12 records changed>

compute in states where population_density != 3 or 2
population_density = 1
< 18 records changed>
```

The recoding is now complete. We can now sort by the new recoded value.

```
list attribute in state population_density

< In database "state" attribute values for
"population_density">
<1   18 records>
<2   12 records>
<3   20 records>
<no missing values>

sort result by population_density
<50 records in result>

replace state with result
<50 records changed>
```

In this exchange, commands are given one line at a time and often must be used in combination to get the desired effect. Note also that many database systems work by performing their operation on a temporary working set (called `result` in the example) that must be placed back into an existing database when necessary. Many DBMSs use menus, variations on different query languages, and different keywords and commands to accomplish the same results.

Before we leave searching by attribute, consider the problems of using these tools to do even a simple search by distance. Merely to find the distance of each record from one point, we would have to do two subtractions, a multiplication, and a squaring, and then a summation and a square-root operation, assuming plane coordinates. Obviously, these database tools, although of immense use, are going to be only humble assistants in our geographical searching needs.

5.3   SEARCHES BY GEOGRAPHY

When we considered searching attributes, we looked at the following search and retrieval commands: *show attributes*, *show records*, *generate a report*, *find*, *browse*, *sort*, *recode*, *restrict*, and *compute*. Moving over to the spatial data within a GIS, some of the operations possible are just spatial equivalents of these, while some are more complex. We discuss the simple retrievals first.

In the map database our records are, instead, features. There are some special attributes specific to the spatial data, and those relate to the coordinates and their measures,

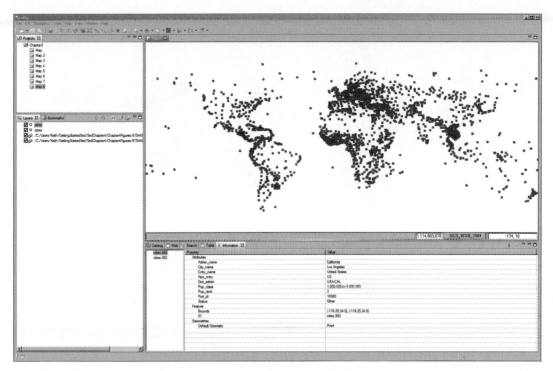

Figure 5.5: Using the uDig GIS to select the city of Los Angeles from a global cities database.

plus the characteristics of the lines and polygons. Showing attributes, then, consists of examining the new spatial attributes, such as the actual coordinates themselves, the lengths of arcs, and the areas of polygons. Note that these are already valuable. They could have been the source of the areas used in the population density calculations in the example in Section 5.2. Using the attribute search functions on these attributes now has spatial consequences. For example, we could find all arcs greater than a certain length, or polygons more than 1 hectare in area.

Show all records in a spatial sense becomes either show all attributes or display all features on a map. Generating a map, which allows us to search for information visually, is a spatial retrieval operation as far as the GIS is concerned. If we wish to generate a report, the spatial equivalent would be to produce a finished map to cartographic standards, including labels, metadata, legend, and so on.

Browsing works on a map by highlighting. We may color-code a specific feature or features. Some GISs allow a displayed single feature to be blinked on and off for visual effect. Finding becomes what many GIS packages call *identify* or *locate;* that is, use a pointing device of some kind such as a mouse to point to a feature. Indicating its selection successfully can then retrieve that feature's attributes from the attribute database (Figure 5.5).

This spatial searching—browsing the map and picking features—is a very powerful GIS capability indeed, especially if the GIS is on a portable computer and the feature located is the one you are standing in front of as located by your portable GPS receiver. On the map we can also search by indicating a single feature, all features within a rectangle

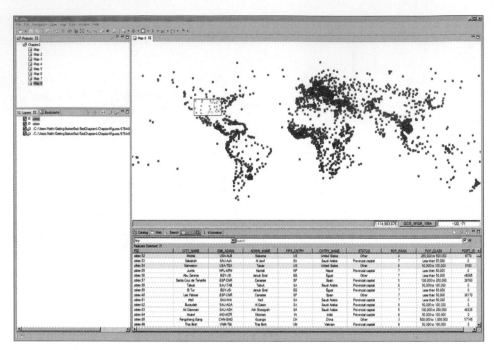

Figure 5.6: Using the uDig GIS to select from the world cities by drawing a search rectangle.

that we drag out, or all features within an irregular area that we sketch on the screen with a drawing tool. This search function is illustrated in Figure 5.6.

Sorting has less spatial meaning and is usually given a GIS context by examining the spatial pattern that results from a sort by attribute. For example, we could sort tornado occurrences in the U.S. by the number of fatalities associated with them (Figure 5.7). Similar to this is recoding by attribute. In the example in Section 5.2, we converted the density values into high, medium, and low values by recoding. We could use this recoded attribute directly to make a choropleth or shade-tone map of the population density, as in Chapter 7.

Operations that perform attribute manipulations, recode, and compute in a spatial sense merely change which features are displayed and how. For example, we could compute and display a new attribute, such as population density in Section 5.2. However, each operation has an equivalent spatial operation. Recoding features spatially, that is, changing the scope of their attributes, is equivalent to a spatial merge. Removing isolated pixels by assigning them to their enclosing or most dominant neighboring region is an example.

Compute can be given a spatial context as distance, length, area, or volume computations and transformations. For example, we could generate a new map that contains the distance to the nearest point feature, the cumulative downstream distance along a stream, or the travel distance along a road network. These new maps could be displayed or used in conjunction with the GIS layers to perform more complex operations, just as we combined the attribute *query commands* in the preceding section to achieve an end result.

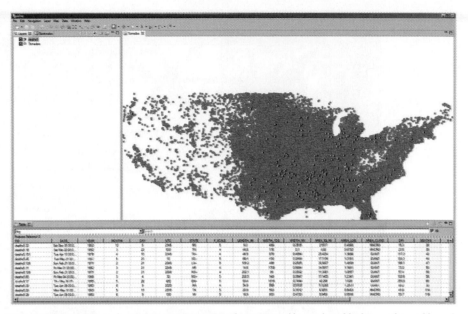

Figure 5.7: Sorting. Tornado occurrences 1950–2008 in the USA sorted by number of fatalities within a table, and the red highlighting of those with more than five fatalities on the map. Example using uDig.

Select and join are the remaining operations. *Select* means to extract specific attributes and to reduce the width of the database. By first selecting, we could use only certain themes or layers in a GIS retrieval operation, or we could change the map scale or extent. Picking a subregion, merging quadrangles into a county, aggregating land cover categories from level 2 to level 1 (from seven classes to one class for urban land, for example), picking only major rivers from a full stream network, or generalizing lines on the map for depiction at a broader scale are all examples of a geographic select.

The form of select used most frequently in GIS operations is the buffer operation, that is, to select only those parts of a map or those features that lie within a certain distance of a point, a set of points, a line, or an area. Examples are to restrict a search for malaria victims in a West African nation to those villages more than 10 kilometers from a health clinic, or to limit our search for a summer cottage to one within 200 meters of a lake. Buffers around points form circular areas, around lines they form "worms," and around polygons they form larger regions (Figure 5.8).

A *join* operation is the cross-construction of a database by merging attributes across flat files, as discussed in the pizza example. In the geographic sense, this is termed a map *overlay*. In a map overlay, a new map is created that shares the space division of both source maps. Every new polygon created on the map has a new attribute record in an expanded attribute database associated with the map overlay. This means that the overlay map is searchable by either of the sets of regions used to create it. Examples are city health districts overlain with zip codes, so that we can use data assembled on health, accidents, and so on with data from mailing lists on population, ethnicity, income, and other variables. By joining the map layers with a map overlay, we can compute, for example, the average income of people suffering from heart disease, or determine by age groups those people at risk from certain age-related diseases. For example, in Figure 5.9 we show the

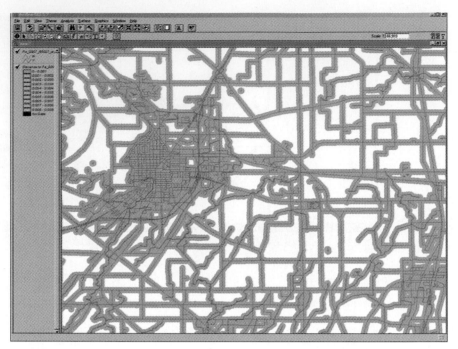

Figure 5.8: Use of ArcView 3.2 to create buffers around the roads and streams from the census' TIGER files in part of Dodge County, Wisconsin.

same buffer as Figure 5.8 (areas within 100m of a road or a river) overlain with areas over 246m in elevation. Creating the overlay has necessitated building several queries that produced intermediate results, as can be seen on the legend to the left.

The power of the attribute database manager comes from the use of multiple operations in sequence. The same is true of GIS retrieval operations. For example, we could take the overlain census data in Figure 5.9 and again multiply it with a computed distance map from hospitals represented as points, perhaps geocoded from the Yellow Pages of the telephone book. This would allow us to display a map showing, perhaps in red, where large numbers of people live long distances from hospitals in a particular congressional district. Some geographic queries don't fit a clear definition of attribute query operations. A map display can be zoomed into and out of to change the equivalent resolution with comparative ease. Some geographic queries can search by geographic properties and topology.

An entire suite of geographic searches are searches and tests by relations of points, lines, and areas. For example, we can select all points enclosed within one or more regions. Join then allows us to assign attributes from the point features to the area features, say weather statistics from weather stations to administrative districts. Typical GIS searches are point in polygon, line in polygon, and point distance to a line. If the points are oil storage tanks and the lines are rivers, this can have great analytical value. We can also weight layers in an overlay, perhaps building a composite layer called *land suitability* in the same way that early planners did with overlay mapping. Another popular GIS layer to construct by weighting is a cost surface, built from a combination of layers and distance calculations. Low spots on this map may be good for business locations.

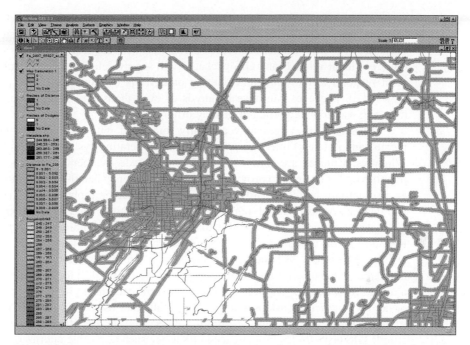

Figure 5.9: Same area as Figure 5.8, showing a map overlay. Areas shown in purple are within 100m of a road or a river, and are above 246m in elevation.

Finally, some highly specific geographic computations are possible. Examples are line-of-sight calculations, computing a viewshed or region that is visible from one place on a map using a base of terrain; maps showing slope and aspect or direction of slope, which might be useful in assessing development suitability or flooding potential; traffic volumes measured on a street network, used to predict traffic jams; or maps showing merged outputs of models or predictions with data, to predict earthquake hazards.

5.4   THE QUERY INTERFACE

Both database management and geographic information management share the fact that the user must somehow interact with the data in an appropriate way. The first generation of DBMS and GIS both used only batch-type interaction with the data, usually closely linked to working with the operating system, the physical management of disk, and so on. This type of interaction dates from the punched card, in that all processes had to be thought out in advance and a file (or stack of cards) produced that could execute the different commands one at a time.

When interactive computing became commonplace, the command line took over as the vehicle for data query. Commands were typed into the computer one at a time, under the control of the DBMS itself, and the software responded by performing the computations one at a time while the user waited for the command to be completed. Many GISs still use this type of interaction, or permit it to allow the use of macros. *Macros* are files containing commands to be executed one at a time. If an error is detected in a macro, the execution can be stopped and the file modified to correct the mistake.

In the typical form, a command consists of a keyword for the operation such as IMPORT, OVERLAY, SELECT, and a set of optional or required parameters. Parameters may be file names, numerical values associated with the task, names of options, or any other pertinent value. Many GIS packages will supply defaults for any parameters left out, most respond to the command without parameters by giving a list of parameters. The largest GIS packages can have hundreds or even thousands of commands, giving more than one way to do a particular task by different routes.

Most GIS packages now are fully integrated with the WIMP (windows, icons, menus, and pointers) interface specified by the operating system, such as Windows or X-Windows. Choices are now most commonly made by menu, with message windows popping up for the user to provide essential parameters when they are required. Values can also sometimes be set by sliders, widgets, and by screen tools such as dials, choice lists, and buttons. An example of a query being composed in a window is shown in figure 5.10.

Another fairly recent trend is that most GISs also contain a language or macro tool for automating repetitive tasks. Examples are ArcGIS's Visual Basic for Application (VBA), MapInfo's MapBasic, and the legacy Arc/Info's AML. In some cases, these languages can interact with the graphical user interface (GUI) tools, presenting a choice as a menu, for example. Thus any GIS user can now become a programmer, establishing his or her own particular task as a query tool for all other users. In large GIS operations when training and employee time are limited, GIS analysts are often employed to automate simple GIS tasks for less experienced GIS users, such as routine queries or simple database updates.

Finally, there have been efforts to provide a suite of database interaction commands that all users can assume as standard interface to relational databases. The result, the Structured Query Language (SQL), has found a broad acceptance as a much-used tool in

Figure 5.10: Example of a query, using ArcGIS 9.3. Figure courtesy of Indy Hurt.

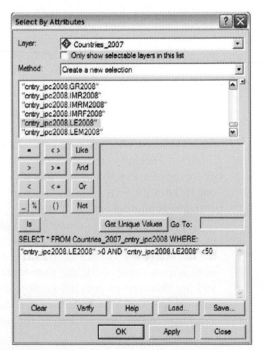

regular database management, although less used in GIS. Some have argued that almost all GIS operations are possible in SQL, while others have sought to extend its capabilities into the spatial domain. Given the differences between DBMSs, this is a welcome effort. SQL is supported by almost all GIS packages, although sometimes in flavors with slight differences, and usually through a menu interface or wizard tool.

Most GISs have often lagged somewhat in their attribute database capabilities behind commercial systems for more routine data applications. The more recent interfaces, however, and the broad support for macros and windowing systems have led to a fair amount of convergence in the "look and feel" of GISs, although we are still a long way away from all GISs functioning in much the same way. Given the rich variety of GIS applications, and the rapidly expanding nature of the field, the standardization of GIS operations and query mechanisms is still many years away.

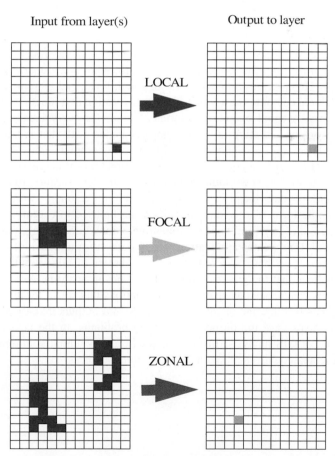

Figure 5.11: Map operations in Dana Tomlin's Map Algebra. GIS transformations of input data (in raster) can lead to an output cell value that is a function of: (1) values across layers at the same place (local); (2) values in adjacent cells across layers (focal); or (3) values in regions or patches, perhaps in the same class across layers (zonal). Incremental operations create a new value based on spatial form, e.g., topographic slope at each point in the grid from elevations.

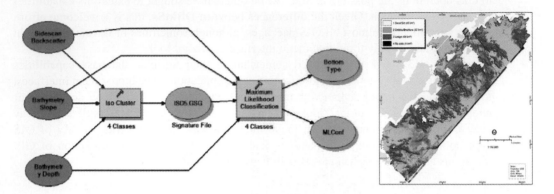

Figure 5.12: Example of a query script. ESRI Model builder schematic of multivariate analysis, showing input data (depth, slope, acoustic backscatter) and processing steps used to create map. Source: U.S. Geological Survey Open-File Report 2005-1293 High-Resolution Geologic Mapping of the Inner Continental Shelf: Nahant to Gloucester, Massachusetts by Walter A. Barnhardt, Brian D. Andrews, and Bradford Butman (2006).

Nevertheless, a very simple and frequently used classification of GIS operations or queries comes from the work of Dana Tomlin, who likens operations in GIS to verbs in language. He describes four types: local, zonal, focal, and incremental (Figure 5.11).

A local computes new values for every location as a function of one or more existing map layers. A zonal operation computes values for every location as a function of the values from one existing layer that are associated with that location's zone on another existing layer, e.g., polygons. A focal operation computes new values for locations as a function of existing values, distances, or the directions of neighboring locations. Lastly, an incremental operation computes a new value for every location to characterize the size or the shape of that location's unique portion of a one-, two-, or three-dimensional cartographic form. These four operations can be used to classify an immense variety of GIS processing steps and their combinations. Many GIS packages allow steps to be saved as a model, for repeated application to new datasets or updated information (Figure 5.12). At this stage, we have covered most of the basics of GIS: input, storage, and retrieval. Many believe that the greatest power of GIS lies in analysis. As we have seen, most GIS operations can be seen as spatial transformations that relate location and attributes. They have inputs that are one or more map layers, and outputs that are new maps. In analysis, the new maps have the purpose of solving problems. In the next chapter, we cover the essentials of GIS analysis. In the real world, the potential solutions to spatial problems using the analytical functions of GIS are almost endless.

5.5   STUDY GUIDE

5.5.1   Quick Study

CHAPTER 5: WHAT IS WHERE?

O *A GIS can answer the question: What is where?* O *Getting information about features from the database or map is called retrieval* O *The secret to effective retrieval is data organization* O *A data model is a theoretical construct that supports data storage and retrieval* O *GIS data organization requires an attribute data model and a map data model* O *GISs often include or share a database management system* O *DBMSs heritage lies in computer and information science* O *A DBMS contains a data definition language, which defines the data dictionary* O *A GIS data catalog allows management of multiple files, layers, projects, and versions* O *All DBMSs support data entry and checking* O *DBMSs support queries; attribute queries include sorting, reordering, subsetting, and searching* O *Query is the way that GIS users interact with the maps and the attribute data* O *First generation database managers were structured hierarchically, but geography has many cases where this does not work well* O *Relational database management has become the standard for database management* O *A relational database can have multiple flat files that can be related together when necessary through a key attribute that is unique across all records* O *Having separate relational files means they can be managed, e.g., updated separately* O *Relational database managers have sets of common commands, standardized using SQL* O *Some commands are necessary for reassembling or reorganizing the database, for example join and select* O *DBMSs have a module to create formal or standardized reports* O *DBMSs contain the ability to create and calculate values for new fields, records and databases* O *SQL and other DBMS operators are often poor at spatial queries* O *GISs contain spatial equivalents of the relational queries: identify, select, recode, and merge* O *A spatial merge is called an overlay* O *A spatial select is often a buffer operation* O *Some geographic query uses topology, e.g., adjacency and spatial properties such as distance* O *GIS search can be relative to points, lines, and areas, or to values in fields* O *Some spatial queries cover the whole geographical space, such as searching a DEM for visible areas* O *GIS usually support query via command lines, menus, or wizards* O *Much power in GIS comes from queries made in sequences and across layers* O *GISs also contain tools to repeat queries, for example in programming or scripting languages* O *Most GISs can use SQL directly* O *Tomlin classified GIS operations and queries into local, focal, zonal, and incremental* O *GIS query is essential for analysis* O *GISs often allow the same query result to be obtained in a variety of ways* O

5.5.2 Study Questions and Activities

Basic Database Management

1. Make a table listing the component parts of a DBMS. What particular task or tasks does each section of the DBMS do? Add a column to the table giving a brief summary of the role of that component. For example, the data-entry module could have the entry, "Permits user to enter attribute data into the database." Add another column that lists parallels between attribute databases and map databases, such as "report generation" and "map display."

2. Use a database manager such as Microsoft's Access or MySQL to create a database similar to the pizza example in this chapter. When the tables are complete, use a relational join to create a single flat file. How big were the data files before and after the join, and why?

Searches by Attribute

3. List and define each of the tools that the user of a DBMS has available for searching by attribute. What are the differences between the types of search; for example, *find* versus *browse*?

Searches by Geography

4. What geographic retrieval tools are available for each of the following: point features, line features, and area features? How can these be used in combination to make complex queries to the GIS?

5. Give an example of a GIS query resulting in each of Tomlin's four types of map operations.

The Query Interface

6. What are the major types of user interfaces for queries that the GIS user could face as he or she moves from one GIS software package to another? What are the advantages and disadvantages of each? Which would you prefer as a new GIS user?

5.6   REFERENCES

5.6.1   Chapter Bibliography

Berry, J. K. (1993) *Beyond Mapping: Concepts, Algorithms and Issues in GIS.* Fort Collins, CO: GIS World.

Burrough, P. A. (1986) *Principles of Geographical Information Systems for Land Resources Assessment.* Oxford: Clarendon Press.

ESRI (1995) *Understanding GIS: The Arc/Info Method.* New York: Wiley.

Huxhold, W. E. (1991) *An Introduction to Urban Geographic Information Systems.* New York: Oxford University Press.

Peuquet, D. J. (1984) "A conceptual framework and comparison of spatial data models." *Cartographica,* vol. 21, no. 4, pp. 66–113.

Tomlin, C. D. (1990) *Geographic Information Systems and Cartographic Modeling.* Englewood Cliffs, NJ, Prentice-Hall.

Warboys, M. F. (1995) *GIS: A Computing Perspective.* London: Taylor and Francis.

5.7   KEY TERMS AND DEFINITIONS

attribute: A numerical entry that reflects a measurement or value for a feature. Attributes can be labels, categories, or numbers; they can be dates, standardized values, or field or other measurements. An item for which data are collected and organized. A column in a table or data file.

batch: Submission of a set of commands to the computer from a file rather than directly from the user as an interactive exchange.

browse: A method of search involving repeated examination of records until a suitable one is found.

choropleth map: A map that shows numerical data (but not simply "counts") for a group of regions by (1) classifying the data into classes and (2) shading each class on the map.

compute: Data management command that uses the numerical values of one or more attributes to calculate the value of a new attribute created by the command.

data definition language: The part of the DBMS that allows the user to set up a new database, to specify how many attributes there will be, what the types and lengths or numerical ranges of each attribute will be, and how much editing the user is allowed to do.

data dictionary: A catalog of all the attributes for a data set, along with all the constraints placed on the attribute values during the data definition phase. Can include the range and type of values, category lists, legal and missing values, and the legal width of the field.

data entry: The process of entering numbers into a computer, usually attribute data. Although most data are entered by hand or acquired through networks, from CD-ROMs, and so on, field data can come from a GPS receiver, from data loggers, and even by typing at the keyboard.

data model: A logical means of organization of data for use in an information system.

database: Any collection of data accessible by computer.

DBMS (database management system): Part of a GIS, the set of tools that allow the manipulation and use of files containing attribute data.

default: The value of a parameter or a selection provided for the user by the GIS without user modification.

feature: A single entity that composes part of a landscape.

flat file: A simple model for the organization of numbers. The numbers are organized into a table, with values for variables as entries, records as rows, and attributes as columns.

file: Data logically stored together at one location on the storage mechanism of a computer.

find: A database management operation intended to locate a single record or a set of records or features based on the values of their attributes.

focal: A GIS operation or query that uses data from cells adjacent to the cell in question in one or more layers to create a new cell value.

geographic search: A find operation in a GIS that uses spatial properties as its basis.

hierarchical data model: An attribute data model based on sets of fully enclosed subsets and many layers.

highlight: A way of indicating to the GIS user a feature or element that is the successful result of a query.

identify: To find a spatial feature by pointing to it interactively on the map with a pointing device such as a mouse.

incremental: A GIS operation or query which uses one or more layers to create a new layer in which the value of each grid cell is computed by iterating over the entire grid. Also known as a global operation.

join: To merge both records and attributes for unrelated but overlapping databases.

key attribute: A unique identifier for related records that can serve as a common thread throughout the files in a relational database.

local: A GIS operation or query that uses data from the same cell in question in one or more layers to create a new cell value.

locate: See **identify.**

macro: A command language interface allowing a "program" to be written, edited, and then submitted to the GIS user interface.

menu: A component of a user interface that allows the user to make selections and choices from a preset list.

overlay: A GIS operation in which layers with a common, registered map base are joined on the basis of their occupation of space.

parameter: A number, value, text string, or other value required as the consequence of submitting a command to the GIS.

query: A question, especially if asked of a database by a user via a database management system or GIS.

query language: The part of a DBMS that allows the user to submit queries to a database.

relate: A DBMS operation that merges databases through their key attributes to restructure them according to a user's query rather than as they are stored physically.

relational model: A data model based on multiple flat files for records, with dissimilar attribute structures, connected by a common key attribute.

renumbering: Use of the DBMS to change the ordering or ranges of attributes.

report generator: The part of a database management system that can produce a listing of all the values of attributes for all records in a database.

restrict: Part of the query language of a DBMS that allows a subset of attributes to be selected out of the flat file.

retrieval: The ability of a database management system or GIS to get back records that were stored previously.

search: Any database query that results in successful retrieval of records.

select: A DBMS command designed to extract a subset of the records in a database.

sort: To place the records within an attribute in sequence according to their value.

SQL (Structured Query Language): A standard language interface to relational database management systems.

subsetting: Extracting a part of a data set.

update: Any replacement of all or part of a data set with new or corrected data.

verification: A procedure for checking the values of attributes for all records in a database against their correct values.

zonal: A GIS operation that uses aggregate cells in a single class or patch in one or more layers to create a new cell value.

5.8   PEOPLE IN GIS

Mark Bosworth—Principal GIS Analyst for Portland Metro, Portland, Oregon

KC: Mark, tell us a bit about your job?

MB: I work as a GIS Analyst for Portland Metro, in Portland, Oregon. I recently became a "Principal GIS Specialist," which means I'm on a technical track professionally, and not a management track. I get to spend my day "Thinking about GIS" (thank your Dr. Tomlinson!)

KC: What is Portland Metro?

MB: Metro is a unique government agency that is responsible for "issues of regional concern" in the Portland metropolitan area. Primarily that means land use and transportation planning (along with a number of other interesting domains like solid waste regula-

tion, regional recreational facilities, and the Oregon Zoo). We are the only elected regional government in the country and our jurisdiction encompasses 25 cities and the urban area of three counties.

KC: What kind of software do you use for GIS?

MB: Metro has been an ESRI customer since 1989. In fact, our customer number is in the "low 3 digits." We use ArcGIS for our desktop applications and various other deployments of the ESRI suite of software solutions: ArcReader and ArcExplorer are popular platforms for delivery of view-only kinds of applications. We also utilize a number of different solutions however—including an

increasing utilization use of Open Source platforms—especially in our Internet deployed mapping applications.

KC: What in your educational background prepared you for your job at Metro?

MB: When I was first hired at Metro, they had just purchased Unix workstations, and my experience with system administration was considered key. Since then we have migrated to a more enterprise IT environment and I rely on my experience with spatial data, analysis, and IT integration issues in general. I'm not especially talented or skilled as a cartographer, but good at reasoning and thinking about spatial processes. Fortunately, I work with a gifted group of professionals who fill out our team.

KC: What sort of backgrounds do the other employees have?

MB: We have a variety of backgrounds in our shop, and I think that is really a strength. I tell my students that I consider myself kind of a "one trick pony" in that my bachelor's degree and masters degrees are in Analytical Cartography and GIS, and my whole professional career has been focused on GIS. Many people

I teach and that I work with come to GIS with a more varied background, which is great. I work with GIS people who have degrees in Biology, Forestry, Geology, and, one of our staff even has a Ph.D in Medieval Architecture—you'd be surprised how that can come in handy sometimes!

KC: What is an example of a typical project that Metro would do that involves GIS?

MB: Given Metro's varied portfolio, we can be involved in urban land development issues, natural resource protection or restoration, transportation analysis, and so on. Most of our projects often come down to land use and planning. For example I've been working on a project with professionals in the health field to study the impact of growth or zoning changes on livability or density. We're using the GIS for to analyze existing conditions and for modeling possible futures.

KC: What is Metro's growth boundary, and what does Metro do with GIS to keep it?

MB: The urban growth boundary (UGB) is the primary tool at Metro's disposal for the regulation of the overall urban form of our region. It is an administrative boundary that separates city-style development from farm and forestland. It has been in place in more or less the same configuration for 30 years—which has helped define development patterns and contain sprawl in our region. Our GIS program was designed specifically from the ground up to support the monitoring and measurement of the effect of the UGB; we maintain an inventory of vacant lands and local zoning regulations so that we can understand the capacity of our region to absorb growth.

KC: Could you give an example of a project where you did spatial analysis?

MB: Most recently we have been studying the impact of the built environment on health and physical activity. The specific analysis we are doing is a longitudinal study of a group of people over almost 20 years - doing regression analysis to identify correlations between the physical health and mobility of subjects and their immediate spatial environment. The GIS is helpful in developing land use characteristics and quantifying the environment using variables like the distance to transit, parks, and other amenities, as well as population and urban services and overall urban form.

KC: What sorts of unique means of visualization do you have?

MB: We have always had a talented team of cartographers who utilize our GIS data in creative ways. One popular product is a 3D enhanced wall map that depicts the region in a color scheme that exploits a color palette that works specifically with 3D glasses. It's cheesy, but engaging. More recently we have been exploring the various visualization possibilities in LiDAR data—specifically the 3D realm. The 3D display of both the built environment—buildings and infrastructure—and the natural environment—trees, vegetation, topography—are all exciting new datasets to exploit in our visualization products.

KC: Do you use open source software for GIS? How?

MB: We have number of open source based applications in production. Specifically, we use MapServer software for many of our web deployed map services. One interesting application that utilizes MapServer and the Google Maps API is www.bycycle.org, an application that recommends bike specific routes, similar

to MapQuest using a mash up of our local bike route infrastructure with Google Maps' cartography.

KC: What sorts of things might Metro be doing ten years from now?

MB: It has certainly been the trend in my career that access to spatial information and geoprocessing tools are becoming more transparent and accessible. What used to be the domain of GIS specialists is becoming more available to the non-specialist. This is a good thing for GIS and geography in general, I think. I believe that in ten years we will have increasingly more compelling and elaborate display and visualization technologies at our disposal. Immersive and multi-dimensional visualizations will become the norm as a way of interacting with spatial data. and GIS will continue to engage and enlighten.

KC: Thanks very much for the interview.

MB: Thanks Keith, this was fun.

Author's Note. Portland Metro's website is: http://www.oregonmetro.gov/

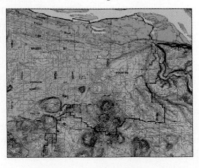

CHAPTER 6

Why Is It There?

"I skate to where the puck is going to be, not where it's been."

Wayne Gretsky (when asked the secret of his success).

6.1 DESCRIBING ATTRIBUTES

As we have seen in the chapters leading up to this point, a GIS has at least two parts: the attribute part and the map part. The attribute data, managed by their regular database manager, are little different from any other type of statistical information when it comes to analysis. In this chapter we move away from the construction and management of data in a GIS to actual use of the information. To best understand information in the form of numbers, we must describe geographic data in methodical and quantitative ways, that is, with well-understood statistics. If this was as far as GIS went, however, there would be few advantages to GIS compared to any of the major computer statistical packages available for information analysis.

What makes analysis within a GIS different is that the attribute data have established links to maps. Any statistic we can think of to describe the data then automatically has geographic properties and as a result can be placed on a map for visual processing. As we will see later in this chapter, the situation is more fruitful than that, because we can use the geographical properties described in Chapter 2 for statistical inquiry as well. This means that in addition to answering the question "Where?" as far as the features are concerned, we can also ask "Why is it there?" We can come up with some definitive answers to these questions and display the answers to the questions and analyses as maps. As this chapter shows, this can give the user an amazing amount of power when GIS analysis is brought to bear on a problem.

The chapter begins by covering how attribute-type data are described. The visual description of the histogram and the number descriptions of the average and difference from the average are covered. In addition, these simple measures are described in terms of their spatial attributes when the two spatial dimensions or coordinates are used as the numbers under description. As shown, map statistical description leads to an initial ability to place onto a map what the numbers demonstrate. The average (mean) and difference from the average (variance) both have visual and geographical meaning.

6.1.1 Describing One Attribute

To revisit the basic structure of a database, review the beginning of Chapter 2. As we saw, all data can be thought of as structured in a table. The rows of the table are records, and the columns are attributes. Each record for each attribute contains a value, and that value falls into a set of types of data, text, numbers, and so on. For example, the values of an attribute called "Date" for the record "357" could be "7/7/2009". The value here is really three numbers (a month, a day, and a year) but for the purposes of the database it is often considered as text. An additional condition is that to be a GIS database, at least one of the attributes must be a link to a map. At the most basic level, the eastings and northings of points can fit as two attributes. We start with this simple case, but as we have also already seen, geographic data can be for features that are points, lines, and areas and combinations of them.

In this chapter, we'll choose a geographical phenomenon, tornadoes, as a means to examine each of the stages by which we examine, query, test, and attempt to explain the pattern of tornadoes in the U.S. and how they cause death and injury to humans. Bringing a research question or problem statement to a data analysis is an important first step in any inquiry. We might be interested in understanding why deaths and injuries occur in order to devise plans to save lives and injury. We might be concerned with building a tornado warning system, or in determining whether there were long term trends in tornado frequency attributable to global warming. In each case, the analysis begins with the same process: describing the data both visually and statistically. This can help in formulation or stating the problem to be solved, and even indicate how to go about solving the problem.

The data to be used come from the National Oceanic and Atmospheric Administration's tornado database. You can find these data on-line at: http://www.ncdc.noaa.gov/oa/ climate/severeweather/tornadoes.html. The format of the data is designed to allow calculation of tornado severity based on the ground area impacted. More information about the magnitudes and the classes they fall into is available at: http://www.nssl.noaa.gov/edu/ safety/tornadoguide.html.

DATE_	YEAR	MONTH	DAY	UTC	STATE	F_SCALE	LENGTH_M	WIDTH_YD	WIDTH_MI	AREA_SQ	AREA_LOG	AREA_CLA	DRI	DCATHS	INJURIES	TDLAT	TDLON	LIFTLAT	LIFTLON
1/3/1950	1950	1	3	1700	MO*	3	9.500	149	0.08470	0.80430	-0.08460	MESO	3.20	0	3	38.770	-90.220	38.830	-90.030
1/3/1950	1950	1	3	1755	IL	3	3.600	129	0.07330	0.26890	-0.57880	MESO	1.10	0	3	39.100	-89.300	39.120	-89.230
1/3/1950	1950	1	3	2200	OH	1	0.100	9	0.00050	0.00050	-3.29330	TRACE	0.00	0	1	40.880	-84.580	0.000	0.000
1/13/1950	1950	1	13	1125	AR	3	0.600	16	0.00910	0.00550	-2.26520	DECIMICRC	0.90	1	1	34.400	-94.370	0.000	0.000
1/26/1950	1950	1	26	130	MO	2	2.300	299	0.16890	0.39870	-0.40810	MESO	1.20	0	5	37.600	-90.680	37.630	-90.650
1/26/1950	1950	1	26	300	IL	2	0.100	99	0.05620	0.00560	-2.24980	DECIMICRC	0.00	0	0	41.170	-87.330	0.000	0.000
1/26/1950	1950	1	26	2400	TX	2	4.700	133	0.07560	0.35520	-0.44960	MESO	1.10	0	2	26.880	-98.120	26.880	-98.050
2/11/1950	1950	2	11	1910	TX	2	9.900	359	0.22670	2.24440	0.35110	MACRO	6.80	0	0	29.420	-95.250	29.520	-95.130
2/11/1950	1950	2	11	1949	TX	3	12.000	999	0.56760	6.81340	0.83320	MACRO	27.30	1	12	29.670	-95.050	29.830	-95.000
2/12/1950	1950	2	12	300	TX	2	4.600	99	0.05620	0.25870	-0.58710	MESO	0.80	0	5	32.350	-95.200	32.420	-95.200
2/12/1950	1950	2	12	555	TX	2	4.500	66	0.03750	0.16870	-0.77280	MESO	0.50	0	6	32.980	-94.630	33.000	-94.700
2/12/1950	1950	2	12	630	TX	2	8.000	833	0.47330	3.78640	0.57820	MACRO	11.40	1	8	33.330	-94.420	33.450	-94.420
2/12/1950	1950	2	12	715	TX	1	2.300	233	0.13240	0.30450	-0.51640	MESO	0.60	0	0	32.080	-98.350	32.100	-98.330
2/12/1950	1950	2	12	1210	TX	2	3.400	99	0.05620	0.19120	-0.71840	MESO	0.60	0	0	31.330	-94.550	31.570	-96.550
2/12/1950	1950	2	12	1757	TX	1	7.700	99	0.05620	0.43310	-0.36340	MESO	0.90	0	32	31.860	-94.260	31.880	-94.120
2/12/1950	1950	2	12	1800	MS	2	0.100	9	0.00530	0.00050	-3.29330	TRACE	0.00	3	2	34.600	-89.120	0.000	0.000
2/12/1950	1950	2	12	1800	MS	1	2.000	9	0.00530	0.01020	-1.99020	MICRO	0.00	0	0	34.690	-89.120	0.000	0.000
2/12/1950	1950	2	12	1800	TX	3	1.900	49	0.02780	0.05290	-1.27680	MICRO	0.20	3	15	31.800	-94.260	31.860	-94.180
2/12/1950	1950	2	12	1830	AR	2	0.100	99	0.05620	0.00560	-2.24980	DECIMICRC	0.00	0	0	34.480	-92.400	0.000	0.000
2/12/1950	1950	2	12	1900	LA	4	82.600	99	0.05620	4.64620	0.66710	MACRO	23.50	18	77	31.970	-94.000	33.000	-93.300

Table 6.1: First 20 entries in the Tornadoes database. Source of the data is the National Oceanic and Atmospheric Administration (http://www.ncdc.noaa.gov/oa/climate/severeweather/tornadoes.html).

Table 6.1 is a listing of the first 20 of the 49,252 data records of tornadoes in the United States between 1/1/1950 and 12/31/2006. There are both non-spatial and spatial attributes. For example, non-spatial attributes are the date and time of occurrence, the tornado class and magnitude, and the number of deaths and injuries that the tornado caused. The spatial attributes are the state and county where the tornado touched down, the latitudes and longitudes of the touchdown and liftup locations of the funnel, and data on the length and width of the impacted area.

Figure 6.1 is a map of all of the tornadoes' touchdown locations, using the uDig GIS. The map shows a huge amount of data with only one small point symbol, so it is hard to see what is going on, although there is clearly an east-west gradient to the map. A first stage in analysis is often selection or sampling. In our case, dealing with all 49,252 tornadoes is unnecessary if we want to examine the impact on humans. It is probably a good idea to focus on only those tornadoes that led to deaths or injuries. This reduces the size of the data set to 6,625 records, much more manageable for visualization and analysis.

Figure 6.2 is a map with only death or injury tornadoes showing both the magnitude and one basic division of the data, tornadoes during the tornado season and those outside of it. Even the most cursory glance shows that this complex data set shows highly structured time and space patterns. Can we use spatial analysis to reveal the hidden structure? Obviously first we need to carefully examine the form of the data, both statistically and visually, to form hypotheses about the form and structure that we can test by measurement.

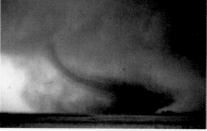

Figure 6.1: Left: Locations of all tornadoes in the US 1950–2006. Map using uDig GIS. Above: A rope tornado (Source: NOAA)

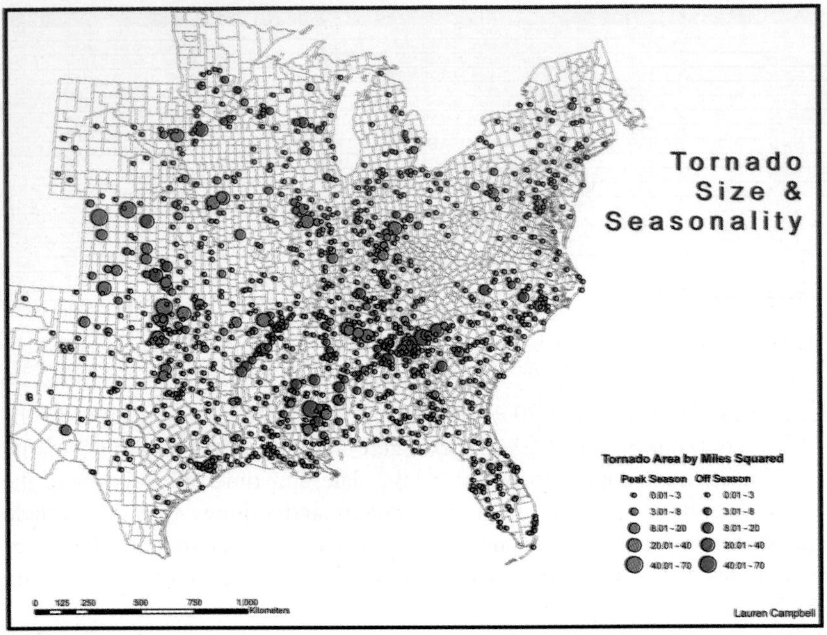

Figure 6.2: Location, magnitude of impacted area, and seasonality of tornadoes in the U.S., from 1950 to 2006 that resulted in deaths or injury. Map by Lauren Campbell.

6.2 STATISTICAL DESCRIPTION

Before we get started on the use of statistics for examining the data, it is worth doing some exploratory data analysis using various graphics tools. Over time, a large set of methods for the visual display of non-spatial data has been created, and in many cases these can be drawn inside popular spreadsheet programs such as OpenOffice.org Calc, and by stand-alone and statistical packages such as the open source programs ViSta and R, and SAS or SPSS. Few GIS packages can create all (or even any) of these plots directly. A few software applications, such as GeoVista Studio, are available expressly for exploring geospatial data interactively.

6.2.1 Statistical Graphics

Figure 6.3 shows some of the tornado data plotted to show different attributes, one at a time. Note that some data are categorical (e.g., tornado severity), others are numerical, and in some cases the values are cyclical, such as month. Different types of plot are suited to each of these. A careful examination of Figure 6.3 reveals that for tornadoes resulting in deaths and injuries, tornado season peaks in April, that the most dangerous tornadoes are small rather than large in area, that tornadoes are most common at midnight GMT (about 6pm local time), and that most tornadoes are in the lowest severity categories. Of these visual methods, only the scatter plot compares one variable against another. A more thorough examination would obviously look for relationships among all of the variables. A spatial analysis would then look for geographical properties among these relationships.

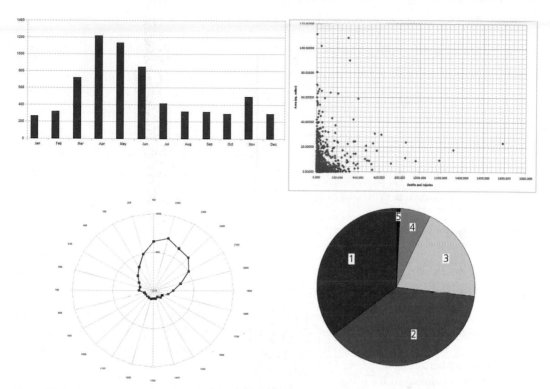

Figure 6.3: Statistical graphics for the tornado data. Upper left: Histogram—frequency of tornadoes by month. Upper right—Scatter plot-area of tornado touchdown vs. deaths and injuries. Lower left: Radar plot—Death and Injury tornadoes by time of day (GMT, subtract about 6 hours for local time, and note time shown counter–clockwise). Lower right: Pie Chart—Tornadoes by severity class.

6.2.2   The Box Plot

One of the oldest forms of non-map statistical graphic is the box plot, invented by John Tukey (Tukey, 1977). The box plot graphically depicts groups of numerical data through their five descriptive numbers. These are: (1) the smallest observation; (2) the lower quartile (3); the median (4); the upper quartile; and (5) the largest observation. A boxplot can indicate which observations, if any, might be considered outliers from the remainder, and whether a distribution is skewed in one direction or another.

These values simply reflect the distribution of the data in the attribute from the highest to the lowest values. The two tick marks show the maximum and minimum values. The higher the central bar, the greater the range. The box is centered on the median value. If the data are ranked based on their values, half of the data would lie above and half below this line. This value should not be sensitive to any abnormally high or low values. Lastly the "box" is a rectangle whose height is given by the bounds of the first and fourth quartile of the data. In other words, if we ranked all the data values above the median, the upper quartile would again divide the group in half, and similarly with the lower. So half the data lie within the box based on their value. A typical box plot is shown in Figure 6.4, along with a box plot from some of the tornado variables. Quite clearly, the distributions are not very even or random, and much structure remains to be explored.

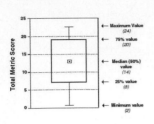

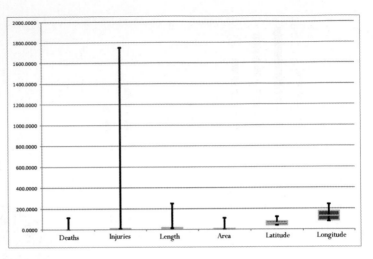

Figure 6.4: Box plots. Above, a typical box plot for normally distributed data (Source: EPA). Right: Box plots for some of the tornado data. generated in Excel. It is possible to generate box plots directly within some GISs using scripts.

6.2.3   Mean, Variance and the Standard Deviation

The median, introduced in the last section, is simply the value of the middle record of an attribute once the values are ranked. When there is an odd number of records, this gives a unique record, but for an even number, we need to average the two middle values. Take the number set {*13, 3, 7, 6, 15, 1*} for example. When ranked these are {*1, 3, 6, 7, 13, 15*}. With 6 values, we have to average the "*6*" and "*7*" to get 6.5. Note an advantage of the median is that it is not sensitive to outliers. If we changed the 15 to 115, for example, the median remains the same. The median is termed a "measure of central tendency," since a single number chosen to represent the whole set.

Another measure of central tendency can be computed without having to rank the numbers first. This is the mean. To calculate the mean or average, we just add up the values of the attribute and divide by the number of values. For example, there were 4783 deaths for the 6625 tornados, giving a mean of 0.721962. So the mean number of deaths per tornado is 0.721962, and similarly the mean number of injuries was 12.3645 for those tornadoes that caused death or injury. Obviously the mean is affected by outliers. The median numbers of deaths and injuries were 0 and 3. In Figure 6.5, the deaths and injury data are plotted as histograms. A histogram is simply a set of graphical bars with axes. The heights of the bars are the number of attributes that fall within a class, and the bars themselves are a set of classes into which the data have been divided. Both distributions (fortunately) are skewed toward zero. In this case, neither the median nor the mean is a good measure of central tendency.

Now let's look at another of the tornado attributes, the day of the month of the occurrence. In this case, the median is 15 and the mean 15.5093 and there is only a minor difference between them, the median is an exact value (because 31 is an odd number), and the mean is probably a more representative value than the median. There is one additional element to note. On the histogram of these values (Figure 6.6), there is an obvious drop in the frequency of tornadoes on the last few days of the month. From the familiar rhyme, we know that "30 days has September … ", meaning that five months have no 31st day, and one has no 30th or 31st, and no 29th 3/4ths of the time.

Figure 6.5: Histograms of the number of deaths and injuries associated with the 6625 tornadoes in the US from 1950–2006 that caused death or injury. Note that these are skewed distributions.

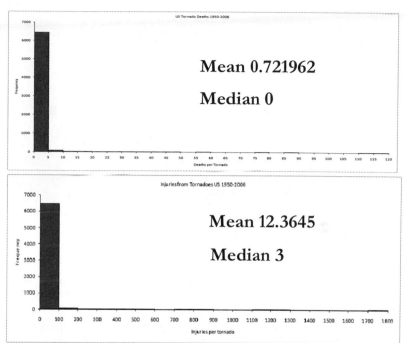

This is a much more even distribution across the range of values. Other than the drop at the end of the month, we might expect that something as random as a tornado might be equally likely to occur on any day of the month. Yet there is an unevenness to the distribution. There are 6625 tornadoes and 31 days, so a first guess at how many there would be on each day is an average frequency of 6625/31=213.709. The third day of the month seems to get more than its share, while the 17th seems to get less. A question to ask is, are these highs and lows what would be expected due to chance, or are they important differences worth further analysis? To answer this question we need to know not only the central tendency of the distribution, but also how much variation to expect within it. In

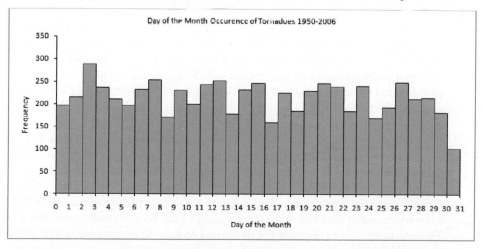

Figure 6.6: Day of the month of the occurrence of tornadoes in the U.S. that caused death or injury 1950–2006.

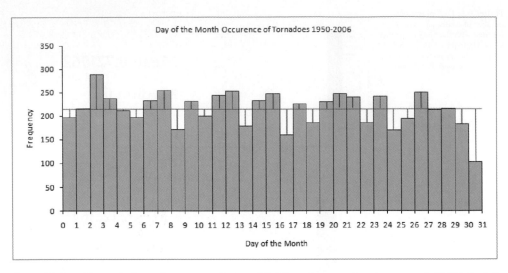

Figure 6.7: Day of the month of tornado occurrence. Horizontal black line is the expected number per month (213.7). Vertical lines are the difference from the mean. Note that some are above, and some below the line.

short, we need to know the average variation from the average. This value is called the standard deviation. It is computed by first calculating the mean, then subtracting it from each value. This is equivalent to the lengths of the solid vertical lines shown in Figure 6.7.

Note that there are approximately as many daily counts above the mean occurrence as there are below, another feature of randomness. A way to deal with both negative and positive values when we subtract the mean from each value is to square the result. A negative value times itself is always positive. These squares are then averaged (sum of the squares, divided by the number of values) to give the mean of the squares, and then we take the square root to get back to the same units we started with. This can be done in a spreadsheet, or using standard statistical software (Figure 6.8). Along the way we also tabulated another key statistic, the sum of the squared deviations, which is called the total variance.

The value for the days of the week data is then 36.115, rounded to three decimal places. We can both add and subtract this amount from the mean of 213.709, getting a range from 177.595 to 249.824. We can add these levels to the histogram (Figure 6.9). Note that when we do so, only a few days of the month fall outside of this range. If we use two standard deviations instead, we get a range from 141.48 to 285.939. When we use two standard deviations, there are only two days of the month that fall outside of this range, the two being the 31st and the 3rd.

We already have an idea why the 31st is unexpectedly low, because there are only 7 of 12 months that have a 31st day. The 3rd is abnormally high, which could, of course, be due to randomness. It could equally be because of some unknown fluke of nature that makes tornadoes more harmful on the third of every month. While the value for the 31st is well below the mean minus two standard deviations line, the value for the 3rd is very close to the mean plus two standard deviations line. To see whether this means anything, we need some way to estimate the likelihood of tornado occurrence at any given level.

Day of the Month	Frequency	Freq. - Avg.	Squared
1	197	-16.70968	279.2134057
2	216	2.29032	5.245565702
3	288	74.29032	5519.051646
4	237	23.29032	542.4390057
5	211	-2.70968	7.342365702
6	197	-16.70968	279.2134057
7	232	18.29032	334.5358057
8	254	40.29032	1623.309886
9	171	-42.70968	1824.116766
10	231	17.29032	298.9551657
11	199	-14.70968	216.3746857
12	244	30.29032	917.5034857
13	253	39.29032	1543.729246
14	178	-35.70968	1275.181246
15	232	18.29032	334.5358057
16	247	33.29032	1108.245406
17	160	-53.70968	2884.729726
18	225	11.29032	127.4713257
19	186	-27.70968	767.8263657
20	229	15.29032	233.7938857
21	246	32.29032	1042.664766
22	240	26.29032	691.1809257
23	186	-27.70968	767.8263657
24	241	27.29032	744.7615657
25	169	-44.70968	1998.955486
26	194	-19.70968	388.4714857
27	249	35.29032	1245.406686
28	212	-1.70968	2.923005702
29	216	2.29032	5.245565702
30	183	-30.70968	943.0844457
31	102	-111.70968	12479.05261

Sum	6625	Sum of Squares	40432.3871
Mean	213.7096774	Mean square	1304.270552
		Standard deviation	36.11468609

Figure 6.8: Using a spreadsheet to calculate the standard deviation of the day of the month tornado occurrence data.

6.2.4   Statistical Testing

A property of the standard deviation is that when we use a distribution that is known, we can calculate the proportions of the area under the distribution's curve that falls between any value and within the mean plus or minus different numbers of standard deviations. We use a standard distribution curve called the normal distribution, well described by its shape "the bell curve" (Figure 6.10). This curve was first described by Abraham de Moivre in 1733.

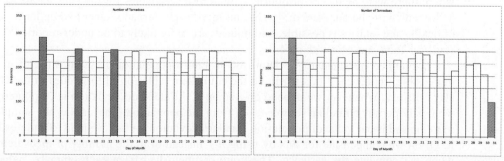

Figure 6.9: Histograms of the day of week of harmful tornado occurrence 1950–2006. Left: Plus and minus one standard deviation. Right: Plus and minus two standard deviations.

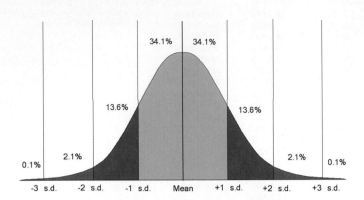

Figure 6.10: The normal curve, also called the bell curve, and the Gaussian distribution. Percentages shown are the percent area beneath the curve between each of the standard deviations. Within one s.d. of the mean is 68.27%, within two is 95.45% and within 3, 99.73%.

If we assume that the distribution we are analyzing can be thought of as a normal distribution, then we can use the equation of the curve and the area under the curve to estimate the probability that a value lies within or outside of the curve at random. Take, for example, the 3rd day of the month, whose frequency of tornadoes was 288. The difference from the mean is 74.29032. This amounts to 2.0571 standard deviations, a value called its *Z-score*. Using tables, or an on-line calculator (e.g., http://www.danielsoper.com/ statcalc/calc02.aspx), this gives a cumulative area under the curve of 0.980157. The probability of a value being greater than this by chance then is 1.0–0.980157, or 0.019843, equivalent to less then 2%. Small, but not beyond the bounds of belief, especially when thousands of tornadoes are involved over 56 years. On the other hand, the 31st day of the month has a *Z-score* of -3.09319, corresponding to an area of 0.0009908. This is less than one tenth of a percent; in other words we can be 99.9% sure that the value is not due to random variation around the mean. Something is clearly wrong with this value, and the shorter months is a valid explanation for the discrepancy.

Two last items to note about statistical testing. First, the areas shown in Figure 6.10 can be used to test whether a value is different from the mean, or whether it is greater or less than the mean. All we do is add the areas differently. Secondly, we need to question whether data we are analyzing can be compared to the bell curve. A perfect normal distribution is only to be expected when the number of measurements is large, when errors are purely random, and when a sample is representative of the overall "population," or all possible values of the data. In the case of tornadoes, we threw out many thousands of cases where there were no deaths or injuries. Some of the variables for the data set are skewed, as we saw with the numbers of deaths and injuries per tornado. Over and under-reporting of deaths and injuries is possible, and injuries are more likely to be under-reported. Use of statistics then, must be with care.

Finding the standard deviation of an entire population is often difficult or even impossible. In our case, this might be the equivalent table for all tornadoes in the U.S. in all of history. Whenever we depend on a sample, the standard deviation is computed from what is hopefully a random sample taken from the population. The smaller the sample, the more biased is our value of population standard deviation; in fact we tend to underestimate it. Often we compensate for this by using *n-1* instead of *n*, the number of measurements, when we compute the variance. These are distinguished from each other by name, the sample standard deviation and the unbiased estimate of the population standard deviation.

In summary, while the mean and median are good descriptors of central tendency, we also need to measure the amount of variation from the mean. An advantage of calculating the standard deviation is that we can compare the value against the theoretical normal curve, and so associate probabilities with values. If we then choose a probability level, such as 95%, we can test a value to see whether it is explainable by chance, or is significantly different from the rest of the values, and worthy of further analysis. Such values may consist of errors (as with the 31st day example), or they may be worthy of further scrutiny. If the data are within a GIS, we can employ spatial analysis and ask where these unusual values lie in geographic space.

6.3   SPATIAL DESCRIPTION

In the preceding section we looked at how to describe a single attribute statistically. The first and most significant factor in dealing with spatial data is that there are at least two spatial measurements, an easting and a northing. We could summarize spatial description, as describing two attributes simultaneously.

In the simplest and most basic way, we can duplicate the attribute descriptions above for the locational data to give spatial descriptions. In this case, we can treat the two separate parts of the coordinates, the eastings and the northings, as if they are each a single attribute, which indeed they are. Just as we began the discussion of describing the values of a single attribute by discussing the concept of a minimum and a maximum value for an attribute and the concept of a range, when the attributes describe coordinates, a first point is described by the minimum easting and the minimum northing, and a second point describes the corresponding maxima. The two points define a rectangle, whose two side lengths are the ranges in easting and northing, respectively, and that encloses all the points.

This is called the bounding rectangle of the points, a concept we met earlier. It can be found by simply sorting the records by easting, and taking the first and last record, and then repeating for the northing. Note that care must be taken with bounding rectangles

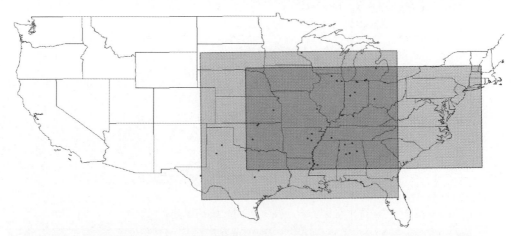

Figure 6.11: Bounding boxes as spatial descriptors. Red box encloses all locations of tornadoes causing 20–39 deaths. Purple box encloses those with more than 40 deaths.

when data are in latitude and longitude, or are projected. The size of the bounding box is the extent of the data. Different extents can be compared by showing bounding boxes for different data sets. A bounding rectangle for the tornados that caused the highest numbers of deaths is included in Figure 6.11.

6.3.1 The Mean Center

In the previous section, we calculated the mean of an attribute by summing the values and dividing by the number of values. We can do this for two attributes at the same time independently, that is for the easting and northing of the coordinates or for their latitudes and longitudes. In doing this for the tornado data, we have two points of interest, the touchdown and liftup points. Note that in the data, many tornadoes have the first but are missing the second. In this case, we simply leave them out of the calculation. When we do this, we get a mean touchdown point at Pascola, Missouri (*36.275484, -89.839066*), and a mean liftup point at Parma, Missouri (*36.609759, -89.779371*). Both places are close to the famous New Madrid loop, and are 37.492 km apart, (computed from their USNG coordinates of 16SBF4499118227 and 16SBF5139655168 respectively) with a bearing of 9.84 degrees from grid north. The fact that the mean tornado contact vector points just east of north is a good overall description of general storm tracks in the middle United States. The two points are shown in Figure 6.12.

The two means (*x, y*) make a point, with both a real geographic location and a special geographic name, the *mean center*. This point is also sometimes called a centroid, a point chosen (in this case statistically) to represent a geographic distribution. Using the mean center is one way to select a centroid to represent a set of points. Points, lines, and area features can also have a centroid, selected in any one of several ways. Note that a mean center is more influenced by outlier points than other representative points, such as the median center.

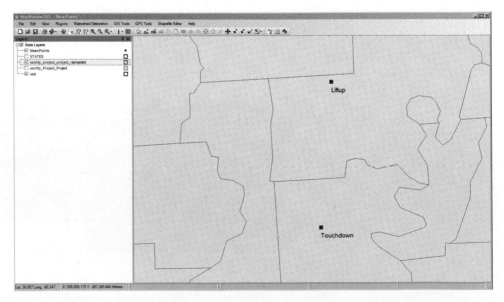

Figure 6.12: Mean touchdown and liftup points for the tornado dataset, shown in the MapWindow GIS.

 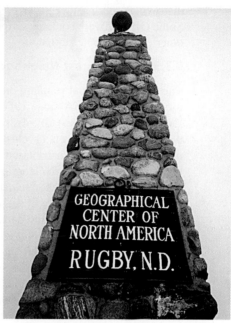

Figure 6.13: The monument at the supposed Geographical Center of North America in Rugby, North Dakota. Photographs by Colette Flanagan.

Figure 6.13 shows photographs of a place in Rugby, North Dakota, that claims to be the geographic center of North America. Although this is a fascinating monument and the nearby diner is probably heavily dependent on its visitors for its food business, it should be quite obvious that, unlike for a set of points, an entire continent could have any number of centroids! For example, this could be the point farthest from any coastline, the center of all of the points making up the coastline, the center of the bounding rectangle, or the center of the largest circle that can be drawn inside North America.The World Almanac lists the geographic center of North America not in Rugby, but in Pierce County, North Dakota, 10 km west of Balta (48° 10' N, 100° 10' W). The mean center calculation would also change depending on the map projection, datum, and ellipsoid. Judging from the flags seen on a visit to the site, it is not even clear whether Rugby's definition of North America includes Mexico, Alaska, Greenland, or Hawaii. Obviously, one place is as good as any for this type of monument.

6.3.2   The Standard Distance

Just as we calculated the standard deviation for the attributes in the last section, we can also calculate the standard deviations of x and y. In doing so, we have to take into account that distances only make sense when projected. We apply a simple correction to the length of a degree for degrees of longitude, using the cosine of the latitude. This ignores the fact that the earth is not a sphere, but it does simplify the math. A degree of latitude and longitude at the equator is 111.319 km. Again we can compute the differences from the mean, square these values, then take the averages across the 6625 records. We add the x and y mean squares together, then take the square root of the result. When this value is complete, we arrive at a *standard distance*, not deviation, of 839.9km. This is quite a large distance, implying that the overall distribution is quite scattered about the mean center. Some GIS software contains scripts to compute the mean center, the standard distance, and other centrographic statistics.

6.3.3 The Nearest Neighbor Statistic

The mean center and standard distance measures tell us that the distribution of dangerous tornadoes in the USA since 1950 is centered on Eastern Missouri, oriented North–North East, and is fairly broadly distributed about that center. But what is the nature of the distribution? Is it even or clustered? One way to tell is to compute a measure that characterizes spatial distributions, the nearest neighbor statistic. This measure has been quite extensively used in Geography and other disciplines.

The nearest neighbor statistic, called R, is a ratio of two densities. First, a polygon must be specified that will be used to intersect the points and used for the density calculations. For the tornado data, an outline of the lower 48 states was used, and 6532 of the points fell within it. Some algorithms available in GIS instead use the polygon that fully encloses all of the points instead, the convex hull.

Next, we computed the expected or average density of points in the space, as the area divided by the number of points. To convert this to a distance, we take the square root and divide by two, because the average spacing is in both directions. This value forms the denominator of the nearest neighbor statistic. For the numerator, we calculate the observed nearest neighbor distance. That is, for each point we determine which of the remaining points is closest, then add up all of these distances and divide by the number of points. So R is equal to the observed mean nearest neighbor distance divided by the expected distance. If the observed distances are small because the points are close together, then the top of the formula approaches zero and the R value is small. When the points are as far apart as possible, the R takes its maximum of 2.15. When the points are in a regular grid, R is 2.0, and when the distribution is random, the value is 1.0. This range is illustrated in Figure 6.14.

Either using GIS scripts, such as the nearest neighbor analyst extension for ArcGIS (http://arcscripts.esri.com/details.asp?dbid=11427), or using the basic GIS tools, it is possible to compute the value of R for a point distribution. Note that over large areas, there is a conflict between the desire for an equal area projection and one that preserves local distances and directions. Using this script, a value for the 6532 points that fell inside the

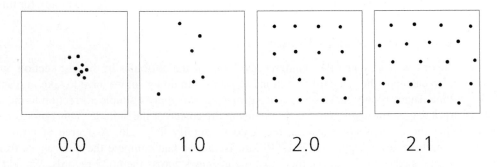

Figure 6.14: Limit (not exact) values of R the nearest neighbor statistic.

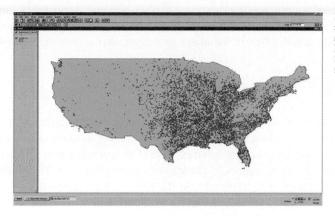

Figure 6.15: ArcView 3.2 with the nearest neighbor script loaded. The script requires a polygon and a point layer, and computes R and other values. For the tornadoes leading to deaths or injuries shown, $R=0.722$.

U.S. outline was computed as $R = 0.72218$. This value is somewhat random, but does not test significantly different from clustered, so the data might be best described as clumped. The points and the map used are shown in Figure 6.15.

6.3.4   Geographic Features and Statistics

In Chapter 2 we met the idea that geographic features can be classified into points, lines, and areas by their dimensions on the map. Describing each of these can lead to measuring spatial properties directly from the digital files containing the geocoded representations of the features. We started Chapter 6 with a set of points, the tornado locations, because points are the easiest type of feature to describe. Although we have so far used quantitative measures to describe geographical features, many arrangements of features are described verbally.

For example, points are clumped, sparse, uneven, random, regular, uniform, scattered, clustered, shotgun, or dispersed. Patterns are regular, patchwork, repetitive, or swirling. Shapes are rounded, oval, oblong, drawn-out, or resemble Swiss cheese. The challenge is to find numbers that say the same thing. The bounding rectangle, the mean center, and measures such as the standard distance and nearest neighbor statistic can provide excellent descriptors of points, although more complex measures are obviously needed for the higher-dimension features.

Lines have a number of points, a line length, a distance between the start and end points (or nodes), the average length of one of the line segments, and a line direction. A useful description of a line could be the ratio of actual line length divided by the start-to-end node length, what could be called a straightness index. For a straight line, this measure would be 1. For the Mississippi River this would be a far larger number. The direction could be taken as the clockwise angle bearing from north (the overall "trend" of the line), although this value would have a big variance along a curved or wiggly line, too.

Areas are even more difficult to describe. Simplest to measure with a GIS are the area in square meters, the length around the boundary, the number of points in the boundary, the number of holes, and the elongation, taken as the length of the longest line axis divided by the axis at 90 degrees to it. We can also divide the area of the bounding rectangle into the area, a space-filling index with a maximum value of 1. If the area has neighboring areas, we could count them or determine the average length of the area's boundary shared in common with a neighbor. Not all of these numbers are easy to compute with a GIS. Sometimes a multiple-step process must be used and information created or computed in the attribute database and passed back to the map for display. Almost every GIS has a compute command, often in the query processor, that allows the performing of math operations like:

```
COMPUTE ATTR5 = (ATTR2 + ATTR3) / ATTR4
```

Each new measure, however, can be an intermediate step in the computation of another statistic. For example, we could measure the lengths of streams by district in kilometers, measure the area of the district in square kilometers, and then create a new attribute in the database of the stream density in meters of stream per square kilometer. This might be an interesting data value to then map by district in the GIS. Computations are so common using areas of polygons that many GIS packages compute and save areas as attributes when they are created, whether they are wanted or not.

6.4 SPATIAL ANALYSIS

Numbers that describe features are useful, but as we noted in Chapter 2 the purpose of geographic inquiry is to examine the relationships among geographic features collectively and to use the relationships to describe the real-world phenomena that the map features represent. The geographic properties we noted in Chapter 2 were size, distribution, pattern, contiguity, neighborhood, shape, scale, and orientation.

Each spatial relation begs three fundamental questions: (1) How can two maps be compared with each other?; (2) How can variations in geographic properties over a single area or GIS data set be described and analyzed?; and (3) How can we use what we have learned using the analysis to explain and therefore predict past, current, or future maps of the geography in question? The third question may be as simple as selecting the best route from A to B on a map, or as complex as modeling the future growth of cities based on their size, shape, and development over time. GIS gives us the capability of doing both, and anything in between. In terms of comparing maps, a simple way is to bring multiple maps into coregistration and then merge their themes to make a composite. This is what is meant by map overlay analysis. An example of map overlay will follow a discussion of spatial models and how GIS adds to their construction, examination, and use.

6.4.1 U.S. Tornado Deaths and Injuries: An Analysis Example

A full set of descriptive statistics for all of the properties listed here is beyond the scope of this book. Instead, two geographic analysis problems will be covered, starting with a simple geographic distribution, and ending with some speculations about prediction.

Let's return to the research questions we posed at the start of the chapter. So far we have described the distribution of deaths and injuries due to tornadoes in the United States between 1950 and 2006. While the information we have discovered so far is useful, it does not really answer the question "Why is it there?" On one hand, we could learn more about the causes of tornadoes, about atmospheric disturbance, severe storm systems, and ocean temperatures. We could investigate whether global warming is increasing the severity of storms, and so increasing the incidence of tornadoes. If we chose this direction, we would still find GIS invaluable, but we would end up with information about why and where

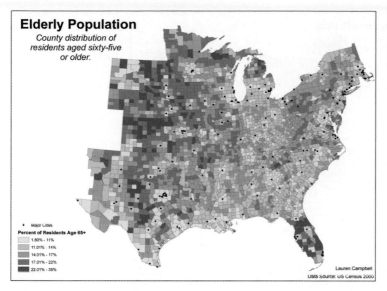

Figure 6.16: Percent Population Aged 65 and over from the 2000 census. Also shown are the locations of major cities. Map by Lauren Campbell using ArcGIS 9.3.

tornadoes occur. Instead, we will examine why they lead to deaths and injury. For this, we will use data from the U.S. Census Bureau about people and where they live.

Two problems arise immediately when we begin to analyze the human dimensions of tornado hazards. First, information about humans is captured by the census bureau for distinct geographical units, such as states, counties, metropolitan statistical areas, tracts and blocks. Thus for every tornado with a point of touchdown (and liftup), we have a polygon containing social data. We immediately have to make a decision, that the touchdown point alone will be used to find which polygon we deal with for social data. We also choose to use the county level census data. Of course in reality, tornadoes cross county lines. Such simplifying assumptions are almost always necessary in any kind of data

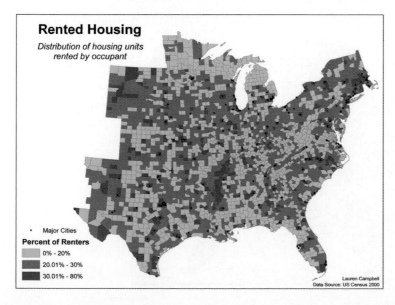

Figure 6.17: Percentage of houses occupied by renters, form the 2000 census. Map by Lauren Campbell using ArcGIS 9.3.

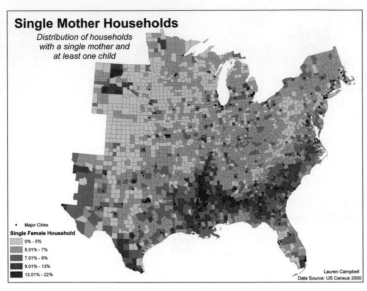

Figure 6.18: Percentage of households headed by single mothers by County. Data from 2000 Census. Map by Lauren Campbell using ArcGIS 9.3.

analysis. We need to be careful to document and explain the reasons for the decisions, because they will have small but possibly significant impacts on our results later.Next we face a temporal problem. The U.S. census takes place every ten years on years ending in zero. So the period covered by our data includes the 1950, 1960, 1970, 1980, 1990, and 2000 censuses. Dealing with all of these is a problem, so instead we select from our data only those tornadoes causing death or injury between 1990 and 2006. We can then use the 2000 census data, although it will be a little out of date. Again, these simplifying assumptions are a necessary part of analysis, and need to be justified.

What factors might explain deaths and injuries from tornadoes? Since 1990, early warning systems have been in place in most places where tornadoes are common. These

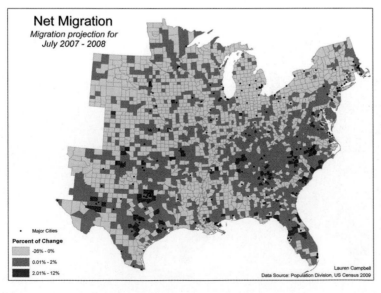

Figure 6.19: Net migration as a percentage of County population. Census Bureau forecasts for 2008–9. Map by Lauren Campbell using ArcGIS 9.3.

include the emergency broadcast network; radio, television, and other media warnings; weather forecasts; and a network of sirens. While many people can respond to these warnings, many cannot. What social groups are least able to respond? We hypothesize that tornado deaths and injuries will be highest where large numbers of people have limited information access. This could be because they do not own radios or televisions, because their first language is not English, or because they are elderly. Some census variables that are indicators of these patterns are the percentage of population aged 65 and over, the number of children in households headed by females, the proportion of recent immigrants, and the proportion of houses that are rented. Because the raw numbers also have an effect, we also collected data on numbers of household, and total population. Maps of these variables were made using ArcGIS (Figures 6.16–6.19). An important step in the analysis was using the join one-to-many function, to create a table that was structured by County (from the Census data) but also contained the tornado information. This join was possible because the GIS software could determine using a point-in-polygon test which county each tornado fell into. The final data set had 2749 tornadoes. Since we want continuous geographical coverage, we include some counties with no tornadoes. We excluded states with no deaths or injuries from tornadoes in the time period.

Each geographic property suggests a question about the data, with a few of these listed below.

SIZE Most of the distributions cover the whole affected area, with lows as gaps within states. Should the analysis be repeated state by state or done once for the whole area?

DISTRIBUTION There is some association of the distributions with city locations. Should the analysis be structured as rural vs. urban? There are also regional peaks, e.g., rental housing in the Mississippi Valley and New England. Should these be analyzed separately?

PATTERN Each map has a general clumping but with regional patterns, e.g., migration is negative in the northern and western zones.

CONTIGUITY There appear to be some donut patterns around cities. Should distance from city centers be included in the analysis?

NEIGHBORHOOD There are distinct urban to rural gradients, and some statewide trends. Again, should distance be included?

SHAPE In some cases state interiors tend to be lower than their edges. Should distance to the state boundary be included?

SCALE There are different patterns around cities vs. within states vs. across the whole continent. Would it be possible to use different units (e.g., States, Census tracts) and do the same analysis?

ORIENTATION Are there connecting lines, perhaps following major highways, between the highs and lows on the map? Are these relevant to an analysis?

Many possible different statistical models could be used to analyze the tornado and the human patterns of risk. In the following, we will use a method called multiple regression modeling. This method can be performed in many GIS, statistical analysis and even some basic spreadsheet packages. The goal of the model is to hypothesize about how the variables we have chosen may contribute to tornado deaths and injuries, then to test the strength of the explanation and adjust the model until it gives reasonable results. Lastly, we explore what the model does and doesn't tell us about the research questions.

6.4.2   Testing a Spatial Model

Is there a statistical relationship between the number of deaths and injuries caused by tornadoes and the census variables that reflect our hypothesis? As a reminder, we believe that the number of deaths and injuries can be explained by the number of people at risk and their vulnerability caused by lack of access to precautionary measures. Vulnerability will, we believe, be reflected by the proportion of rental housing, the proportion of elderly, and the proportion of migrants and dependent children. If we had to express this mathematically, we could say that the tornado deaths and injuries T are a function $f(\ )$ of population P, rental housing R, dependent children C, elderly E, and migrants M. Then we also have to take into account the severity of the tornado. Tornado severity is measured in two ways, the Destructive Potential Index (D) and the force, using the Fujita Scale of Tornado Severity (F).

$$T = f(P,\ R,\ C,\ E,\ M,\ D,\ F) \tag{6.1}$$

The simplest form that this relationship might take is a linear relationship. You may remember from high school math the formula for a straight line, $y = a + bx$. The y is called the dependent variable, because its value depends on what is computed on the right-hand side of the equation and because it is the one we want to predict. The independent variable, x, is the one for which we have the measurement. The value b is the slope of the line. Finally, a is called the intercept. This is the value of y when $x = 0$.

The difference is that we have six of these relationships and not one. Nevertheless, the linear model is the same. Regression is a means of computing the best-fit line through the data for a linear model. It uses the method of least squares; that is, it selects the linear equation that minimizes the sum of the squared differences between the data values and the linear model in the dimension of the dependent variable. For a two-variable straight-line case this is vertical differences from the line. For many variables the linear model is extended into multiple dimensions, but the least squares principle remains the same.

There is, however, a problem. As we saw above, hypothesis testing about statistical relationships is usually only possible when we assume that the sample data we use are selected at random from normally distributed populations. We saw from the preliminary examination of the tornado data that this is not the case. It is also probably true of the new variables we have introduced. So first, we need to examine each of our variables to see whether they approximate a normal distribution, i.e., follow a bell curve. Some of the histograms of the variables are shown in Figure 6.20. There are several metrics to test for a normal distribution of a variable: the skew, kurtosis, and Kolmogorov-Smirnov test. Another way is to create a *P-P* plot (probability-probability), which plots the cumulative

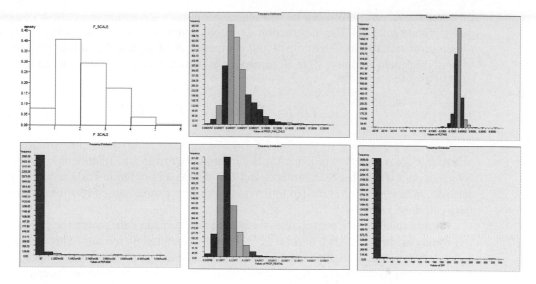

Figure 6.20: Histogram plots for the tornado data. Shown from upper left are F (Fujita tornado force), proportion of household headed by females with children, percent recent migrants, census 2000 population, proportion of homes that are rental, and the destruction potential index.

distribution of the actual distribution against that expected from the normal curve. Probably superior, however, is to examine the histograms carefully.

Distributions that are not normal will violate the assumption of normally distributed errors that the regression model assumes. A solution is to apply a transformation to the variables to draw out one end or other and force a more normal pattern. Of course this runs the risk of making the model a bit harder to comprehend. Typical transformation that can be applied are root, logarithm, or sine/cosine. In this case, the following transformations were applied to the distributions. A demonstration of the impact on the histogram of the transformations of the variables is given in Figure 6.21.

$$sqrt(T) = f(ln(P),\ R,\ C,\ E,\ M,\ sqrt(D),\ F) \qquad (6.2)$$

The Fujita scale is somewhat like the Beaufort wind scale. It takes on values 0 through 5, with zero having maximum wind speed of 40–72 miles per hour (tree branches and chimneys broken) up to five with winds speed of 261–318 miles per hour (homes lifted off

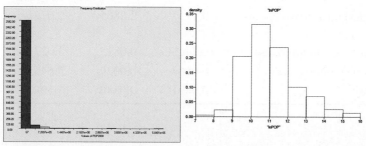

Figure 6.21: Left: original histogram of the value for county population in which tornado struck. Right: variable after transformation to a natural logarithm.

of foundations, cars thrown up to 300 feet) (see: http://www.nssl.noaa.gov/users/brooks/public_html/tornado). The destruction potential Index is a more complicated value (Thompson and Vescio, 1998), involving computing the length and width of the tornado ground track, and summing over all tornadoes in one outbreak. The DPI is given by:

$$DPI = \sum_{i=1}^{n} a_i \quad (F_i + 1) \tag{6.3}$$

where n is the number of tornadoes, a is the tornado damage area (path length multiplied by mean path width) for each tornado, and F is the maximum Fujita-scale rating for each tornado. Since the value is related to F, we should expect some autocorrelation, or linkage between these variables.

From the database perspective, the point-based tornado data have to be joined with the county-wide census data to form a single table structured by tornado. This is a further source of error, because different spatial units are used depending on where the tornado falls (county, ground path, point). In some cases, the join could not be made because there was not a unique identifier on the point. In all, 606 tornadoes survived as the sample representative of the set.

The open source statistical program Smith's Statistical Package (http://www.economics.pomona.edu/StatSite/SSP.html) was used to compute the multiple regression. Input files were created by reading the DBF file created by ArcView 3.2 into OpenOffice.org *calc*, then saving a comma-separated file (*csv*) for input to SSP.

$$sqrt(T) = -2.8935 + 0.2956 \ln(P) - 3.0334 R + 13.0316 C + 0.1717 E + 0.2466 M$$
$$+0.0164 \; sqrt(D) + 0.9491 F \tag{6.4}$$

Variable	Coefficient	Standard error	t-value	One-sided P value
intercept	-2.8935	0.8994	3.2171	0.0007
ln(P) (population)	0.2956	0.0691	4.2766	0.0000
R (rental housing)	-3.0334	1.4540	2.0862	0.0187
C (female-head of household with children)	13.0316	3.5894	3.6306	0.0002
E (elderly, 65 and up as a percent of population)	0.1717	2.2865	0.0751	0.4701
M (proportion of recent migrants)	0.2466	0.0815	3.0271	0.0013
D (destruction potential index)	0.0164	0.0029	5.6465	0.0000
F (Fujita scale)	0.9491	0.0848	11.1981	0.0000

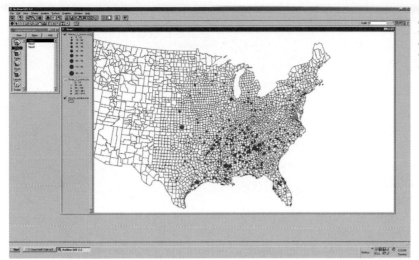

Figure 6.22: Distribution of deaths and injuries due to tornadoes 1990–2006, based on the multiple regression equation 6.4. Map made in ArcView 3.2.

The overall regression results are that equation 6.4 has standard error of estimate of 1.855. with a corresponding coefficient of determination or *r-squared* of 0.3271, which becomes 0.3192 when adjusted for the degrees of freedom. The coefficient of determination is the proportion of the variance that is accounted for by the regression equation. This needs to be adjusted for the degrees of freedom (a function of the number of variables and cases) because the addition of each variable could add to the explanation purely by chance. The modeled distribution of deaths and injuries when re-transformed back to actual values is shown in Figure 6.22.

Obviously statistical results need interpretation. First, the contribution of the variables to the explanatory power of the equation is given by the sign of the coefficient, and the standard error. The *t*-test value is the coefficient divided by the standard error. Values over 2 are good, and the P value is the proportion of the one-sided bell curve that is outside the value, a confidence test as discussed above. There are three items of note in these results: first, that the elderly population's inclusion in the regression does not add to its strength (*t*-value of 0.0751). Secondly, our belief that deaths and injuries will be higher in areas of rental housing is wrong. In fact the opposite is true: in areas with higher proportions of people in rental housing there are fewer deaths, and this result is significant at the 5% (but not at the 1%) level of confidence. Third and last, the variables we have proposed do show a statistically significant relationship, but they account for a mere 33% of the overall variance. Clearly our model leaves much unexplained about tornado deaths and injuries. Running the model again and leaving out the two variables E and R actually decreased the adjusted *r*-squared, to 0.2794.

6.4.3   Residual Mapping

A common way of seeking a deeper understanding of a spatial relationship is to examine the amount that each record deviates from the current model under analysis. In a simple linear regression, if we plug the values for the independent variable (x) into the linear equation $y = a + bx$, we end up with an amount above or below the line in the up-down (y or dependent variable) direction on the scatter diagram (Figure 6.23). If we add all these

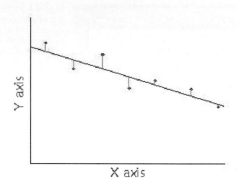

Figure 6.23: A linear regression line derived from observations. The residual comes from using known values for the independent variables, calculating the predicted y value, then subtracting that from the actual value.

together, they sum to zero, just as when we examined the deviations from the mean in the attribute description in Section 6.1.

These amounts are called *residuals*. Each record has a residual, just as each record has a geographic extent. Again, we can use the compute command or its equivalent in our GIS, spreadsheet, or database manager to calculate the residual for each instance. In a multiple regression, the value is computed straight from the regression formula, as shown in equation 6.4. The results are subtracted from the actual values. Since we transformed the dependent variable T (tornado deaths and injuries), we must first untransform the predicted and actual values, then do the subtraction. The result is the map in Figure 6.24.

The map again takes a little interpretation. The first thing to notice is that the negative residuals (where the model overestimated the number of deaths and injuries) are clearly outnumbered by the positive, when the model underestimated the number of deaths

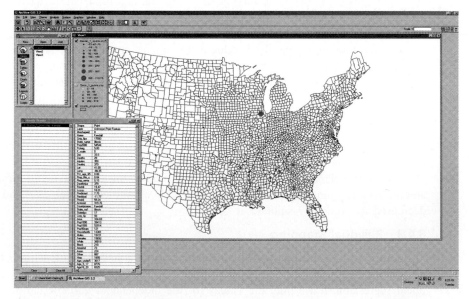

Figure 6.24: Map of regression residuals for the tornado deaths and injuries multiple regression. Highlighted case is for Kendall, Illinois, with 29 deaths and 350 injuries on 8/28/1990, underestimated significantly by the regression model.

and injuries. There could be two explanations for this: first, because the dependent variable is skewed (many more small numbers of deaths and injuries than large), the model may be better at the low end of the death and injury scale than the upper end. Secondly, the *r-squared* implied that the model only explained a third of the variance, so there is one or more additional factors that increase the number of deaths and injuries. As yet, this is unmodelled variation. It may be contained in some of the additional factors considered in Section 6.4.1. Of course it is also possible that tornadoes are so unpredictable that no model could capture a majority of the variance in their prediction.

The example we have followed here has a few lessons to offer. First, spatial analysis follows the same path as much of scientific inquiry. We begin by displaying the attributes that we are interested in and looking at their aspatial (e.g., the histogram) and spatial (the map) characteristics. We try to see whether geographic properties have influenced the form of the distribution and if they are able to explain it. Then we formulate a model of the geographic relationship. In the case above this was a multiple regression model relating the tornado deaths and injuries to social and tornado force factors.

We then formulate a way to test the model, usually a measure of goodness of fit between the model and the data that we have. In statistical terms, we suggest a hypothesis about the relationship, propose a counter or null hypothesis, and then devise a test to accept or reject the hypothesis based on the results. This often involves a probability from the normal distribution; in statistics, cutoff probabilities such as 95% or 99% are used to accept or reject a hypothesis. In this case, while the model was statistically significant, it underpredicted deaths and injuries, and further investigation would be necessary.

Spatial analysis then goes further. We seek to explain geographically why the model does or does not fit. If the fit is inadequate, we can choose another model, change the geographic scope of the problem (e.g., the scale or the extent), or expand the model to include more attributes, that is, build a more complex model. Good science dictates that a simple model is preferred over a complex model, but that when a complex model explains a data set successfully, it is acceptable.

Lastly, any analysis is left with the need to justify the choices and parameters chosen. In the tornado case, data had to be left out (including the highest incidence of death, 5/3/1999 with 36 deaths and 583 injuries in Oklahoma). Information had to be fused from different sources (NOAA and Census), and different geographical units chosen and compared. Transformations were chosen, and maps were interpreted. Even so, the model can only account for one-third of the deaths and injuries from tornadoes. So what is the use of such a model? The answer, in a nutshell, is the power to predict.

6.4.4 Prediction

The final stage of a model's use is to predict in addition to explain. Ideally, the geographic properties themselves, and their mapping, have some explanatory power and can point to a process in action. For example, if we analyzed a disease and found a high concentration in a single district, we could speculate that there was a clustered distribution with a single "source" and an outward diffusion from the source. Proof of this model would be to find a sample of people who had the disease and to show that they all contracted the disease in the single district. Prediction would then follow. An all-out attack on the disease in the one district would be the best strategy to eliminate the disease.

We can return to the tornadoes example to take a closer look at prediction. We can generate a map of expected tornado deaths using the regression equation just as we did in Figure 6.22. This would obviously be extremely difficult to test by experiment. We could easily, however, use different data to examine the model. Using the data for an earlier or later time period could be tried to see whether the same model did a good job for data from the 1950s, for example. Similarly, we could test the model in parts of map that we excluded from analysis, or for Canada and Mexico. Clearly, the model has geographic constraints upon its predictive powers, and these can be tested and described. If the model is geographically variant, this too is of great interest. Virtually every phenomenon on earth, even gravity, varies in some way over geographic space. If this were not the case, geography would not exist as a discipline, and GIS would be no more powerful than a spreadsheet program! There is value in mapping data even if no spatial analysis is intended, as a visual description of a geographic pattern.

As GIS use becomes more commonplace, and as geographic explanations become of more value in managing resources, more of the geographic analyst's time will be spent in searching GIS data for spatial relationships. This geographic detective work can go a long way toward dealing with all the phenomena that have yet to be examined rigorously. This explains why GIS is being so rapidly accepted in disciplines such as archeology, demography, epidemiology, and marketing. In these fields, GIS allows the scientist to look for relationships that could not be seen without the lens of cartography and the integrative nature of GISs turned toward information and data like a high-powered telescope. The eye-opening moment of seeing the data visually for the first time is a common GIS converts' experience. Just as the explorers of the last centuries mapped out America and the world, so the GIS experts of today are mapping out new geographic worlds, invisible at the surface, but visible and crystal clear to the right tool controlled by the right vision.

In the search for spatial relationships, the GIS analyst is largely alone, however. The tools for searching are only now being integrated into GIS. In cartography, a process called the design loop is often used in map design. The map is generated to the specifications, and then the tools of digital cartography are used to make slow incremental improvements until the optimum design is reached. Obviously, not all of the improvements lead to success. Many serve to show what not to do; nevertheless, trial and error are important steps in the map's improvement.

Such a process is often used in analysis with GIS. A typical analysis is the result of extensive data collection, geocoding, data structuring, data retrieval and map display, followed by the sorts of description and analysis highlighted here. The next steps, prediction, explanation, and the use of these in decision making and planning, are the ultimate goal. Without seeing spatial relationships, however, these last prizes will be lost, and the GIS will not have reached its potential as an information management tool. The search for spatial relationships is in turn driven by posing compelling questions to pursue by analysis.

The GIS spatial relationship search loop consists of the steps outlined above, data assembly, preview, hypothesis design, hypothesis testing, modeling, geographic explanation, prediction, and examining the limitations of the model. It is the automated nature of the GIS and its flexibility that allows effective use of trial and error, and exploratory visualization. In this process the GIS components most used are the database manager, for selecting and reorganizing attributes (e.g., with relational *joins* and *relates*); the map display module, for display of intermediate and final analyses; and the computational or statistical tools available as part of (or beyond) the GIS. Some of these are sophisticated, although in many GIS packages they are rather simplistic. In many cases the right spatial analysis tools are user-contributed scripts, add-ons, or extensions, so the issue becomes finding and installing these.

One capability now being added to GIS is that of allowing examination of how distributions and relationships vary over time. Time is not particularly simple to integrate into a GIS, because each attribute data set and its maps are best interpreted as a single time snap-shot. Nevertheless, both the attributes and the map are undergoing constant change, and the geographic phenomena they represent are indeed very dynamic. Even an apparently stable attribute such as terrain is affected by mining, erosion, and volcanoes. Human systems are virtually all in a constant state of change. Even simple geographical coverages, such as satellite imagery, are often pieced together with tiles or maps created at different times.

We can simply compare two time periods for which we have data. Many human and social data are collected only once a decade, for example by the census, meaning that changes happening more rapidly than this will go unnoticed. Comparing two time periods allows only a single measurement, by value or map, of change. We can map, for example, on a satellite image all those areas that have changed between the time periods shown in the two images, assuming the same geographic extent at similar map scales. This can give us a direction of change, the addition or loss of wetlands, for example, and the amount of addition or loss, but not a rate of loss or gain. To measure a rate of change needs a minimum of three images or maps (Figure 6.25).

The most effective tool for time-sensitive geographic distributions is animation (Peterson, 1995). Animation allows the GIS interpreter to see changes as they take place. It is a key part of scientific visualization as far as a GIS is concerned, because usually more can be learned by examining the dynamics of a geographic system than its form. Imagine seeing glimpses of a chess game, for example, at only three stages in a game. The forms

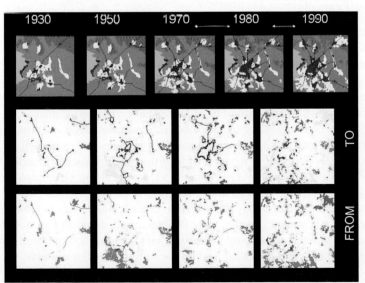

Figure 6.25: Hypothetical land use data (test data for the SLEUTH land use change model), showing land use classes over time (top) and then change between data sets, with changes colored by the class changed from and to.

Figure 6.26: Animation sequence of world earthquakes. Source: USGS (http://earthquake.usgs.gov/eqcenter/recenteqsanim/ world.php.

remain largely unchanged, although some pieces have moved around and disappeared from the board.

As we get more "frames" in the sequence, we arrive at the stage where every move is visible and the rules of movement of chess pieces become discernible. Finally, a great number of frames allow us to see not only the rules but the players themselves, their strategy, and the drama of the game. Just as there is an appropriate geographic resolution for map data, so is there a suitable time resolution. Like enlargement and reduction in space, time can be made slower or faster than "real" time to reach this appropriate resolution. Often frames are added between data points to smooth the appearance of change, a process called "tweening." More information about cartographic animations can be found in Slocum et al. (2009), Chapter 21.

Figure 6.26 shows a few individual frames from an animation built with a GIS by the USGS, in this case the global distribution of earthquakes. As is obvious, a static textbook cannot bring across the dynamic nature of the sequence. For this, use the animation on the World Wide Web at http://earthquake.usgs.gov/eqcenter/recenteqsanim/ world.php.

6.4.5 Map Overlay Example

One of the oldest analytical methods used in GIS, and indeed the simplest, is map overlay. Map overlay is the set of procedures by which maps with different themes are brought into geometric and scale alignment so that their information can be cross referenced and used to create complex themes, or to show spatial linkages. We have met the method already several times, and should recall that the maps to be overlain must be of the same spatial extent, on the same map projection and datum, be at comparable granularity (that is, the spatial units, whether pixels or polygons, should be about the same average size), and if the layers are to be used with map algebra, at the same raster grid size and resolution.

The power of the GIS is in handling the geometry of the overlay process. Handling and preparing the themes is up to the GIS analyst. Under the simplest possible configuration, GIS layers are all converted to binary maps, and an overlay then sifts the map space to leave open the areas that satisfy the selection criteria in use. This is the case in the simple overlay analysis we met in the early chapters, and duplicates in a GIS methods that were worked out using transparent overlay maps and blacked out areas on the transparencies. Many of these methods date back to the turn of the 20th century.

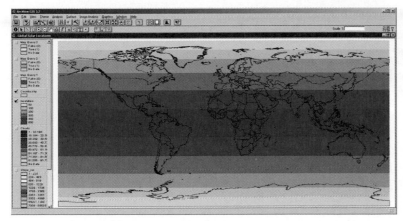

Figure 6.27: Overlay exercise for solar power. Global insolation, using ArcView 3.2. Derived from a figure in *Geosystems* by Robert Christopherson.

One means of map overlay is to intersect all of the layers involved to generate a set of most common geographic units. In map algebra, the raster plays this role. The attributes are then inherited or passed down to subsetted areas, and the attribute table gets longer and longer as more and more units are created. We have already seen the many problems with vector map overlay, including sliver polygons. Blind map overlay will happily assign attributes to very small sliver polygons, and use them in further analysis. A solution to this problem is to first process each layer to reduce the number of solution classes that will find their way into the final map. A selective query from each layer is one simple way to do this. A third overlay method is to find some common unit into which all values can be transformed. One GIS project that the author worked on solved the apparent incompatibility of the GIS layers for an ocean GIS by converting all of the themes into dollars, and adding them together to yield a composite. This is not always possible, of course, and far more common is to weight the overlays involved by some prechosen set of values reflecting the relative, not the absolute, importance of the layers.

For example, Figures 6.27–6.30 show world maps of the factors considered for large scale siting of global solar power production. Information was searched for on the World Wide Web, downloaded, and then processed into ESRI's ArcView 3.2 so that map layers could be registered and converted to a common map projection using ArcView's projection wizard extension. The themes used were those thought to relate to the possible supply and

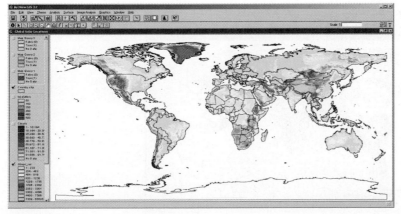

Figure 6.28: Overlay exercise for solar power. Global topography. Source: GTOPO30 USGS.

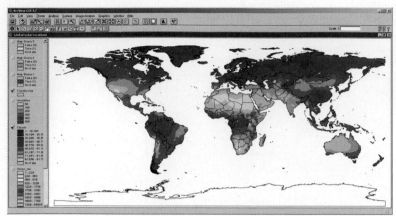

Figure 6.29: Map overlay exercise for solar power. Global average cloud cover. Source: UNEP Grid.

demand for solar-generated power. On the supply side, layers were the maps of incoming solar radiation, the expected average cloud cover, and global topography. On the demand side, a global population layer was used to buffer the solution set and limit it to those areas located close to major population agglomerations.

The overlay exercise consisted of creating a query to the GIS that essentially converted each of the layers into a binary map, a thresholding operation. The query requested insolation levels greater than 200 Watts per square meter from Figure 6.27, three levels of cloud coverage on average less than 65, 70, and 75 percent from Figure 6.29, and elevations lower than 5000 feet (1524m) from Figure 6.28, since although incoming solar power increases as the atmosphere becomes less dense with height, large scale construction favors lower and flat topography. These were combined by selecting those areas with population densities greater than 50 people per square kilometer (Figure 6.30). The resultant map shows the largest contiguous areas suitable for large scale solar power generation include the Southwestern United States, Northern Chile, South Africa, the edges of the African Sahara, the Arabian peninsula, and Pakistan (Figure 6.31).Having done the analysis with three different thresholds as far as cloud cover is concerned, two issues related to overlay analysis are shown. First, the different criteria are subjective, and need to be weighted to reflect their relative importance in the GIS solution. For example, it may be that total insolation is far more, say 10 times, more important than elevation in deciding where solar power generation is located. This can be accommodated by

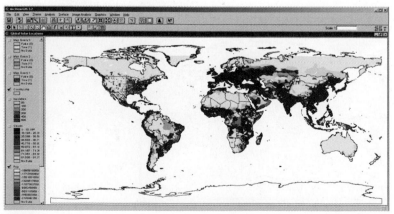

Figure 6.30: Map overlay exercise for solar power. World population density. Source: NCGIA.

multiplying the weights by the binary layers, each factored so that the sum of the weights is one, and then summing them across the layers. The final map will reflect the critical factors and their importance. However, selecting the weights can be a very complex process.

Secondly, in this case the solution area is highly influenced by only one of the layers, the cloud cover. This is why three levels of cloud cover were used in the final map. This "most sensitive" or critical layer in the map overlay process often marginalizes the contribution of other layers, perhaps even eliminating them from the final solution space altogether. This layer sensitivity is especially important to understand. Often a small amount of testing can reveal the critical layer. Some research suggests treating the layers as fuzzy, combining the factors together as a smooth field and creating margins of error on the solution map. Another important factor that affects analysis is how the error, for example caused by different generalization on each input layer, impacts the final decision by propagating through the overlay analysis and influencing the result.

Map overlay remains one of the most common forms of GIS analysis. With the use of buffers and distance transforms, some very sophisticated analysis can be done. This method is extensively applied in planning, but is also increasingly used in all GIS applications, from fire modeling to habitat suitability mapping.

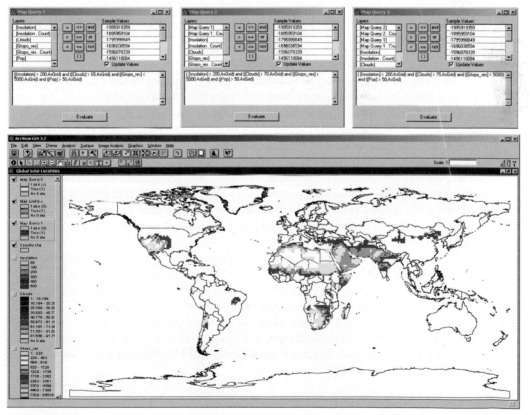

Figure 6.30: Solution set for the map overlay exercise. Global suitability for solar power production. Above the map are the three query windows that generated the solution set. Figures 6.27–6.31 by Jeff Hemphill and Westerly Miller.

6.4.6 GIS and Spatial Analysis Tools

In the early days of GIS, much criticism was made of the fact that GIS software rarely came with any true analytical options. As we have seen, the basic tools of description are those of arithmetic and statistics, and the tools of modeling involve allowing the encoding of a model or formula into the system. Many models work on network flows, dispersion in two or three dimensional space, hierarchical diffusion, or probabilistic models based on weights determined by buffers, and so on. This sort of model is manageable in a GIS using the tools of retrieval: overlay, buffering, and the application of spatial operators. Even a simple model, however, can become a quite lengthy sequence of steps for the GIS's user interface.

Almost all GIS packages allow operations to be bundled together as macros or as sequences of operations as part of a model. ESRI's ArcGIS, for example, contains a set of tools for building graphical models of repetitive procedures, the model builder. Although this goes a long way toward routine analysis, exploratory GIS data analysis is still something of an art. Many operations can be performed in the database manager only, and often GIS users move the data from the database manager into a spreadsheet package such as Microsoft Excel, or a standard statistical package such as SAS (Statistical Analysis System) or SPSS (Statistical Package for the Social Sciences) for analysis. As we have seen, open source and shareware tools such as R are increasingly popular for use with GIS analysis.

Most GIS analysts use statistical and GIS tools in tandem during the analysis stage of GIS operation. The ability to produce nonspatial graphics—for example, a scatter plot or a histogram—is often far easier this way. Given the broad acceptance of statistical packages, and the large number of scientists and others trained in and familiar with their use, a compromise solution seems best. GIS packages can avoid duplicating the many functions necessary for statistical analysis by making two-way data movement between GIS and statistical software easy.

In summary, one of the greatest strengths of a GIS is that it can place real-world data into an organizational framework that allows numerical and statistical description and permits logical extension into modeling, analysis, and prediction. This important step, along with examining and thinking about one's data, is the bridge to understanding data geographically. In the beginning of this chapter we saw that by exploring GIS data visually, through exploratory statistical graphics and maps, a great deal of information and knowledge can be gained.

This understanding is enhanced by GIS because many phenomena simply cannot be fully understood, and certainly cannot be predicted, without an understanding of the geographic forces at work and their expression among the map's features as imprints of the principal geographic properties. Unfortunately, most GIS packages contain only rudimentary tools for spatial analysis. However, GIS practitioners have filled the gaps with scripts and standard statistical software, and great strides are now being made as GIS contributes to the new models that are resulting in a host of different applications beyond the traditional scope of geography.

6.5   STUDY GUIDE

6.5.1   Quick Study

CHAPTER 6: WHY IS IT THERE?

O *Spatial analysis examines, queries, tests, and tries to explain geographical phenomena* O *Analysis begins with a question, then hypotheses, then data; and with non-spatial statistical and spatial graphic description* O *Visual exploration helps formulate research problems* O *Exploratory visual tools for attributes include the box plot, histogram, scatter plot, radar plot, and pie chart* O *Box plots use a display of five numbers* O *Measures of central tendency are single values to summarize many* O *The median is the middle member of a distribution when ranked* O *The mean is the sum of the record values divided by the number of records* O *Measures of central tendency are impacted by outliers* O *The standard deviation is the average difference from the average* O *The sum of the squared differences from the mean is the total variance* O *The Z-score is the number of standard deviations a value lies from the mean* O *The theoretical normal distribution follows a bell curve, and can be used to compute the probability of values being beyond a given Z-score* O *A perfect normal distribution means random error and a large population* O *The smaller the sample, the more we underestimate standard deviation* O *Spatial description summarizes two variables x and y, simultaneously* O *The bounding box is formed by the maximum and minimum x and y* O *The mean of x and y form the mean center, and the normalized standard deviations form the standard distance* O *The mean center is also a centroid, but not necessarily vice versa* O *The nearest neighbor statistic for point distributions is the observed mean nearest point separation divided by the expected separation* O *The value R ranges from 0 (clustered) to 1 (random) to 2 (square) to 2.15 (triangular); usually computed in GIS with a script* O *Other measures are available for lines and areas; many require multi-step calculations* O *Issues of spatial relations include map comparison, spatial variation, and spatial modeling and prediction* O *Spatial analysis invariably involves simplifying assumptions, which must be justified* O *A spatial model formulates a geographic relationship that is statistically testable* O *A simple spatial model is a multi-variable linear model, testable using multiple least-squares regression* O *The method assumes that variables are normally distributed, so some may need transformation* O *Regression minimizes the sum of the squared deviations from the linear model in the dimension of the dependent variable* O *A model is testable by the significance of its variables and its goodness of fit* O *A model gives a predicted value, that can be subtracted from the data to give residuals, which can be mapped and examined* O *Spatial models should be examined for their assumptions, errors, and lack of fit* O *Examining errors can lead to model improvement* O *Variation over time is hard to examine with GIS; with two time periods change can be measured and with three, rates of change* O *Animation is effective for examining change* O *Overlay analysis is a simple form of spatial analysis* O *Layers can be equal or weighted in their importance; complex sequences of variables can be used in analysis* O *GIS software often needs to be supplemented with database, spreadsheet and statistical software tools, and results sent back to the GIS for display* O *Exploratory spatial analysis is something of an art* O

6.5.2   Study Questions and Activities

Describing Attributes

1. Write out step-by-step instructions for a child to calculate a mean and a median from a list of 10 numbers. Modify the instructions for 11 numbers. Write a one-paragraph explanation of what the numbers mean.

2. Why is spatial analysis different, though related to, statistical analysis? Give examples.

3. Copy from a map such as a 1:24,000 series USGS map a group of objects, including a set of points with known elevations, some rivers, and forested areas. List as many measurements as you can devise to characterize the basic geographical properties of each feature, classified by dimension. Which measures are the easiest to calculate and why?

Statistical Description

4. Draw a flow diagram of the stages of scientific inquiry surrounding analysis in a GIS. What is necessary before analysis can be conducted? What does a successful analysis lead to? What might prevent the ease of movement through the flow in the diagram?

5. For any data set of your choice, make a box plot, scatter plot, and a histogram. What tools had to be introduced from outside of the GIS you are using?

6. How might the length of a line and the area of a polygon be calculated (a) in a vector GIS and (b) in a raster GIS? Why might they be expected to give different results?

7. For any attribute you may choose, calculate the minimum, maximum, median, mean, and standard deviation. Now repeat for a second attribute and compare the values with your initial results.

Spatial Description

8. Use your GIS to plot the bounding rectangle and mean center of a set of points.

9. Download and apply a nearest neighbor statistic script and use it to test point distributions that you enter by hand into your GIS. How low and high can you make the value go?

10. Use your GIS to compute a new attribute based on a transformation of two or more others. Map the result.

Spatial Analysis

11. Design a model that might account for the risk of wildfire for a GIS data set consisting of layers for vegetation type and condition, soils, streams, topography, and wind direction. How might the model be tested?

12. Use a GIS to overlay a map of topography with polygonal districts such as counties. Using any method available, compute and then map the variance or the standard

deviation of the elevation values within each district. Explain the distribution on the map.

13. Download the tornado data from NOAA and repeat the analysis in this chapter. Can you build a more explanatory model of tornado deaths and injuries?

14. Repeat the chapter's solar energy analysis for your state. Where is the best place to generate solar power? How large an area is available in a single contiguous zone, say for a solar power plant?

6.6 REFERENCES

Ashley, W. S. (2007) Spatial and temporal analysis of tornado fatalities in the United States: 1880–2005. *Weather Forecasting*, 22, 1214–1228.

Campbell, J. (2000) *Map Use and Analysis*, 4 ed. Boston, MA: McGraw-Hill.

Earickson, R. and Harlin, J. (1994) *Geographic Measurement and Quantitative Analysis*. New York: Macmillan.

Peterson, M. P. (1995) *Interactive and Animated Cartography*. Upper Saddle River, NJ: Prentice Hall.

Slocum, T. A., McMaster, R. B., Kessler, F. C. and Howard, H. H. (2009) *Thematic Cartography and Geovisualization*. 3ed. Upper Saddle River, NJ: Pearson Education.

The World Almanac and Book of Facts. New York: Pharos Books. Published annually.

Thompson, R. L., and Vescio, M. D. (1998) The Destruction Potential Index—a method for comparing tornado days. Preprints, 19th Conf. Severe Local Storms, Amer. Meteor. Soc., Minneapolis, 280–282.

Tukey, J. W. (1977) "Box-and-Whisker Plots." In *Exploratory Data Analysis*. Reading, MA: Addison-Wesley, pp. 39–43.

Unwin, D. (1981) *Introductory Spatial Analysis*. London: Methuen.

6.7   KEY TERMS AND DEFINITIONS

analysis: The stage of scientific inquiry when data are examined and tested for structure in support of hypotheses.

attribute: An item for which data are collected and organized. A column in a table or data file.

bearing: An angular direction given in degrees from zero as north, clockwise to 360.

bell curve: A common term for the normal distribution.

bounding rectangle: The rectangle defined by a single feature or a collection of geographical features in coordinate space, and determined by the minimum and maximum coordinates in each of the two directions.

box plot: A simple way of graphically depicting groups of data through five summary numbers (minimum, lower quartile, median, upper quartile, and sample maximum).

centroid: A point location at the center of a feature used to represent that feature.

compute command: In a database manager, a command allowing basic arithmetic on attributes or combinations of attributes, such as summation, multiplication, and subtraction.

data extremes: The highest and lowest values of an attribute, found by selecting the first and last records after sorting.

dependent variable: The variable on the left of the equals sign in a formula model, whose values are determined by the values of the other variables and constants.

error band: The width of a margin plus and minus one standard error of estimation, as measured about the mean.

expected error: One standard deviation in the units of measure.

goodness of fit: The statistical resemblance of real data to a model, expressed as strength or degree of fit of the model.

histogram: A graphic depiction of a sample of values for an attribute, shown as bars raised to the height of the frequency of records for each class or group of value within the attribute.

hypothesis: A supposition about data expressed in a manner to make it subject to statistical test.

independent variable: A variable on the right-hand side of the equation in a model, whose value can range independently of the other constants and variables.

intercept: The value of the dependent variable when the independent variable is zero.

least squares: A statistical method of fitting a model, based on minimizing the sum of the squared deviations between the data and the model estimates.

linear relationship: A straight-line relationship between two variables such that the value of the dependent variable is a gradient times the independent variable plus a constant.

mean: A representative value for an attribute, computed as the sum of the attribute values for all records divided by the number of records.

mean center: For a set of points, that point whose coordinates are the means of those for the set.

median: The attribute value for the middle record in a data set sorted by that attribute.

missing value: A value that is excluded from arithmetic calculations for an attribute because it is missing, not applicable, or is corrupted, and has been signified as such.

model: A theoretical distribution for a relationship between attributes. A spatial model is an expected geographic distribution determined by a given form such as an equation.

multiple regression: A least squared method that examines relationships among multiple variables.

nearest neighbor statistic: The measured mean separation between points and their nearest neighboring points, divided by the expected mean separation.

normal distribution: A distribution of values symmetrically about a mean with a given variance.

normalize: To remove an effect biasing a statistic; for example, the influence of the size of the sample.

null hypothesis: The state opposite to that suggested in a hypothesis, postulated in the hope of rejecting its form and therefore proving the hypothesis.

Pearson's product moment correlation coefficient: A measurement of goodness of fit computed as the square of the sum of the cross-variance of two variables divided by the sum of the variance in the independent variable. When squared (the coefficient of determination or r-squared) the value is the proportion or percentage of the variable "explained" by the model under test.

population: The total body of objects from which a sample is taken for measurement.

prediction: The ability of a model to provide information beyond that for which measurements are available.

r-squared: A common term for the coefficient of determination.

random: Having no discernible structure or repetition.

range: The highest value of an attribute less the lowest, in the units of the attribute.

record: A set of values for all attributes in a database. Equivalent to the row of a data table.

residual: The amount left when the observed value of the dependent variable has subtracted from it that predicted by a model, in units of the dependent variable.

sample: A subset of a population selected for measurement.

sort: To place the records within an attribute in sequence according to their values.

standard deviation: A normalized measure of the amount of deviation from the mean within a set of values. The mean deviation from the mean.

standard distance: A two-dimensional equivalent of the standard deviation, a normalized distance built from the standard deviations of the eastings and northings for a set of points.

table: An arrangement of attributes and records into rows and columns to assist display and analysis.

units: The standardized measurement increments for values within an attribute.

variance: The total amount of disagreement between numbers. Variance is the sum of all values with their means subtracted and then squared, divided by the number of values less one.

Z-score: The number of standard deviations that an attribute's value is from the attribute's mean.

6.8   PEOPLE IN GIS

Anne Girardin, Database Management Specialist, AIMS, Afghanistan

KC: Hi Anne. Which one is you in the photograph?

AG: That's me in the center, showing local staff how to use a total station to lay out a property parcel.

KC: What is your job, and how do you use GIS in your work?

AG: I actually have several hats. I am originally a French Chartered Land Surveyor and have been doing mostly geographical database management. I have been working for different types of companies, in archaeology, mobile telephony, as a geomarketing consultant, in navigation for TeleAtlas, now Tom-Tom, as a process engineer and database quality project manager. I've worked in France on the surveying and mapping of sewer networks. Recently in Afghanistan, I've worked on land formalization as part of the Pilot Urban Street Addressing Project and USAID/LTERA (Land Titling and Economic Restructuring in Afghanistan), a project to improve land tenure security and support economic growth. Most recently I'm working on the E-Governance of infrastructures (AIMS, Afghanistan). For all these activities, GIS was the basis of operation: location, location, location. Location of people, location of infrastructure, linkage of these located elements to all types of information in order to better manage land and offer services to people.

KC: What is the Afghanistan Information Management Service, and what has been your role in the program?

AG: The AIMS project is building information management capacity for the Afghani government and delivers information management services to organizations across Afghanistan. AIMS seeks to build appropriate skills in government to manage information management systems. AIMS goal is to exit from these activities once capability exists within the government to manage the specific information systems concerned. I am a database management specialist and my main task is to organize the many types of data into a Data Model. The purpose of our project is to develop software for the monitoring and evaluation of infrastructure such as road construction, school repair, and hospital maintenance. When all data are integrated into the database and the software developed, we can produce maps of our assets and other kinds of information such as population density, poverty and also make analyses of the economic growth.

KC: What education did you receive in GIS?

AG: I actually have a land survey and mapping Master's Degree from l'Ecole Superieure des Geometres et Topographes, Le Mans, France. The program at ESGT is divided approximately into: 20% general topics such as Maths, Physics, French, English; 20% land surveying (geodesy, topographical survey, photogrammetry, GIS); 20% urban and rural planning; 20% law (civil, land tenure, regulations, court systems …); and 20% computer science. GIS study was divided between computer science and land surveying. We received instruction both in how to use GIS software (it was MapInfo) and how to

model data (not only geographical). We had to use GIS in some urban and rural planning practical work.

KC: What led you into GIS after your Master's degree?

AG: I really discovered the world of GIS when I attended the ESRI conference in San Diego in 1998 when I was still at school. It opened my mind when I saw all the capabilities of GIS: from mapping the transportation in New York City to observing the condor's habitat in South America.

KC: Give an example of the day-to-day tasks that you are involved with.

AG: I analyze the types of projects and information that the different institutions are dealing with. I transfer the data into a Conceptual Data Model. I help develop mapping standards and adapt them to Afghanistan and the different customer requirements. We are releasing the Data Model to the software developers. And we ensure that data are correctly implemented and integrated into the software.

KC: What advice would you give a student just getting started with GIS?

AG: It depends what the student wants to achieve. To becomes a technician I would definitely advise to focus on the functionalities of the GIS sotware. To become an engineer, then focus on the modeling aspect (without neglecting knowing the functionalities of the GIS software, of course).

KC: Thanks Anne.

CHAPTER 7

On the Surface

I prefer winter and fall, when you feel the bone structure of the landscape—the loneliness of it, the dead feeling of winter. Something waits beneath it, the whole story doesn't show.

Andrew Wyeth

7.1 FIELDS AND FEATURES

In earlier chapters we met the two different GIS models of geography, on one hand the world of features, points, lines, and polygons, and on the other hand the geographical field. A field differs from the discrete models of geographic space, which assume a world of objects set against a blank "void" or background. It is as if the world can be fully described as the objects it contains and their attributes, and these can be cartographically depicted with points, lines, and areas. The alternative model is necessary when a single property, a theme or field, can be evaluated or measured at every point in the geographical space of interest. The attribute is simple, a value in a set of known units, but our knowledge of the field distribution becomes our set of measurements of the space.

A tangible example is surface temperature. It would make very little sense to show temperatures only at weather stations, or to show average temperatures by county in the U.S. Instead, we assume that surface temperature is a field variable, and that it varies smoothly over space. We make measurements at specific locations, such as at weather stations, and then fill in the remainder of the map by making the assumption that temperature changes slowly and predictably as we go from one place to a nearby place. Imagine two weather stations, one recording 10°C and another 20°C. Now imagine that you start at the first point, and drive straight toward the second. You might expect that the temperature increases by the lower temperature subtracted from the higher and divided by the distance from the first point. At points along the way, the actual temperature is determined by the surrounding temperatures and the distances to where those measurements were taken.

Now imagine hundreds of weather stations. An easy best guess at the temperature anywhere would be the average of all the temperatures. Obviously all of the measurements contribute in some way to the value at any particular point. However, clearly the places nearby would be more influential than those further away. At some distance, the influence of a single point's temperature would reduce to effectively no influence on other places. The temperature in London, for example, is not directly influenced by the temperature in Los Angeles at all. This property, that nearby things are more similar than distant things, is called spatial autocorrelation. The property seems intuitive, but was first stated by Geographer Waldo Tobler, and has become called "Tobler's first law" (Sui, 2004).

One last theoretical point about fields before we move onto digital terrain. Returning to our example of temperature; a detailed map of temperature in a house might show general spatial autocorrelation, from room to room in the house or across a room as one moves away from the heating vents, for example. Yet in certain small (often square or rectangular) zones the temperatures are either very much higher than the surrounding points (the stove, water heater, and furnace) or very much lower (the air-conditioned room, the refrigerator and the freezer). Imagine the thermal map of a whole city, where breaks in temperature would occur at the edges of different land uses (Figure 7.1). These sudden changes are discontinuities, and while Tobler's first law applies, the rule is locally changed. These discontinuities might be thought of as tears or folds in the smooth uniformly varying field. They are places where the field model works worse than the

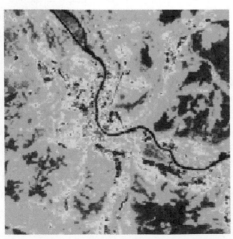

Figure 7.1: Image of a city from the thermal band (6) on the *Landsat 7* satellite. Warmer areas are red, colder are blue. Source: NASA.

feature model. In the world of terrain, these places are cliffs, ridgelines, streambeds, overhangs, and caves.

Terrain or land surface topography is the geographical field on which we live our lives. Much of what is geography is directly or indirectly structured by topography. Mountains form our political boundaries, rivers divide cities, watersheds provide our water, and the slope of the landscape determines whether we can grow crops, or even build a house. Topography is when the "top-down" two dimensional nature of a map becomes three dimensional. With digital terrain, GIS is capable of an almost unlimited number of calculations that inform us about the lay of the land, from slope analysis to intervisibility.

7.2   TERRAIN DATA STRUCTURES

In Chapter 3 we examined the various means by which a map can be stored digitally inside the GIS. Terrain is a little different, because we are dealing with the uppermost surface over a region, a continuous or field data type. In terrain, we use the surface elevation, not the elevations of holes or caves below the surface, for example. The discontinuities we mentioned above, and the smoothness with which we represent the terrain surface, or any other surface for that matter, are very much a function of how we sample the surface, and then how we store the data in the GIS. In Chapter 3 we also examined how all geographical data can be stored in the computer as numbers, noting that the feature model captures geographical objects well, and that the raster or grid model was better for fields. What about terrain specifically? In this section we'll cover the structures that are used to store and represent terrain. Some of these we've touched on already, but now we will consider their implications for dealing with digital topography.

7.2.1   Points or Postings

The simplest way to imagine a field is to record a data value at a point. In the temperature example above, you could use a GPS receiver to record your location, while using a thermometer to record temperature. The GIS representation is very straightforward—you have a measured attribute and a location, geocoding at its most basic! In fact the GPS receiver is already measuring your elevation or height along with your position. This point takes the form (x, y, z), where x and y are the coordinates and z is the elevation. Clearly we need to measure z with respect to some reference level. The GPS usually uses a known datum, such as WGS84, and calculates heights with respect to this model.

With only a few points, not much is known about the surface. Most GISs have the capacity to do interpolation. In interpolation, a sampling frame is established, such as a grid bounded by latitude and longitude values, and a procedure used to "interpolate" values at the grid intersections (or centers) based on the known points (Figure 7.2). The reasons why the point data are sparse may be many: bore holes may be expensive to drill, samples difficult to find such as rare plant species, or the data may just be that way, such as finds in archeological digs. The intesections of the sampling frame to which we interpolate are called postings. Almost any configuration of postings is possible, but regular postings, such as in grids, allow for complete coverage of an area. Sometimes the postings are actual sample points, where the measurements were taken, and we have little control over their distribution. This is usually the case in remote sensing. When the density of points is high, we oversample the surface and need to select or subsample the "point cloud" to select regular postings or to average adjacent values.

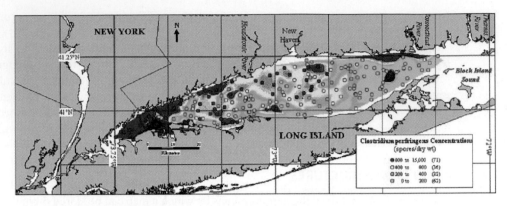

Figure 7.2: Concentration of sewage-related spores in ocean sediments in Long Island Sound. Note the surface has been interpolated from points taken at the sample locations indicated. From USGS Open-File Report 00-304.

There are many possible ways to interpolate data with a GIS. Many GIS packages include a popular method called inverse distance weighting. In this method, a posting is selected and a search conducted for all points that fall within a neighborhood, often defined by a given search radius. The points are then weighted by the inverse distance, often raised to a power, to the posting. The interpolated value at the posting is then the sum of the inverse weighted elevations within the neighborhood (Figure 7.3). The idea is that nearby points have the most influence over the interpolated point value. This method can lead to isolated peaks at high points within the data.

An alternative method uses statistical theory to optimize the interpolation. This method, called *kriging*, was originally applied to ore bodies in gold mining and is based on the mathematical theory of *regionalized variables,* which breaks spatial variation down into a drift or structure, a random but spatially correlated part, and random noise. Kriging involves a multistep approach to interpolation. First, the drift is estimated. Next the

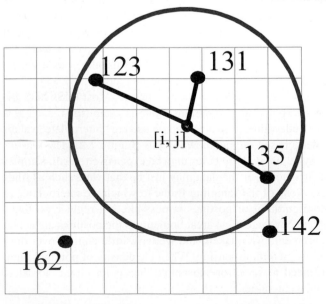

Figure 7.3: Interpolation. In this operation, known height sample points within the neighborhood (the red circle) are used to estimate the value at a location, such as grid cell *[i,j]* where the field value is unknown. In inverse distance weighting, the known elevations (123, 131, 135) are weighted by the inverse of their distances from *[i,j],* and the values summed for the new estimate at *[i,j].* This can be repeated for all grid cells.

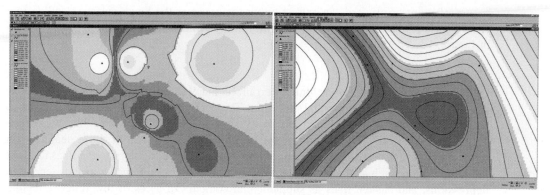

Figure 7.4: Different surfaces interpolated from the same elevation points (black triangles) using ArcView 3.2. Left: Inverse squared distance weighting, 5 nearest points used. Rights: Spline interpolation, 12 nearest points, weight 0.1. Note the data range and pattern differences.

relationship between variation in values and distance is computed and replaced with a statistical model, and lastly the model is applied to estimate the postings. Because kriging yields a surface that passes directly though the data points, and because the technique also yields the estimated variance at each interpolated point, the technique is statistically preferred. Different interpolation methods can give vey different results (Figure 7.4).

One type of commonly used terrain data that is based on the point cloud is Light Detection and Ranging (LiDAR). These data can come from an aircraft or from tripod-based instruments. Data are collected by measuring the time taken by tiny light pulses from a laser to leave the instrument, reflect off the ground, and return. Since the light pulses can be very short, a very large number of points can be collected. Some pulses reflect off of a single hard surface, such as a lake or a building roof. Others penetrate through vegetation and other surface cover. The result is a "cloud" of points that literally reflects the surface. The location of the points depends on the pulse frequency of the laser, the viewing geometry of the flight, the topography, and other factors, giving an uneven

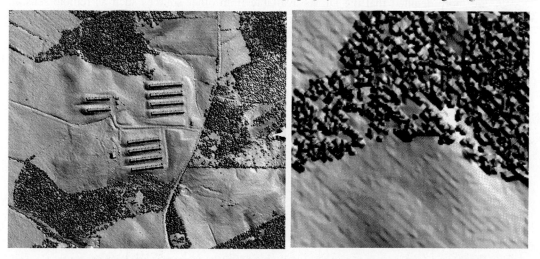

Figure 7.5: Detail of LiDAR image of part of North Carolina. The locations of points in the point cloud can be seen in the detail at right.

scatter of points. Software can extract either the upper or the lower surface, leading to a terrain surface model of great accuracy and fidelity. An example of a point cloud is shown in Figure 7.5.

7.2.2 Contours

One of the oldest ways of representing terrain on maps is to use contour lines. A contour is a line joining points with equal elevation. The contour interval is the vertical distance between successive contours, and is usually a round number, e.g., 2m, 5m, or 20 feet. There are "rules" of contour drawing that aid interpretation, such as those dealing with areas where contours are dense. There are also symbolic modifications to the lines to assist in interpretation, such as thicker contours at index levels, hachure marks to show enclosed depressions, and dashed lines as supplemental contours in areas of low relief. Labeling contour lines is also a skilled task.

Fortunately, many GIS packages have fully automated contouring. Many of the visual aspects (e.g., the level of smoothing, the weight of the lines, etc.) are under user control. Many GISs also allow further modification of the contours, e.g., plotting them over other terrain visualization methods, or filling the areas between contours with color sequences. Some examples are shown in Figure 7.6.

As far as data structures for GIS are concerned, while contours are descriptions of a surface, and represent successive two-dimensional slices of the landscape at regular vertical intervals, logically they are simple vectors consisting of nodes. Since contours do not cross, they have a simple topology and can be drawn with ease. As a terrain data structure they are often not sufficient to embed the critical features of the landscape. They are often used because legacy maps contain them, often paper maps with a contour separation of distinctive color such as brown. These maps can be easily scanned, and the contour lines vectorized. While tagging the lines with their elevations can create problems, often because the drawn lines have gaps to accommodate elevation labels, nevertheless, they can be used to create very reliable and accurate terrain data, especially when the maps are more detailed that the DEMs available from other sources such as remote sensing. Digital terrain data from contour lines sometimes show a statistical bias toward the elevations of the contours used. Among the different terrain storage methods for GIS, they probably produce the smallest file sizes.

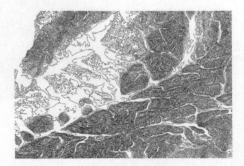 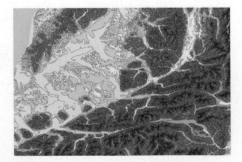

Figure 7.6: Contour map of part of the South Island of New Zealand. Data source: SRTM. Software: ArcView 3.2 and GlobalMapper.

Figure 7.7: A Delaunay triangulation, the underlying basis for a TIN. TIN translation of Santa Barbara, CA 30m DEM using LandSerf.
Source: the author.

7.2.3   TIN

The triangulated irregular network, or TIN, was also introduced in Chapter 3. TINs are often a bridge data structure between point and line methods and gridded Digital Elevation Models. As a logical structure, they are surprisingly robust. This is because the points they include can be chosen to carefully match the discontinuities in the landscape. Points that are at summits, along water features, ridges, and other form lines such as faults are often called "very important points" because they are on the terrain skeleton. This is a network that outlines the peaks, depressions, saddle points, stream channels, and ridge lines that partition the terrain into facets, such as hillslopes and watersheds.

The TIN is a set of points chosen as part of some terrain sampling scheme, for example GPS data collection points. A TIN is created by forming a set of triangles, with their vertices at the points, such that the TIN covers the whole landscape. If there are points along or across the edges of a study area, then the TIN can cover the whole terrain. Any point on the surface is either a node, falls on a TIN edge, or falls inside a triangle formed from triplets of points.

Obviously the full set of possible triangles is large, but TIN selects a particular triangulation called the Delaunay triangulation (Figure 7.7). A Delaunay triangulation is a set of planar triangles formed at points configured such that no point is inside the circumcircle of any triangle that is part of the triangulation. This means that if a circle is drawn that exactly touches the three nodes of a Delaunay triangle, it contains no other nodes. The solution is unique (except for one special case) and maximizes the minimum angle of all the angles of the triangles, so TINs generally avoid skinny triangles. The method was invented by Boris Delaunay in 1934 and is implemented in many GIS packages, sometimes without letting the user know, for example, it is an intermediate stage in creating contour lines.

Once the surface has been partitioned into the TIN, it can be treated logically as a set of triangles. For fitting lines and smooth surfaces, often methods are used to ensure exact fit across triangle edges. Many TIN implementations deliberately include surface points such as streams, ridge lines, roads, and buildings. As a result, both the maps created from a TIN and analyses based on TINs can be quite accurate. TINs are often favored in GIS applications where volumes must be computed, for example in cut-and-fill calculations. They are also good for computing surface visibility, and interact well with computer graphics systems that are designed to rapidly render and shade polygons.

7.2.4 Digital Elevation Models

We also met the digital elevation model in Chapter 3. Several terms are often used interchangeably with respect to gridded representations of a surface. A Digital Elevation Model (DEM) is a regular grid (often square, or with equal spacing in latitude and longitude) of cells, each containing an elevation value. These values are with respect to a reference datum, such as WGS84 or NAD83. They can be integers, in which case the files are smaller, but the elevations are rounded to a whole number of elevation units, or they can be in floating point or decimal format. The latter is sometimes called a grid or a lattice. A trick is to change the elevation units to store better vertical resolution, for example storing decimeters as integers instead of the first decimal place of meters. Note that this means that a DEM has two resolutions, the ground spacing of the side of the cell in ground units, e.g., 10m, and a vertical resolution, e.g., 1m.

The elevation values stored in the DEM are ideally those of the ground surface. Traditionally in mapping (and for good reason, e.g., in figuring surface runoff or flooding) the elevations of features on the surface are removed. This includes natural features such as vegetation, and human features such as buildings. Lakes and water bodies are shown at their water surface elevations. In some cases, we need DEMs that include the actual hard surface of the landscape, e.g., build structure tops and lake depths. Such a model is called

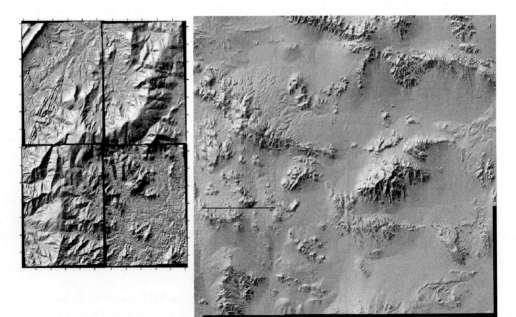

Figure 7.8: Some common problems of mosaicing DEMs. Left: Missing data due to use of UTM coordinates and lat-long quads. Rights: Remnant DEM boundaries visible after mosaicing. Source: U.S. Army TEC: USGS.

a Digital Terrain Model, or DTM. In other cases we also want the actual measured upper surface of the land and its features, e.g., including trees and buildings, for aircraft navigation. Such a model, common in LiDAR mapping as the first reflected return, is called a surface model. These are often used in 3D modeling and simulation.

A DEM has a resolution in the x, y, and z dimensions. Since DEMs can cover projected map space, these resolutions can be constant (e.g., 1km) or varying (e.g., 3 arc seconds). Since the (x, y) space is projected, we also need to take into account the datum for the z direction, and this may change with reprojection. For a DEM we also need to know the extent of the grid, usually stored as the number of columns and rows in the array. Grids may also need metadata, for example the latitude and longitude of the four grid corners. It is important to realize that resampling or reprojecting a DEM can result in pixel loss and duplication. Often a GIS will fill in missing pixels by interpolating (e.g., averaging) from the neighboring cells. This may have unintended consequences, for example smoothing the sharp edges of a lake or coastline.

Another issue with DEMs is mosaicing (Figure 7.8). Depending on the projection and coordinate system, adjacent grids may match exactly, or leave slivers of missing or duplicate data. Ideally, DEM data are "seamless" and can be joined without noticing the effect of the fact that DEMs were mosaiced. In practice, small errors, datum shifts, etc. often mean that seams are visible in DEMs. Another major problem is the registration of DEMs with known landscape features, such as rivers, ridgelines, lake boundaries, or coastlines. Many DEMs were generated prior to exact measurement capabilities of LiDAR, and so are quite generalized. A 30m DEM, for example, is almost guaranteed to miss the actual location of a mountain peak or a stream bed. In some cases, for example with the Shuttle Radar Topographic Mapping (SRTM) mission data, systematic attempts were made to use another source of water body data to "flatten" lakes, estuaries, and ocean areas. This is also sometimes done with rivers and some other features, a process known as "burning." Thus exactly which DEM you use can make an important difference in how well the terrain surface matches with other map layers, such as building features, streams, and water bodies.

There is no shortage of DEM data. World coverage less the polar regions is available at 90m resolution from the SRTM (http://www2.jpl.nasa.gov/srtm/). NOAA makes a large amount of global topography and bathymetry (ocean depths) available through its National Geophysical Data Center (http://www.ngdc.noaa.gov/mgg/bathymetry/relief.html) and supports a DEM discovery portal (Figure 7.9). For the United States, the National Map Viewer (http://nmviewogc.cr.usgs.gov/viewer.htm) contains DEM coverage for the United States, including the older 30m DEMs, the 10m National Elevation Database, and in some cases much higher resolution LiDAR coverages (Figure 7.10). Data sets are available at 6, 1, 1/3, and 1/9 arc seconds of latitude and longitude for all or parts of the country. Spot elevations, depth soundings, and bathymetry can also be found and downloaded from the USGS seamless data server. A distinct advantage of using the various DEM data portals is that downloaded terrain data usually come with metadata, including projection and datum, that the GIS can interpret when the data are imported. This makes any further editing, mosaicing, etc. much more straightforward.

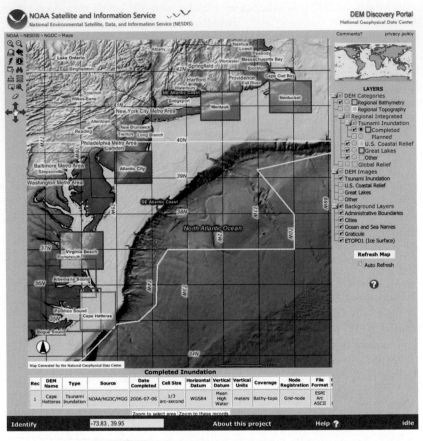

Figure 7.9: NOAA's DEM discovery portal on the World Wide Web at www.ngdc.noaa.gov.

To supplement the SRTM 90m global data, a global digital elevation model was created from approximately 1.3 million individual stereo-pair images collected by the Japanese Advanced Spaceborne Thermal Emission and Reflection Radiometer, or Aster,

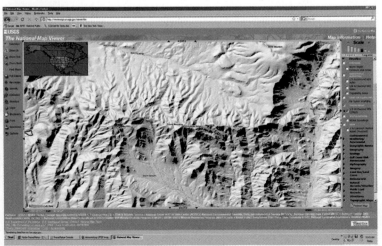

Figure 7.10: Searching for digital elevation data on the National Map Viewer. Yucca Mountain area Nevada, shown with the 1/3 arc second DEM hillshaded.

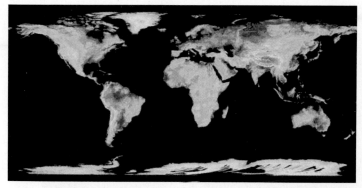

Figure 7.11: GDEM data for the world, from the NASA/METI Aster dataset. Low elevations are purple, medium elevations are greens and yellows, and high elevations are orange, red, and white. See: http://www.nasa.gov/topics/earth/features/20090629.html

instrument aboard NASA's *Terra* satellite. NASA and Japan's Ministry of Economy, Trade and Industry (METI) developed the global DEM (GDEM). The GDEM data expand global land DEM coverage to 99 percent, from 83 degrees north latitude to 83 degrees south at 30m resolution (Figure 7.11). Data users can download one degree tiles from the Aster global digital elevation model at: https://wist.echo.nasa.gov/~wist/api/imswelcome.

7.2.5   Volumetric Models and Voxels

A voxel (volumetric pixel) is a volume primitive, representing a value on a regular grid in three dimensional space, i.e., a cell in x, y, and z. This is the 3D equivalent of a pixel in an image, or a grid cell in a DEM. Voxels are frequently used in the visualization and analysis of scientific data, including within GIS. As with pixels, voxels themselves typically do not contain their coordinates and location is implicit within a wider framework or 3D model.

Voxel terrain is usually used in games and simulations. Most commonly, voxel terrain is used instead of an elevation model because of the ability to represent overhangs, caves, arches, and other 3D terrain features (Figure 7.12). Such concave terrain features would not be possible in a DEM or surface model because they can store only one elevation value at each point on the surface. In many GIS application fields, site engineering, building modeling, geophysics, sub-surface geology, marine science, and ground water hydrology, for example, this means of terrain representation is essential.

GIS contains rather limited functionality for voxels and 3D rendering in general. Many GIS packages allow data to be exported as KML scripts for 3D display in

Figure 7. 12: Terrain with caves and overhangs, generated from voxel data sing the C4 graphics Engine (www.terathon.com/c4engine) and rendered using a triangular mesh, a triplanar projection of the vertex positions, bump texture mapping and ambient occlusion in the crevasses. Image courtesy of Eric Lengyel. Used with permission.

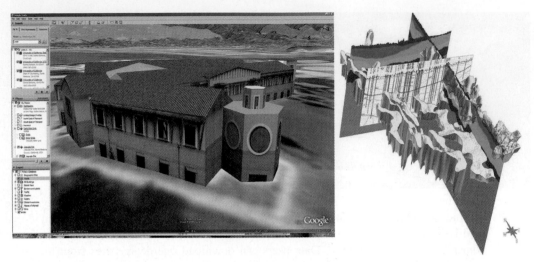

Figure 7.13: 3D data from GIS rendered in different visualization packages. Left UCSB's Kavli Institute for Theoretical Physics in Google Earth (via Google Sketchup). Right 3D Geophysical data Source: USGS

GoogleEarth and other geobrowsers (Figure 7.13). Some specialist GIS-like software allows true 3D rendering for sub-surface data and other 3D information. Add-ons and extensions, such as ESRI's ArcGIS Arcscene, allow interactive and animated 3D rendering of surfaces stored as TINs and DEMs (Figure 7.14). Just as there are standards for posting maps on the web and making them interactive, there is also a way to create 3D models that can be interactively viewed with a web browser and extension. Such a 3D standard is the Virtual Reality Mark-up Language, or VRML. An extension, GeoVRML, allows maps to be viewed, zoomed, panned, and rotated through a web interface. Some GIS software can write GIS data directly into GeoVRML, allowing on-line 3D browsing. An example of a 3D model in a GeoVRML browser is shown in Figure 7.15.

Figure 7.14: Two Mako Shark tracks in the Southern California Bight using ESRI's Arc-Scene 3-D with extruded dive data. Source: http://www.nmfs.noaa.gov/gis/how/inventory/descriptive/stock2.htm.

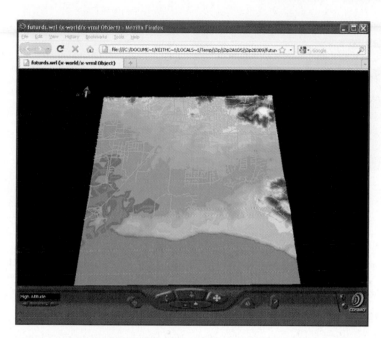

Figure 7.15. GeoVRML 3D model of Goleta, California viewed using the CosmoPlayer addition to the Mozilla Firefox web browser. Source: Author.

7.3   TERRAIN REPRESENTATION

7.3.1   Contour mapping

Just as there are many ways of storing surface data and then getting it into a GIS, so has cartography devised an elaborate suite of methods for depicting and rendering the form of the surface. In general, these can be divided into methods that retain the ability of the map user to retrieve height information, such as the highest and lowest elevation points on the map, and those methods that focus on creating a visual perceptual mental image for the surface form, to aid in overall interpretation. These might be called metric and visual methods. Simplest of the metric methods is the use of contours. Contour lines are lines connecting points on the surface with equal height. Heights chosen for contours usually beging with the base level, and then choose uniform increments at a contour interval sufficient that there are enough contours on the map to see the surface form. Too small an interval and the map drowns in contours. Too big, and the terrain cannot be visualized (Figure 7.16). In addition to the lines, it is possible to fill the areas between the lines with a range of colors. A standard Atlas style color range is shown in Figure 7.16.

Contour lines follow formal rules. A contour line will either wind across the surface until it exits at the map edge, or it will loop back upon itself to form a closed loop. Contour lines do not cross, nor do they suddenly end. When contours surround an enclosed depression, they are annotated with small hatch lines, at 90° to the contour and pointing downhill. Reference contours, often labelled with their elevations, are often drawn with a heavier line than the intermediate contours. In areas of little relief, sometimes supplemental dashed contours are added. Few GIS contouring functions implement all of these conventions, such as labeling. Most, however, allow contour lines to be superimposed with other representational methods, such as hillshading.

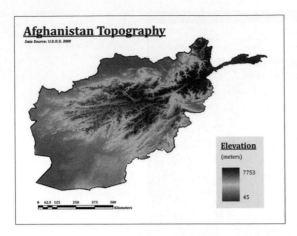
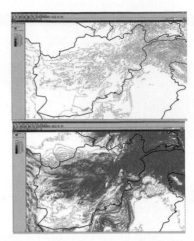

Figure 7.16: Contour representations of terrain in Afghanistan. Map by Michael Titgemeyer. Left: hypsometric tinting, where a sequence of colors fills the space between the contours. Right: Upper 1000m (too few) and Lower: 100m (too many) contour interval maps of the same topography.

7.3.2   Hillshading and Shaded Relief Maps

There is a long tradition in manual cartography of shading terrain to create the illusion of the third dimension. Until the computer, these methods were too manually intensive to be often applied. Now the use of artificial shading and shaded relief on maps is commonplace, largely because the methods are embedded in most GIS packages. In hillshading, the facets of a TIN or the cells of a DEM are used to calculate the degree of incidence with light assumed to be coming from a single direction (azimuth) and elevation (zenith angle). If a

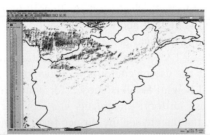

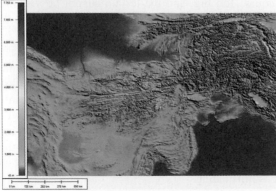

Figure 7.17: Some variants of hillshading and shaded relief. All images using the Afghanistan DEM. Above: Arc-View 3.2 hillshading using azimuths of 315 degrees (left) and 135 degrees (right). Left: Blending hillshading with contrast stretching and color sequencing using Global Mapper.

surface facet has a surface normal (i.e., a vector at the center of the facet which is at 90° to the surface of the facet) that points directly to the light source, then the facet or DEM cell will be bright. If the surface normal points away from or at 90° to the incoming light, then it will be dark. The range of light to dark is often scaled so that the image balances light and dark overall. The result is a hillshaded topographic map (Figure 7.17). Simple hillshading may not be effective. It is possible to have multiple light sources, to employ highlighting effects, to add the ridge lines and drainage to the shading to supply contrast, and to employ custom shading techniques. Other methods use slope-related shading and non-linear shade scaling, or employ color to enhance the effect. These shaded relief maps are common in reference maps, but also in maps where terrain structure is an important part of the message that a GIS map is sending. Custom effects are more common in special purpose terrain mapping packages than in basic GIS. In 3D models, shadows and reflections from the terrain itself can be included.

7.3.3   Perspective Views

Three dimensional viewing allows the creation within the computer of a viewpoint anywhere within or above the terrain surface. Creating such an image view means calculating which parts of the surface from a particular viewpoint are visible and which are hidden. This is often done by ray-tracing, that is starting a virtual ray or vector from every point on the surface and seeing whether any other parts of the surface intercept it before it reaches the camera or viewpoint. In perspective views, often the DEMs, voxels or TINs form a framework for the geometric calculations and if the surface is visible, then a color can be chosen to represent the surface. Popular surface color models are satellite images, air photos, or shaded relief images. Important to the use of perspective views is whether or not the surface fills the visible space or must be depicted with a "base." Also critical is the amount of terrain exaggeration employed. Terrain depicted at 1:1 vertical to horizontal looks surprisingly flat from above, so relief exaggeration is common (Figure 7.18).

7.3.4   Movement, Fly-by, and Fly-through

Once a GIS can create a perspective view, it is possible to create them frame by frame from different viewpoints and then to assemble them in sequence to make a movie or animation. There are many such examples on the World Wide Web, and the idea is embedded in Geobrowsers such as Google Earth, and is even common on television weather broadcasts. Specialist stand-alone software, or GIS extensions such as ESRI's ArcScene, can be used to plot out a flight line, determine view angles, and give equivalent heights above the ground. Different animation sequences include keeping a single view, but zooming in or flying over the terrain; remaining a distant range and view angle, but having the terrain rotate beneath the view around a common point; or moving into the terrain and moving around within it to illustrate a particular view, or to follow a feature such as a fault line.

Creating a 3D sequence can be time consuming with GIS, but the effects can be remarkable. Animations are very effective ways of communicating ideas with GIS.

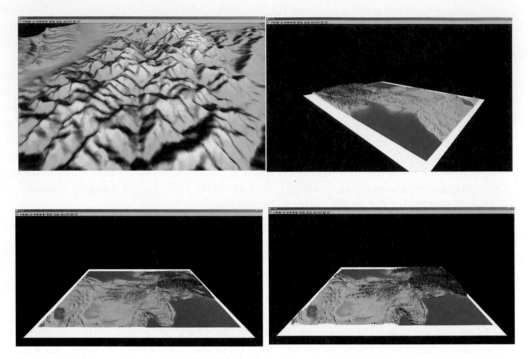

Figure 7.18: Variants on perspective views. Above left: View from within scene, so edges are unnecessary. Above right: Rotation of viewpoint. Lower left: Terrain exaggeration at 1:1. Lower right: Terrain exaggeration at about 10x. Images made using Global-Mapper.

7.3.5 Terrain Mapping in 3D

The use of LiDAR, both from aircraft and from terrestrial scanning, can create point clouds and digital terrain models that are so detailed they become *de facto* 3D models of the landscape. Many states and agencies such as the USGS and FEMA have devoted a great deal of their resources in recent years to updating DEMs with this new and accurate technology using airborne LiDAR. Terrestrial LiDAR has become commonplace in detailed mapping and engineering applications. Figure 7.19 shows a LiDAR point cloud from a terrestrial scan of a building on the UCSB campus. Simultaneous digital photography that is co-positioned with the 3D measurements allows the creation of a virtual representation of the actual building.

Extra work is necessary if the 3D model is to also have an inside. This can come from digitized floor plans, rendered digital photography, or from careful drawing with 3D modeling tools from known measurements. Similarly, 3D models can include subsurface infrastructure such as pipes, sewers, and power lines. This makes 3D modeling, and the interface between GIS, visualization packages, and computer assisted design, a very powerful combination. Another example from the UCSB campus is shown in Figure 7.20.

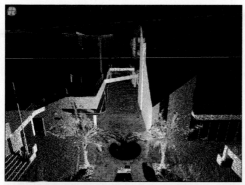

Figure 7.19: LiDAR point scan of the Science and Engineering Building on the UCSB campus. Top Left: Raw point cloud. Top right: Cloud showing color from digital camera. Note moving person. Bottom Left: Data colored by distance range. Bottom Right: Taking the terrestrial LiDAR scan. Photo by Jerome Ripley, scans by Bodo Bookhagen.

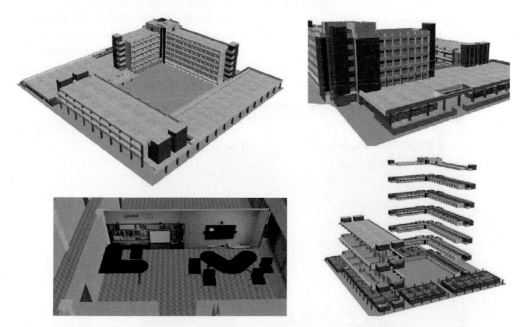

Figure 7.20: 3D renderings of UCSB's Phelps Hall, including interiors built by extruding and registering building floor plans and adding digital photographs. Data from Geography 176C project, Spring 2008, images by Cheyne Hadley, Doug Carreiro, Scott Prindle, and Paul Muse.

7.4   TERRAIN ANALYSIS

While GIS is very versatile at reading in and displaying digital elevation and terrain data, it is important to remember that analysis is a critical part of GIS's capabilities. It is spatial analysis of terrain (and indeed, other types of fields) that provides much of the power of GIS as a scientific method. This will be illustrated with a small sampling of some of the derivative surfaces than can be created once the basic terrain map is at hand.

7.4.1   Gradients and the Profile

As we move across terrain, either along the x or y axis, or along a road or river, we trace out a profile. Many GIS packages allow the creation of cross-sectional profiles to show the surface form along an axis, or along a straight line between two points. You could also imagine plotting a map of a hike or jog showing the elevation and steepness of the path or trail. Such maps are commonly used in geography, geology, archaeology, and elsewhere. An example is shown in Figure 7.21. Gradients can also be calculated from contour maps. The shorter the distance between contours, the steeper the hill.

7.4.2   Slope and Aspect

Just as in mathematics, if we can imagine height as a continuous surface function of x and y, then we can also imagine that the surface has a slope or gradient. For a TIN facet, slope has a magnitude (steepness) and an aspect (direction). In a DEM, which slope and aspect are used and how they are calculated is important. Usually we examine a cell's neighbors in eight directions (i.e., compare a cell with its immediate 8-cell neighbors) and select the maximum slope and its direction as the slope values to use. Almost all GISs have the capability of computing and creating as a new layer a map of the terrain slope and aspect. These values can be depicted as classes or continuously as shades (Figure 7.22). Slope class maps are very common in urban planning and transportation engineering. For example, for further analysis, we could compute in a GIS all areas greater than a 30% slope that might be susceptible to landslides. Similarly, we could find all northward-facing slopes on a mountain that would be last to lose their snowpack in the spring.

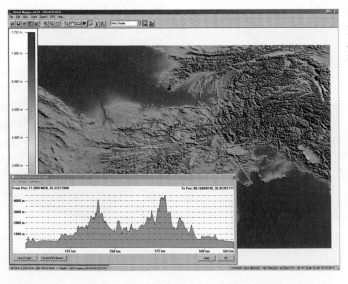

Figure 7.21: Topographic map of Afghanistan in GlobalMapper software showing the creation of a line of a cross-section (in yellow) and the resulting terrain profile.

Figure 7.22: Afghanistan topography. Left: Slope, computed in Quantum GIS. Right: Aspect, computed in ArcView 3.2.

7.4.3   Fundamental Terrain Statistics

Slope and aspect are only two of the many local values that can be computed for terrain. Within GIS another frequently computed value is the slope curvature. Curvature attributes are based on the slope of topographic slope, in mathematics called a second derivative: the rate of change of slope or aspect, usually in a particular direction. The two curvatures most frequently computed are plan curvature, the rate of change of aspect along a contour, and profile curvature, the rate of change of slope down a flow line, i.e., the line of steepest downhill gradient. Profile curvature measures the rate of change of potential gradient, while plan curvature measures topographic convergence and divergence or the propensity of water to converge as it flows across the land. Thirdly, tangential curvature was suggested by Mitasova and Hofierka (1993) as more appropriate than plan curvature for studying flow convergence and divergence because it deals better with flat zones. Tangential curvature is tangential with respect to an inclined plane perpendicular to both the direction of flow and the surface (Wilson and Gallant, 1996).

Another fundamental descriptor of slope is the downslope flow direction (Figure 7.23). Although there are several ways to calculate this, in all cases surface water flow is assumed to begin at the center of the source pixel and travel to an outlet point. Whether the point has to be another cell or can be multiple cells is important for another parameter, downslope flow accumulation. This is the sum at every cell in the DEM of all the flow deriving from all upstream cells. When this value is small or zero, the cell must be at a peak or ridge line, i.e., on the edges of a watershed. On the other hand, when this value is large, we are probably at a point that is part of the stream system on the terrain. The closer we get to a stream's outlet point, the larger the value gets. Examples of the use of GIS to compute these values are given in Figure 7.23. These values set the scene for using GIS to extract the features of the drainage system directly from the DEM, a very valuable transformation.

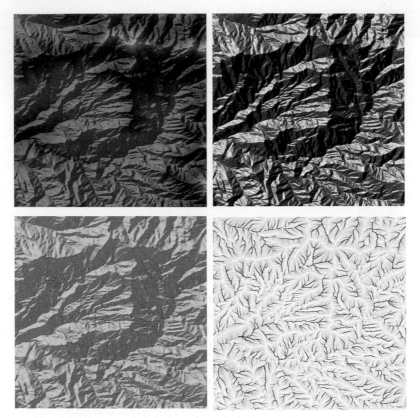

Figure 7.23: Upper Left: Part of the Sequoia National Forest: Digital terrain. Upper right: Downslope flow direction. Lower left: Profile curvature. Lower Right: Downslope flow accumulation. Images created using LandSerf (www.landserf.org).

7.4.4 Extracting Terrain Features

Although when we look at terrain from the ground it appears complex and varied, from a topological point of view, terrain forms a rather simple pattern. At the extremes, terrain has high and low points, called peaks and pits. Lines that follow zero slope gradient points from local peaks are called ridges. The inverse, lines that start at local pits and follow local surface minima, are drainage lines or streams. A saddle point is a place where ridge and drainage lines intersect, i.e., the point is both a local maximum and a minimum. If we connect these lines on a surface, the resulting network is often called the terrain skeleton or surface network. This network was first described in Geography by Warntz (Warntz and Waters, 1975).

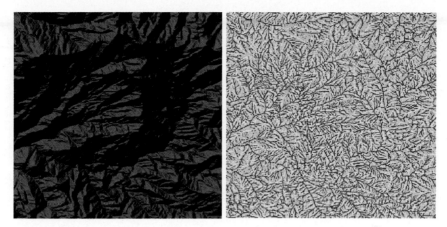

Figure 7.24: Terrain features extracted from the Sequoia National Forest DEM. Left: Shaded relief. Right: Terrain features. Streams are blue, ridgelines yellow, peaks red, and saddle points green.

Using downslope flow accumulation, it is possible to find all values that are close to zero (ridges) and values that are high (streams). Using a GIS, it is possible to connect across the discontinuities that result when these values are computed at high resolution, often by computing the same values for a smoothed version of the surface, or at a lower grid resolution. The result is a raster or vector map showing the terrain features. Also the peaks, pits and saddles can be extracted.

Figure 7.24 shows the process for the same data as Figure 7.23. Different GIS software packages allow the terrain skeleton feature extraction to various degrees. Probably best at this process is GRASS (and hence Quantum GIS). Once the skeleton is extracted, drainage basins and the boundaries of watersheds can also be created. These zones are essential for creating maps and doing analyses related to surface water flow, flooding, erosion, etc. Figure 7.25 shows how these features can be used to derive the expected location of a drainage system, whether for a single basin or a whole country.

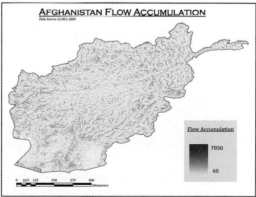

Figure 7.25: Left: Flow direction computed for the SRTM digital terrain data for Afghanistan; overlay is the vector drainage from the AIMS database. Right: Computed flow accumulation, showing the digitally derived drainage. Maps made by Michael Titgemeyer.

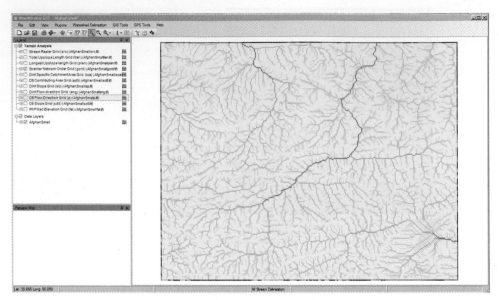

Figure 7.26: Strahler stream order computed for the Afghanistan DEM (subset) using TAUDem within the Map Window GIS. The darker the stream, the higher the Strahler order.

An example of an analysis function that can be performed once the stream network has been computed and extracted is to classify the stream segments by their Strahler stream order. In this system, all streams at their sources are assumed level one. When two streams of the same order meet, their order increases by one. Often drainage systems are analyzed in terms of how the stream order is associated with height, drained area, slope, curvature, etc. The stream order, once computed, can then be assigned as an attribute of the stream segment for further analysis. An example of the extraction of the Strahler order is shown in Figure 7.26.

7.4.5 Intervisibility and the Viewshed

While the number of possible terrain transformations is immense, one set is of particular value. Given any two points in the terrain, at any height, a vector can be drawn from one to the other to see whether that vector intersects the terrain. If the line intersects, then a viewer at one point would be unable to see an object at the other and vice versa. This property is called intervisibility. At some distance, the earth's curvature and other factors would place a limit on the visibility range. If we can compute all intervisible points from a single location, then a polygon can be created called a viewshed. All points within the viewshed are visible from that location (Figure 7.27).

Advantageous places from which to calculate a viewshed are mountaintops, cell phone towers, radio transmitters, etc. The viewshed is then the set of places that can either see a light on the mountain peak, or those that could be served by a transmitter at that location. Viewsheds are used to see how to plan highways and ski resorts, and to see what homes have views of the ocean or a landmark. Militarily, it is advantageous to know where on a road, for example, a vehicle will be seen from a particular location. Viewsheds are often a very important layer in using a GIS for planning or decision-making.

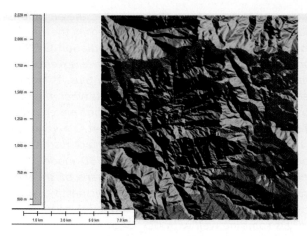

Figure 7.27: DEM from Sequoia National Forest. Shaded in red are all places visible from the right side of the peak at map center. Images created using GlobalMapper.

In the latter discussion we have primarily discussed terrain transformations that use the grid as their data structure. Many of these options are also easily performed on the grid, or to some extent with point clouds. In some cases, such as computing intervisibility, the strain on the computer can be considerable. In this case, often another structure, for example the TIN, is first extracted from the terrain and used to assess whether or not computations are necessary over all parts of the DEM. It has been noted that when exact skylines are needed in terrain views, or for intervisibility, a detailed TIN is better than a DEM. Regardless, this chapter has introduced the immense variety of data structure, display, and analytical options available for dealing with terrain data. While the land surface has been emphasized here, these methods are equally well suited for bathymetry and for any continuous surface computable within GIS.

7.5   STUDY GUIDE

7.5.1   Quick Study

CHAPTER 7: ON THE SURFACE

O *GISs can use the surface or field model for continuous variables* O *In a field, a single attribute is sampled in space and then assumed continuous, e.g., surface temperature measured at weather stations* O *Values on a field are most similar to those in the same vicinity* O *At some distance, the influence of one point on a field's value diminishes to zero* O *Spatial autocorrelation is summarized as Tobler's first law: nearby things are more similar than distant things* O *Terrain and other surfaces can have breaks and discontinuities* O *Terrain discontinuities are cliffs, ridgelines, stream channels, overhangs, and caves* O *GIS can compute many different terrain derivatives that are useful for analysis* O *Most GISs really only deal with the upper surface of terrain, sampled at specific locations* O *A simple terrain model is to record elevations at points, and to interpolate to regular postings* O *If undersampled, we must interpolate; if oversampled, we have a point cloud that must be generalized* O *Two interpolation methods are inverse-distance weighting and kriging* O *LiDAR is common in terrain mapping and creates dense point clouds* O *Contour lines have been used to show surfaces for centuries, they are lines joining points of equal elevation* O *Most GISs support automated contouring ○ DEMs extracted from contours show a bias in the elevations ○ TIN has the advantage that surface critical points can be included in the terrain model ○ TIN links the sample points with a unique Delaunay triangulation ○ TIN is favored for volume computation and contouring ○ DEM grids have extent, and vertical and horizontal resolution ○ A DTM is a DEM of the terrain plus its surface features such as buildings and trees ○ Problems can emerge with DEMs when they are reprojected or mosaiced ○ Low resolution and accuracy DEMs may not match other surface features in maps ○ Much DEM data is available via the Internet ○ A voxel is a 3D grid cell for visualization and modeling ○ Voxels can represent surface discontinuities ○ GIS often needs special purpose extensions for dealing with 3D models ○ A standard for 3D in GIS is GeoVRML ○ Cartography uses many means of showing surfaces: contours, hypsometric colors, hill shading, relief shading, perspective views, animations, and 3D visualizations ○ LiDAR point clouds can be so precise as to be de facto 3D terrain models ○ 3D models combine well with GIS, CAD, and visualization packages ○ Terrain analysis starts with computing 2D cross sections of terrain ○ Most GISs will calculate surface slope and aspect ○ Many can calculate fundamental topographic parameters: slope direction, plan curvature, profile and tangent curvature, downslope flow, and flow accumulation ○ Terrain parameters allow extraction of the terrain surface network, ridges, stream channels, and very important points ○ These can be assembled into drainage networks and watersheds ○ GIS allows the computation of intervisibility and viewsheds on surfaces ○ Viewsheds and watersheds are useful in many other GIS analyses, and for planning ○ The same properties of land surface terrain apply equally to bathymetric and abstract geographical surfaces ○*

7.5.2   Study Questions and Activities

Geographical Fields

1. Make a list of attributes in geography that are best considered as fields rather than features. For example, population density, disease rates and weather data. Decide which terrain data model would be most suitable for working with each attribute in GIS.

2. Brainstorm what spatial discontinuities would be expected in a particular type of geographical field, e.g., air pollution rate. What would be the causes of the breaks? How could GIS deal with them?

Terrain Data Structures

3. Select a particular geographical field variable and consider the advantages and disadvantages of representing that field in any of the several terrain data models discussed in Chapter 7. Which is the best and worst among them for a particular task?

4. What terrain applications and transformations are easier or more effective when performed on a TIN model?

5. What are the advantages and disadvantages of storing terrain data in a point cloud? Consider what happens as the point cloud becomes a more or less dense sample of the actual terrain surface.

Terrain Representation

6. For a locale of your choice (best suited to a mountain or hill) extract a DEM from the USGS's seamless data server at as high a resolution as possible. Use a GIS of your choice to represent the terrain in as many of the methods discussed in this chapter as you can, e.g., contours, shaded relief, perspective view. Which of the methods is best at portraying (1) the metric nature of the terrain (i.e., heights) and (2) the perceptual image of the terrain.

7. Using the digitizing features of your GIS, capture as many points as possible from a digital contour map, and tag them with the point elevation. Then interpolate the surface from this sample of points, using a method such as inverse distance interpolation or kriging. Subtract the result from the original DEM from which you created the contours. How do you explain the differences?

Terrain Analysis

8. Research the various fundamental terrain parameters covered briefly in this chapter. What properties of terrain does each capture? You may have to read both books and research papers to come up with the answers.

9. What applications are there for visibility maps? Make as comprehensive a list of them as is possible. How would height error in the DEM impact each of the applications?

7.6   REFERENCES

Chu, T. H. and Tsai, T.H. (1995) Comparison of accuracy and algorithms of slope and aspect measures from DEM, *in Proceedings of GIS AM/FM ASIA '95*, pp. 21-24, Bangkok.

Jenson, S. and Domingue, J. (1988) Extracting Topographic Structure from Digital Elevation Data for Geographic Information System Analysis. *Photogrammetric Engineering and Remote Sensing*, Vol. 54, No. 11, pp. 1593–1600.

Moore, I. D., Grayson, R. B. and Ladson, A. R. (1991) Digital terrain modeling: A review of hydrological, geomorphological, and ecological applications. *Hydrological Processes*, Vol. 5, pp. 3–30.

Rodriguez, E., Morris, C. S. and Belz, J. E. (2006) A global assessment of the SRTM performance, *Photogrammetric Engineering and Remote Sensing*, vol. 72, no3, pp. 249-260.

Sui, D. Z. (2004) Tobler's First Law of Geography: A Big Idea for a Small World? *Annals of the Association of American Geographers*. 94(2): 269–277.

Warntz, W. and Waters, N. (1975) Network Representations of Critical Elements of Pressure Surfaces. *Geographical Review*, Vol. 65, No. 4, pp. 476–492.

Wilson, J. P. and Gallant, J. C. (2000). *Terrain Analysis. Principles and Applications*. New York: J. Wiley.

7.7   KEY TERMS AND DEFINITIONS

aspect: The direction that a specific point on a slope faces. Often the direction of maximum slope magnitude.

bathymetry: Water equivalent to topography, the study of underwater depth of the third dimension of lakes or ocean floors.

burning: Artificially lowering digital elevations at the location of a drainage feature, such as a stream bed, to more accurately depict that feature.

contour interval: The vertical spacing between successive contour lines on a contour map.

contour line: A line joining points on a surface with equal elevations.

contour map: An isoline map of topographic elevations.

cut-and-fill: The computation and management of information relating to volumes of material to be removed or added to grade or flatten a surface, for example along a road.

Delaunay triangulation: An optimal partitioning of the space around a set of irregularly spaced points into non-overlapping triangles and their edges.

DEM: Digital elevation model. A raster format gridded array of elevations.

digital terrain model: a surface model of the upper surface of land features, including buildings and vegetation.

downslope flow: On a terrain surface, the direction and downslope accumulation of surface water.

drainage basin: The zone where surface precipitation drains downhill into a body of water, such as a river, lake, reservoir, estuary, wetland, sea, or ocean.

elevation: The vertical height above a reference datum, in units such as meters or feet.

field variable: A geographic value that is continuous over space.

fly-by: An animation made from rendered digital terrain models where the viewer's eye position passes across or around a surface.

fly-through: An animation made from digital terrain models where the viewer's horizon is filled by aspects of the surface.

geobrowser: A web browser that permits the user to interact visually with the geographic space of the content, e.g., by panning, zooming, and movement.

geoVRML: An extension to VRML that allows specification of real-world coordinates.

gradient: The constant of multiplication in a linear relationship; that is, the rate of increase of a straight line up or down. See also *slope*.

grid: A logical map data structure consisting of a rectangular array of cells with equivalent size and containing a single attribute value.

grid cell: A single cell in a rectangular grid.

grid extent: The ground or map extent of the area corresponding to a grid.

height: Vertical distance above a reference datum.

hillshading: Use of an apparent shading effect of raised topography so that the land surface appears differentially illuminated, as it would in low sun angles naturally.

interpolation: A method of constructing new data values at user specified postings within the range of a discrete set of known data points.

intervisibility: The property of being able to view one location or feature from another.

inverse distance weighting: An interpolation method using values from a usually scattered set of known points weighted by the inverse of the distance between points and the unknown location.

isoline map: A map containing continuous lines joining all points of identical value.

KML: The Keyhole Markup Language, an XML based language schema for expressing geographic annotation and visualization on Web-based, two-dimensional maps and three-dimensional Earth browsers.

kriging: A group of geostatistical techniques to interpolate the value of a random field at an unobserved location from observations of its value at nearby locations.

lattice: A logical grid with extent over a surface and allowing any value on that surface.

LiDAR: Light detection and ranging.

mosaicing: Dealing with the data fusion issues around splicing together separately compiled but adjacent geographical tiles, especially for DEMs.

peak: A local or global highest point or surface maximum on a topographic surface.

point cloud: A large number of spatial measurements taken at points in a specific (x, y, z) space.

posting: A point in a regular sampling framework onto which surface values are interpolated or sampled.

profile: A vertical 3D cross section of a sample of topography, often to show surface form.

raster: A data structure for maps based on grid cells.

ray-tracing: A technique that traces the path of light through pixels in an image plane, for example to test their intersection with a surface.

reference datum: A base reference level for the third dimension of elevation for the earth's surface. A datum can depend on the ellipsoid, the earth model, and the definition of sea level.

relief: The highest minus the lowest elevation on a section of terrain.

ridgeline: A line following local high points from a peak to a peak, or from a peak to a saddle on a terrain surface.

saddle point: A terrain surface point that is both a local surface minimum and a maximum, in different directions.

seamless: Data that have had the effects of tiling, differential compilation, and generalization removed.

shaded relief: The use of hillshading and other techniques to depict visually the form of a topographic surface.

slope: The constant of multiplication in a linear relationship along a surface; that is, the rate of increase of elevation. See also gradient.

sounding: The bathymetric depth, measured at a point.

spatial autocorrelation: The relationship of the value of a variable with itself across space.

spot elevation: The height or elevation on land, measured at a point.

SRTM: The Shuttle Radar Topography Mission (SRTM) obtained elevation data to generate the most complete digital topographic database of Earth. SRTM consisted of a radar system that flew onboard the Space Shuttle during February 2000.

summit: The surface local maximum, usually of a significant terrain feature, such as a mountain.

surface: The spatial distribution traced out by a continuously measurable geographical phenomenon, as depicted on a map.

slope curvature: The second derivative properties of a surface, measured from a plane tangent to the plane and aligned with the maximum gradient. Plan curvature is the rate of change of aspect along a contour. Profile curvature is the rate of change of slope down a line of steepest downhill gradient. Tangential curvature is measured with respect to an inclined plane perpendicular to both the direction of flow and the surface.

surface discontinuity: A point, line, or area on a surface where spatial autocorrelation is temporarily suspended, e.g., a cliff.

surface facet: Any partitioning of a surface, e.g., a grid cell or TIN triangle.

surface model: An abstract set of assumptions about the properties of a geographical field.

surface network: A geometric network of lines connecting all peaks, pits, and saddle points on a surface.

surface normal: The vector pointing at 90 degrees to the surface at a point, usually to a tangent plane of maximum local slope and in the aspect of the maximum slope.

surface sample: Any set of point measurements taken from a surface and used to represent that surface.

terrain: The third dimension of the land surface form.

terrain exaggeration: The scaling of the vertical or z dimension with respect to x and y used in a terrain representation.

terrain representation: A means of cartographic depiction of topographic relief.

terrain skeleton: A surface network that includes terrain features such as discontinuities, drainage lines, and ridgelines.

TIN: A vector topological data structure designed to store the attributes of volumes, usually geographic surfaces.

Tobler's First Law: Everything is related to everything else, but near things are more related than distant things.

topography: Study or mapping of Earth's surface shape and features and their description.

viewshed: The area or part of a surface defined by all points from which a vector to a viewer location from those points does not intercept the surface.

voxel: A volume element, representing a value on a regular grid in three dimensional space.

VRML: An ISO standard file format for representing three-dimensional models over the World Wide Web.

watershed: The area drained by a single river or stream feature and its tributaries.

7.8   PEOPLE IN GIS

Brian G. Lees, Professor of Geography, Head, School of Physical, Mathematical and Environmental Sciences, UNSW@ADFA (University of New South Wales—Australian Defence Force Academy, Canberra, Australia)

KC: How did your education prepare you for your work in GIS?

BL: Previously I trained as a flight navigator in the RAF. Flight navigator training covered some cartography (including projections and some geodesy), climatology, advanced maths and more. It was a great foundation. When I left the RAF I ended up working in Air Surveys, doing high precision photography for Australia's National Mapping agency, geophysical exploration and early radar remote sensing. When the mining industry slump hit, I went off to university. I have a BA (honours) and a PhD from the University of Sydney. My thesis was on beach sediment dynamics and the PhD was on continental shelf sediment dynamics. From psychology I got a grounding in statistics, a lot of exposure to the mechanics

of perception, and theories of cognition. From physical geography I picked up computing skills as we set out to instrument the surf zone with homebuilt electronic equipment.

KC: How did you get interested in GIS and what areas of research in GIS have you specialized in during your career?

BL: I took a job in the Geography Department at the Australian National University in 1985. There I taught not only geomorphology, which I'd been hired to do, but also remote sensing and GIS. I helped manage the university field station at Kioloa. Courses were first taught using hand calculators and data sets gridded on drafting film. Within a year we were teaching with GIMMS (vector-based cartography software from the University of

Edinburgh) and Dana Tomlin's MAP Analysis Package. I began teaching geomorphology with a lot of GIS, geomorphometry, erosion modeling, and hydrology. It made sense to build teaching datasets and exercises around Kioloa and its land management. This led to a specialization in land cover modeling, decision support, and data fusion. An interest in sources of error in these models led to the closer examination of DEM and DTM error.

KC: What kinds of things do Australian students do with GIS?

BL: GIS is taught in most universities in Australia in Geography, Surveying, Archaeology, Geology, Epidemiology, Engineering, and Environmental Science. There may be a slightly greater emphasis on land cover mapping than in Europe, but the mix of activities in Australia is similar to the U.S.

KC: Please tell us something about the journal *International Journal of GIScience*, which you edit, and its role in GIS research worldwide.

BL: IJGISc was the first and, for a long time, the only academic journal in the field of geographical information science. The journal was founded by Terry Coppock, from the University of Edinburgh, one of the pioneers of computer mapping and GIS. Peter Fisher took over in 1994 and remained Editor-in-Chief until 2007, when I took over. The number of papers we publish has increased steadily and we'll increase it again for 2010 to cope with the increasing rate of article submissions. The journal growth reflects the increased level of activity in the field very accurately. We want the journal to reflect the best research the community produces so that it stays relevant. It's one of the few GISc journals to be truly international; our European editors are as busy as our Asia-Pacific editor, and both are as busy as our American editor.

KC: What has been your work specifically on terrain analysis? What do people who specialize in terrain analysis do in Australia?

BL: With a teaching and research emphasis on land cover and geomorphological process modeling, it was natural to take an interest in geomorphometry and hydrological modeling too. Included in this were land degradation and dryland salinity modeling. This was influenced by the context at ANU, where Ian Moore and Mike Hutchinson also worked and Andy Skidmore and John Gallant studied. My own work, with many students, has been to seek out sources of error in habitat and land cover modeling, dryland salinity prediction and land degradation and to refine the input variables. Since the DEM is fundamental in all of these, improving DEM production and evaluating the use of terrain variables in modeling took up a lot of my time. Given Australia's serious concerns with lack of water, dryland salinity, and so on, these are the natural focus of many other Australian researchers. All of this work left us well equipped to tackle other problems which may seem unrelated, but proved not to be, such as modeling the fear of crime in the inner city.

KC: Do you have any advice for a beginning student just taking their first GIS class?

BL: Learning GISc is different. There is a big entry cost both in time and intellectual effort. The payoff, however, is worthwhile as GIS skills allow you to tackle fascinating problems. You can really only learn by doing, making choices and making mistakes.

KC: Any other comments you care to make?

BL: One of my biggest mistakes was thinking that we had a five-year life running that first GISc course before we began to saturate employment demand. I couldn't have been more wrong.

KC: Thanks Brian.

C H A P T E R   8

Making Maps with GIS

What's the good of Mercator's North Poles and Equators,
Tropics, Zones, and Meridian Lines?
So the Bellman would cry: and the crew would reply
"They are merely conventional signs!"

Lewis Carroll (The Hunting of the Snark)

8.1   THE PARTS OF THE MAP

A *map* can be defined as a graphic depiction of all or part of a geographic realm in which the real-world features have been replaced by symbols in their correct spatial location at a reduced scale. Maps, as we have already seen, are the storehouses of spatial information that we use as sources of data for GIS. They are also the final stage in GIS work, the means by which the information being extracted, analyzed, and reconstructed using the powers of the GIS is at last communicated to the GIS user or the decision maker who relies on the GIS for knowledge. Maps within a GIS can be temporary, designed merely for a quick informative glance, or permanent, for presentation of ideas as a substitute for a picture or a report.

As was necessary in Chapter 2, we return to the cartographic roots of GIS for a discussion of the critical information needed to make maps correctly with a GIS. Regardless of a map's context, a map has a visual structure. Just as a sentence in the English language needs to follow grammar and syntax to be understood, so a map has to follow its own visual grammar. As a starting point, we cover in this chapter what the various parts of a map are called and which of them are essential. This is followed by a description of the methods used by GIS systems for displaying maps and how the choice of a type of map is sometimes made for us by the sorts of attributes and the geographic character of the data in the GIS. We then cover map design, summarizing some of the rules that cartographers have developed for selecting map symbols, such as colors, and then using them to assemble effective maps.

Just as a map has a structure, so that structure can vary according to which media we use for map display. GISs usually use the computer monitor to display a map, rather than the traditional paper. Only now, after many years of computer mapping, are cartographers beginning to understand how map design depends on the display medium. The GIS has been a major reason why this has become an important consideration.

First, however, we should define the terms used in describing maps. Figure 8.1 shows a set of cartographic *elements*. A cartographic element is one of the building blocks that make up a map and from which all maps are assembled. Each element should be present, although in some cases exceptions are possible. The two basic parts of the map are the *figure* and the *ground*. The figure is the body of the map data itself and is the part of the map referenced in ground coordinates. Almost all other parts of a map are located on the map using *page coordinates*, defining locations on the map layout itself rather than the piece of the world it shows. The *graticule* or *grid*, or often a *north arrow*, is the reference link between the two coordinate systems on the final map. Also part of the figure,

Figure 8.1: A map is a collection of map elements. These include a title, neat line, scale, legend, reference data, source, and other key components. Map by Michael Titgemeyer.

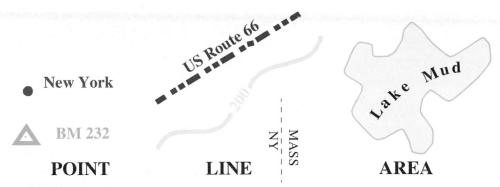

Figure 8.2: Some cartographic label placement conventions. Points: right and above preferred with no overlap. Lines: Following the direction of the line, curved if a river. Text should read up on the left of the map and down on the right. Areas: On a gently curved line following the shape of the figure and upright.

the ground, and the legend are the *symbols*. The *legend* translates the symbols into words by locating text and the symbols close to each other in the page coordinate space.

The *border* is the part of the display medium (paper, window, computer screen, or other medium) that shows beyond the *neat line* of the map (Figure 8.1). In special circumstances, additional information can be provided in this space, such as the map copyright, the name of the cartographer, or the date. The *neat line* is the visual frame for the map and is usually a bold single or double line around the map that acts as a rectangular frame. From a design standpoint, the neat line provides the basis for the page (i.e., cartographic device) coordinate system, in units such as inches or centimeters on the page.

Text information is an integral part of a map, and no map is complete without it. Text is contained in the *title* (whose wording sets the theme and the "feeling" for the map), in *place-names*, in the *legend*, and in the *credits* and *scale*. The scale is a visual expression of the relationship between the ground coordinate space and that of the map page space. Because representative fractions change as the map is projected or rendered onto different display devices or windows, a graphic scale is preferred. Place names follow a strict set of placement rules, both on the figure and related to features and within the map space. Point, line, and area features have different placement rules, and there are also rules for dealing with the overlap of names and feature symbols. Some of the rules are shown above in Figure 8.2.

Finally, an *inset* is either an enlarged or a reduced map designed to place the map into geographic context or to enlarge an area of interest whose level of detail is too specific for the main map scale. An inset should have its own set of cartographic elements, although it is usually highly generalized and many elements may be omitted. To avoid confusing the main map with the inset, the inset should be clearly distinguishable from the figure and ground. Many Americans believe that Alaska and Hawaii are small islands off San Diego!

8.2   CHOOSING A MAP TYPE

Over 3000 years of cartographic history, cartographers have designed numerous ways of showing data on a map. One way to divide up the methods is to look at those that show attributes by their geometric dimension, so that we can have point maps, line maps, and area maps, plus maps that show a three-dimensional view. Many maps show some or all

Figure 8.3: Reference Map. Shown is the CIA World Factbook entry for Afghanistan. Source: https://www.cia.gov/library/publications/the-world-factbook/.

of the types of features at the same time. These are often called general purpose maps. Thematic maps show just one or two themes or layers of information, often coded, colored, or grouped for convenience. In this section we take a look at the breadth of map types available. A basic outline or reference map (Figure 8.3) shows the simplest properties of the map data. An example is a world outline map, with named continents and oceans. A general reference map, usually showing a suite of features including terrain, streams,

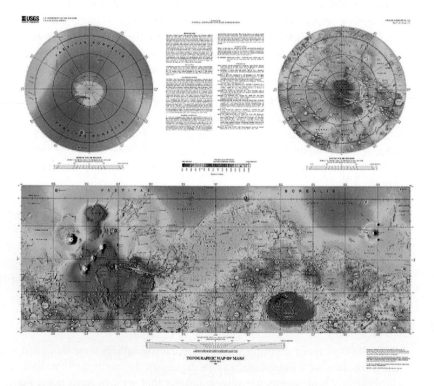

Figure 8.4: Topographic map. Shown is the topographic map of Mars by the USGS. Source: http://wrgis.wr.usgs.gov/open-file/of02-282/. Note the projections.

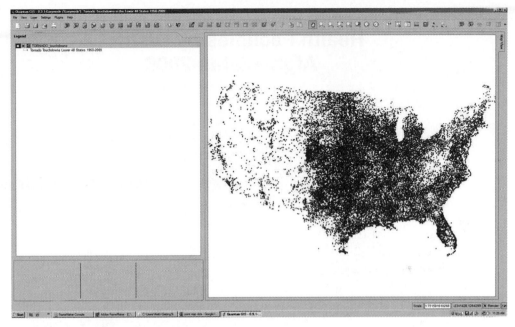

Figure 8.5: Dot map. Example is locations of tornado touchdown points in the lower 48 United States from 1950 to 2006. Map displayed in the Quantum GIS software.

boundaries, roads, and towns, is called a topographic map (Figure 8.4). Topographic maps are often used as reference information behind GIS map layers. A dot map (Figure 8.5) uses dots to depict the location of features and may show a distribution such as population against a base map. A picture symbol map (Figure 8.6) uses a meaningful symbol, such as skulls and crosses, to locate point features such as tornado deaths and injuries. The graduated symbol map (Figure 8.7) is the same, except that the symbol size is varied with the value of the feature. The symbols, in this case the circles, can be grouped into a set of size classes, in which case the map is termed a graded symbol map. Typically, geometric symbols such as circles, squares, triangles, or shaded "spheres" are used.

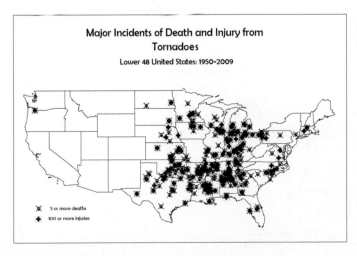

Figure 8.6: Picture symbol or iconographic map. Example created with QGIS software.

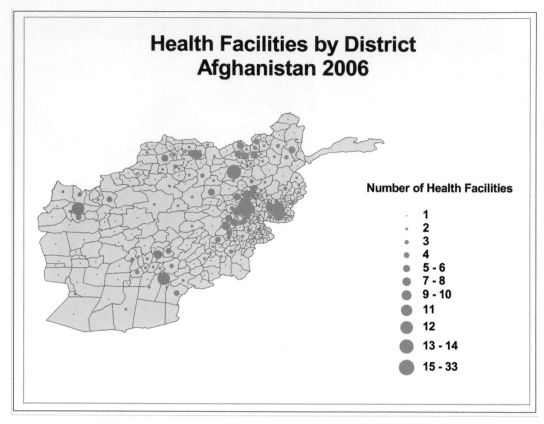

Figure 8.7: Proportional symbol map. Made with ArcView 3.2. Note that circles can be scaled proportionally or in groups (a graduated symbol map). Ideally, smaller circles should overlap larger.

A network map shows a set of connected lines with similar attributes. A subway map, an airline route map, and a map of streams and rivers are all examples. The flow map (Figure 8.8) is the same, but it uses the width of the line to show value—for example, to show the air traffic volume or the amount of water flow in a stream system. Flow direction can be shown using arrows.

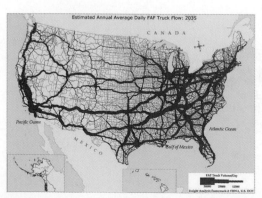

Figure 8.8: Flow map. Forecast of Freight Analysis Framework Freight Traffic in 2035. Source: Alam, Fekpe, and Majed (2007). See: http://ops.fhwa.dot.gov/freight/freight_analysis/faf/faf2_reports/reports7/.

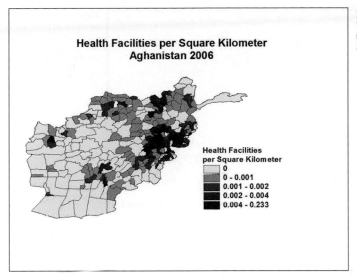

Figure 8.9: A choropleth map. Values that account for areal difference are classed, then shaded. Made using ArcView 3.2.

A choropleth map is the familiar shaded map where data are classed and areas such as states or countries are shaded or colored more or less densely according to their value (Figure 8.9). Most GIS packages will make this kind of map. Note that these maps should be used in cases where the counts or numbers to be displayed have been normalized by conversion to a ratio, or to a density. Otherwise, areas will dominate the graphic purely by size, rather than significance of value. A variation on this, the unclassed choropleth, uses a continuous variation in tone or color rather than the steps that result from classes. An area qualitative map (Figure 8.10) simply gives a color or pattern to an area—for example, the colors of rock types on a geological map, or the land-use classes derived from image classification in remote sensing. .

Another means to portray area data is to actually distort the map so that the area of the feature shown is proportional to the size of the attribute. These maps, called cartograms, are diagrams as much as maps. They can be generated in several ways, for example with the units as discontinuous shapes, or by retaining the topology but distorting

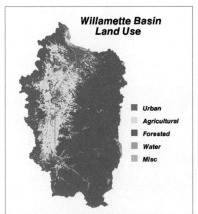

Figure 8.10: Land use map, an example of an area qualitative map. Colors are assigned to different types of areas. Source: USGS NAWQA program, based on GIRAS land use data.

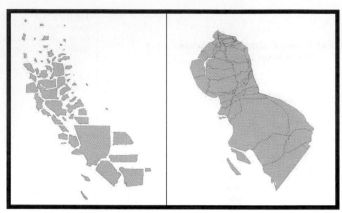

Figure 8.11: Cartograms. California by county, with area proportional to population. Left: non-contiguous cartogram. Right: contiguous cartogram. Maps by Steve Demers. Source: Cartogram Central (NCGIA, USGS).

size (Figure 8.11) Cartograms usually can be made by using special stand-alone software packages, or by using scripts with a GIS. Several such scripts are available. One information source on cartograms is the Cartogram Central website at the NCGIA's website (http://www.ncgia.ucsb.edu/projects/Cartogram_Central).

Volumetric data can be shown in several ways. Discontinuous data are often shown as a stepped statistical surface (Figure 8.12), a block-type diagram viewed in perspective.

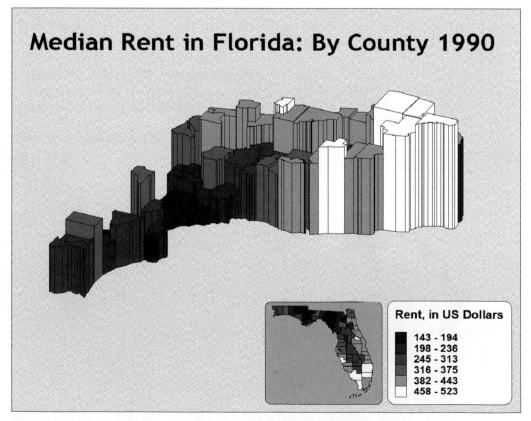

Median Rent in Florida: By County 1990

Rent, in US Dollars

- 143 - 194
- 198 - 236
- 245 - 313
- 316 - 375
- 382 - 443
- 458 - 523

Figure 8.12: Stepped statistical surface map. In this apparently 3D method, heights of surfaces are proportional to the data values shown. Map produced by Golden Software MapViewer 5.

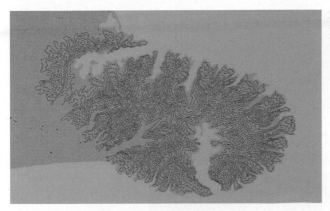

Figure 8.13: Isoline (contour) map. Area shown is section of South Island, New Zealand, from the SRTM digital elevation data. Map made in ArcView 3.2.

The standard isoline map (Figure 8.13) is a map with lines joining points of equal value. Surface continuity is assumed, meaning that sharp breaks are usually smoothed. The terrain equivalent is the contour map, with its characteristic datum and contour interval. A variant is the hypsometric map in which the space between contour lines is filled with color using a sequence designed to illustrate variation. Image maps and schoolroom topographic maps use this type.

Three-dimensional views of surfaces rendered in perspective can be either a gridded fishnet (Figure 8.14), where a grid is distorted to give the impression of three dimensions, or a realistic perspective (Figure 8.15), when an image or shaded map is draped over the surface rather than a grid. The latter technique is often used in animations. Map views of terrain are often represented using simulated hill shading (Figure 8.16) where illumination of shadowing is simulated by the computer, and a gray scale or a colored map is used to show the surface. A variant is illuminated contours, in which the shading algo-

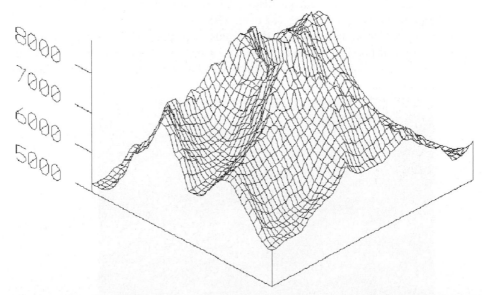

Figure 8.14: Three dimensional fishnet perspective view. Mount Everest summit area. Source: Analytical and Computer Cartography, 2ed. by the author.

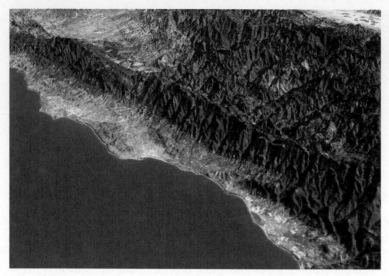

Figure 8.15: Realistic perspective view. Image is Landsat Thematic mapper data draped over the USGS DEM data. Image by Martin Herold and Jeff Hemphill. Used with permission.

rithm is applied only to the contours themselves. The penultimate map type considered here is the image map (Figure 8.17), in which a value is depicted as variation in tone on a color or monochrome grid. Most raw and false-colored satellite image maps fall into this category, as does the orthophoto map.

Last but not least, probably the most versatile and useful map does not show a single attribute at all, but combines reference features of different dimensions, points, lines, and area, plus symbols for volumes with place names and feature labels. Different themes are separated by color or symbol type. This is the familiar topographic map. Increasingly, many of the design features of different map types are being incorporated into topographic maps. Examples are image maps with shaded relief (Figure 8.18). Today, almost all topographic maps are produced using GIS.

So far we have covered the various map types. The GIS user should think of these as a set of possible methods, to be used when the GIS data to be shown have a given set of characteristics. Earlier in the book we classified features on a map into those that are points, lines, areas, and volumes. Obviously, the nature of the map data in the GIS is dif-

Figure 8.16: Relief hillshading. In this case, color is added to the gray-scale simulated hillshading. Generated with Globalmapper software.

Figure 8.17: An image map. Section of the San Francisco, CA, color digital orthophoto quadrangle. Source: USGS.

ferent for each of these. A three-dimensional location, for example, usually needs latitude, longitude, and elevation. In addition, the type of attribute information determines what mapping methods can be used.

Figure 8.19 places the mapping methods covered in this section into a framework of a division of the number of dimensions of the GIS features being shown. Similarly, the types of maps make certain assumptions about the nature of the attributes themselves, not just their graphic representation. For example, a reference map that shows cities has point information and text attributes—the names of the cities. The proportional circle map requires that for every point the attribute must be an integer or a floating-point number. A choropleth map requires a floating-point number that has been grouped into shade categories. These data requirements are also given in Figure 8.19.

Particularly common mistakes in the choice of mapping method are choosing choropleth maps for data that are simply "counts," such as population totals, rather than values, rates, or percentages (instead, proportional symbol mapping should be used at the

Figure 8.18: Topographic map. Point, line, area, and volume maps are shown together, along with text. Note the incorporation of an image map into a standard USGS topographic map.

	Feature Present	Categorical Attribute	Numerical Attribute
Point	*Dot map Picture symbol map*	*Graduated symbol map*	*Proportional symbol map*
Line	*Network map*	*Symbol hierarchy map (e.g., major/ minor roads)*	*Flow map*
Area	*Area qualitative map*	*Choropleth*	*Cartogram*
Volume	*Relief shading Perspective view Image map*		*Stepped statistical surface map Isoline map*

(A "Reference map" arrow spans the Point through Volume rows in the Feature Present column.)

Figure 8.19: Types of maps sorted by dimension of features and type of attribute.

area centroids) because larger areas look like "more" simply because of their geographic size; choosing methods that classify attributes when it is unnecessary; or using too many classes for symbolization (Figure 8.20). Most other problems are problems of design and are covered in Section 8.3.

8.3   DESIGNING THE MAP

The last stage in the mapping process is the conversion of the GIS data into a map design. Note that for any map type we can have an almost infinite number of choices or symbols, fonts, colors, line thicknesses, and so on. Selecting the "best" design can make an enormous difference in the effectiveness of the map. If a map has taken a large amount of work to generate, it is well worth the GIS user's effort to make doubly sure that the design is sound.

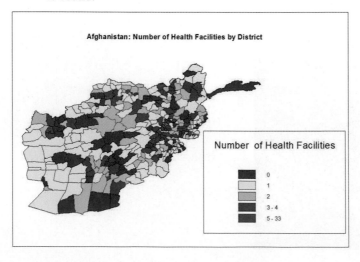

Afghanistan: Number of Health Facilities by District

Number of Health Facilities

- 0
- 1
- 2
- 3 - 4
- 5 - 33

Figure 8.20: Map symbolization errors common in GIS. Title is too small; color scheme varies hue, not saturation or intensity; data are raw counts, unweighed by area or population; legend too big; no scale or reference graticule.

8.3.1 Basics of Map Design

Some characteristics of the design are predetermined by the type of map you choose. Primarily, the design stage consists of devising a balanced and effective set of cartographic elements to make the map. A trial-and-error interaction between a map design and a set of symbols or colors comes into play, called the design loop. The GIS makes this process possible in the first place by supplying the tools to create, modify, and re-create the map.

It is important to place the map elements correctly. Placement of the elements is usually in one of three ways: first, by having the GIS draw a map, then passing it to a graphic design program and interacting with the map in a design loop; and second, especially in a GIS, by editing a set of macro-like commands that move elements to specific places in the map space. This technique is less efficient and involves many traverses through the design loop. Thirdly, many GIS software packages support a layout mode with various graphical editing tools available, although the suite of tools is not the same as that available in the professional design software.

Most cartography texts state that the cartographer should aim for harmony and clarity in the composition—visual balance and simplicity (Figure 8.21). This comes from experience and an aesthetic sense that can take years to perfect. For the beginner in GIS, MacEachren (1994), Dent (1996), Slocum (2009), and Krygier and Wood (2005) give summaries of the design experience of professional cartographers.

Text is an important design element. Map text should be clear, correctly and tersely worded, and the words should be positioned as the graphic elements they are. It is easy to make a map title or legend labels either too small or too big, unnecessarily grasping the map reader's attention. Map text should be edited carefully. Many a map in final form has retained a typographical error that should have been eliminated at first glance, or has misspelled a foreign name that should have been checked.

Facts to bear in mind to balance the map elements are (1) that the "weight" of the elements can change when a symbol set (line widths, colors, text fonts, etc.) is chosen; (2) that the elements act in concert with each other in a visual hierarchy, that is, some of the elements naturally stand out from or "above" others, and that using deliberately exaggerated contrast to enhance this hierarchy is usually most effective; and (3) that the combined

Figure 8.21: Creating visual balance by layout choice. Overall map layout should be visually balanced, and symetrical left-right and top-bottom.

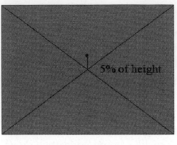

Figure 8.22: The map visual center. Map readers will favor maps centered or focused slightly above the geometric center of the map.

Landscape   Portrait

effect of all the elements is to draw the eye to the center of gravity of the elements. Theory implies that the "visual center" of the map be placed 5% of the map height above the geometric center (Figure 8.22).

8.3.2   Pattern and Color

The symbolization aspect of design has been studied by cartographers in detail, and more than a few rules of thumb exist. Some symbolization methods are simply not suitable for certain types of maps and certain map data configurations. For example, a frequent misuse of color is on choropleth maps, especially when the computer gives access to thousands of possible colors. Choropleth maps usually establish value by shading, pattern, or color intensity, but rarely by color as such. Thus a sequence from light yellow to orange with a slight color change looks right, but a sequence from red to blue across the rainbow makes the map look like a decorated Easter egg! Color changes are appropriate to distinguish between opposites on the same map, such as a surplus/deficit, above/below a statistical average, or two-party election results.

When only monochrome is to be used, the equivalent applies. Shade sequences should be even, flowing from dark to light, with dark usually being high, and light being low. Don't forget that white or blank can be a shade tone, leaving the map looking less

Figure 8.23: Visual variables influencing design and visual combination of features. Left to right: line weights, pattern, shading, hue, and outline.

cluttered as a result. Another issue is pattern. Combinations of crosshatching, dot patterns, and so on can be confusing to the map reader. Combining unmatched patterns can create undesirable optical illusions. Some of these visual variables are shown in Figure 2.23.

Even on general-purpose maps, color balance is essential. Computer displays use pure color, to which the eye is not usually subjected. Less saturated colors, if available, are more suitable for mapping. In addition, cartographic convention should be followed. Ground colors are usually white, gray, or cyan, not black or bright blue. Contours are frequently brown, water features cyan, roads red, vegetation and forest green, and so on. Failure to follow these conventions is particularly confusing to the map reader. Imagine, for example, a globe with green water and cyan land! Map colors can also look completely different on a white rather than a black background, and even on different monitors and plotters.

Color is a complex visual variable. Colors are often expressed as red, green, blue triplets (RGB) or sometimes as hue, saturation, and intensity (HSI). These values are either determined by the hardware device (8-bit color allows a total of 256 colors from any of 256 x 256 x 256 combinations of individual values of RGB) or are decimal values of HSI between zero and one. For example, in RGB, a mid-gray would correspond to [128,128,128]. It is possible to translate directly between the RGB and HSI color representations. Whereas RGB values are simply the degree to which the respective colored phosphors of the monitor emit light, HSI is closer to the way in which people perceive color.

Hue corresponds to the wavelength of light, going from red at the long-wave end of the visible light spectrum to blue at the other end. Saturation is the amount of color per unit display area, and intensity is the illumination effect or brightness of the color (Figure 8.24). Cartographic convention dictates that hue is assigned to categories and that saturation or intensity is assigned to numerical value. When several hues appear in juxtaposition on a map, the colors are perceptually altered by the eye, a phenomenon known as simultaneous contrast. Thus maps that use several hues, even as background and line color, should be designed with caution. In addition, the eye's ability to resolve contrast varies significantly with hue, highest in red and green and lowest in yellow and blue. To assist in making color decisions, especially when depicting variation in quantitative attributes such as in choropleth mapping, a very useful aid is ColorBrewer, an online tool

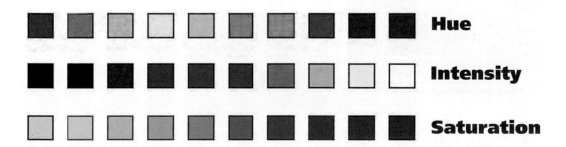

Figure 8.24: The dimensions of color. Most GIS packages allow control over hue, saturation, and intensity of color, but the number of levels varies by application.

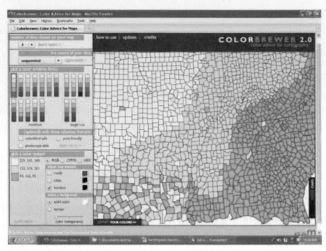

Figure 8.25: Screen capture from ColorBrewer, a color design advisor for GIS maps on-line at http://www.personal.psu.edu/cab38/ColorBrewer/ColorBrewer.html. ColorBrewer is the work of Cynthia Brewer at Penn State University. (Used with Permission).

created by Cynthia Brewer at Penn State University (Brewer, 2003). This tool is especially useful because it allows color combinations to be seen and colors chosen accordingly (Figure 8.25).

Lastly, in most cases when GIS has compiled data, another software tool is often used to create the final map, whether it is to be printed or displayed on the Internet. Software tools include those specific to GIS (e.g., Avenza, ArcPress, and Star-Apic's Mercator) and tools that allow GIS layers to be introduced into a professional graphical design package such as Macromedia Freehand, Adobe Illustrator, or Adobe Photoshop. Other packages or languages are necessary if the intended output is an animation or 3D visualization, e.g., Macromedia Flash. While these packages are beyond the scope of getting started with GIS, eventually the limitations of the GIS interface for map design and manipulation become apparent, and one or more of these solutions become necessary (Figure 8.26).

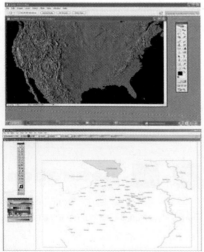

Figure 8.26: Selected tools to add to GIS capabilities for map publishing and editing.

8.4   SUMMARY

The design of a map is a complex process. Good design requires planning, achieving visual balance among map elements, following conventions, employing the design loop, and correctly using symbols and map types. Without consideration of design, and certainly without having all the required map elements, however impressive it may look on a computer screen, the product is just not an effective map. If the map is the result of a complex GIS process, good design is even more important to the person who will have to interpret the map. As we have seen, the relationship between cartography and GIS is a close one. While making a map is often given little thought in the GIS process, it is nevertheless an important stage because it is using maps that particularly distinguishes GIS as being a different scientific approach, and it is the map that has the primary visual impact on the GIS user or decision maker using GIS. Just a little extra care and attention to detail at this final stage can lead to immense improvements in the finished GIS product and to the perception that the entire information flow used in the GIS process is professional and complete.

8.5   STUDY GUIDE

8.5.1   Quick Study

CHAPTER 8: MAKING MAPS WITH GIS

O *A map is a graphic depiction of all or part of a geographic realm in which real-world features have been replaced with symbols in their correct spatial location at a reduced scale* O *GIS maps can be temporary, used to check a result, to answer a query; or a permanent fully featured cartographic product* O *GIS maps should reflect knowledge about map design* O *A map has a visual grammar or structure for the best map design* O *Selection of a map type is determined by properties of the data and the attributes* O *A map has a set of basic cartographic elements: neat line, scale, border, figure, ground, labels, insets, credits, legend, and title* O *The figure is shown in ground coordinates rather than page coordinates* O *Map text, especially labels, follows placement rules specific to the feature dimension and map properties* O *Maps are thematic and general purpose, and of different dimension* O *Types of point maps found in GIS software are dot maps, picture symbol maps, and graduated symbol maps* O *Types of line maps in GIS systems include network maps and flow maps* O *Some types of volume maps that GIS can produce are isolines, hypsometric maps, gridded fishnet, realistic perspectives, and hillshaded maps* O *GIS map design uses a design loop* O *Good map design has map elements in a balanced arrangement within the neat line* O *Visual balance depends on the symbols' "weight," the visual hierarchy and the location of the elements* O *Symbols are subject to the constraints of cartographic convention* O *Color is a complex visual variable specified by RGB or HSI values* O *GIS design errors include incorrect selections of map type and symbolization errors* O *Software packages are available to take GIS layers to publishing-quality maps* O *On GIS maps from complex analytical processes good design is essential for understanding* O *The map is what distinguishes GIS as an approach to information management; extra care should be taken to improve the final maps that a GIS generates* O

8.5.2   Study Questions and Activities

The Parts of the Map

1. Using a map that you have found in a newspaper or magazine, identify the map elements listed in the caption of Figure 8.1 and label them on the map. Are any of the elements missing? Could the map have been improved by adding any of the elements listed in the figure?

2. Using a USGS topographic quadrangle or any other general reference map such as a wall map, a road map, or an atlas map, copy onto a diagram examples of label placement for point, line, and area features. Are there any examples where the "rules" of text placement have been violated? How has the cartographer dealt with the problem that in dense label areas, feature names would overlap?

3. Name six items that could legitimately be found within a map's border and outside the neat line (not coffee stains!).

Choosing a Map Type

4. Make a list of the different types of maps listed in Section 8.2. Verify that the classification of feature dimension is correct. Which map types cross categories? Can you find examples of cartographic methods that cross the boundaries of these types?

5. Make a set of conditions for data to be suitable for display on a choropleth map.

Designing the Map

6. Give three simple rules for a complete GIS novice to keep in mind when using a GIS to produce a map.

7. What design issues should be kept in mind when making a choropleth map?

8. Annotate Figure 8.21 with each of the mistakes labeled in the caption. Can you see any other mistakes? Can you find any mistakes in other figures in the book?

9. Carefully read the documentation for a GIS to which you have access, and compile a list of what map types the software is capable of producing, compared to the map types listed in Section 8.2. Is the subset of map types suited to a particular dimension of data attributes, such as areas?

10. Open up a new map inside your GIS and allow the system to take all the defaults in creating a map layout. Print this map and critique its design using knowledge you have gained from this chapter.

11. Use your GIS package to draw a simple choropleth map. What tools are available within the GIS to assist in the choice of classes for the choropleth data? Does your GIS allow you to make choropleth maps using values other than ratios or percentages? Is there any guidance for choice of colors, shades, or a map layout? How might the documentation for the system be improved to guide the new cartographer better?

12. Make two different maps with your GIS of the same data, one in which you choose a design to enhance differences in the data and one where you try to hide them. Show the maps to some friends or colleagues and ask them about the distributions. Can their opinions about the data be shaped by the choice of symbols for a single map type? Repeat the task using the same data and two sets of symbols, say gray tones versus shading, or red hues versus green hues.

13. Using a topographic map or any map you choose, perhaps from the documentation for your GIS, analyze the placement of labels on the maps. Check a cartography textbook for the conventional cartographic rules of label placement. Can the GIS change placement of the labels?

8.6 REFERENCES

Brewer, C. A., Hatchard, G. W. and Harrower, M. A. (2003) ColorBrewer in Print: A Catalog of Color Schemes for Maps. *Cartographic and Geographic Information Systems*. vol. 30. no. 1, pp. 5–32.

Dent, B. D. (1996) *Cartography: Thematic Map Design*, 4th ed. Dubuque, IO: Wm. C. Brown.

Imhof, E. (1975) "Positioning names on maps," *The American Cartographer*, vol 2, pp. 128–144.

Krygier, J. and Wood, D. (2005) *Making maps: A visual guide to map design for GIS*. New York: Guilford.

MacEachren, A. M. (1994) *SOME Truth with Maps: A Primer on Symbolization and Design.* Washington, DC: Association of American Geographers Resource Publications in Geography.

Robinson, A. H., Sale, R. D., Morrison, J. L., and Muehrcke, P. C. (1984) *Elements of Cartography*, 5th ed. New York. Wiley.

Slocum, T. A., McMaster, R. M., Kessler, F. C. and Howard, H. H. (2009) *Thematic Cartography and Geovisualization* (3 ed.) Upper Saddle River, NJ: Prentice Hall.

8.7   KEY TERMS AND DEFINITIONS

area qualitative map: A type of map that shows the existence of a geographic class within areas on the map. Colors, patterns, and shades are generally used. Examples are geology, soils, and land-use maps.

border: The area between the neat line and the edge of the medium or display area on which a map is being displayed. Occasionally, information can be placed within the border, but this area is usually left blank.

cartographic convention: The accepted cartographic practice. For example, water is usually cyan or light blue on a world map.

cartographic elements: The primitive component part out of which a map is assembled, such as the neat line, legend, scale, titles, figure, and so on.

choropleth map: A map that shows numerical data (but not simply "counts") for a group of regions by (1) grouping the data into classes and (2) shading each class on the map.

clarity: The property of visual representation using the absolute minimum amount of symbolism necessary for the map user to understand map content without error.

color balance: The achievement of visual harmony between colors on a map, primarily by avoiding colors that show simultaneous contrast when adjacent to each other.

contour interval: The vertical difference in measurement units such as meters or feet between successive contour lines on a contour map.

contour map: An isoline map of topographic elevations.

credits: A cartographic element in which the sources, authorship, and ownership of the map and the map attributes are cited, often including a date or reference.

design loop: The iterative process in which a GIS map is created, examined for design, improved, and then replotted from the modified map definition until the user is satisfied that a good design has been reached.

dot map: A map type that uses a dot symbol to show the presence of a feature, relying on a visual scatter to show spatial pattern. Most often used where point features are the GIS data, but dots can be scattered at random throughout areas.

figure: The part of a map that is both referenced in the map coordinate system rather than the page layout coordinates and that is the center of the map reader's attention. The figure is contrasted against the ground, or background.

flow map: A linear network map that shows, usually by proportionally varying the width of the lines in the network, the amount of traffic or flow within the network, and optionally flow direction by arrows.

fonts: A consistent design for the display of the full set of English or other language characters, including special characters such as punctuation and numbers.

graduated symbol map: A map type that varies the size of a common geometric symbol to show the amount of an attribute at points or at centroids of areas. For example, cities could be shown with circles of area proportional to population, or census tracts could have a proportional circle divided as a pie chart at a representative point inside the tract.

gridded fishnet map: A map of a three-dimensional surface showing a set of profiles, often parallel to the x, the y, or the viewer's axis so that the surface appears three-dimensional, as a raised fishnet viewed in perspective.

ground: The part of the body of the map that is not featured in the figure. This area can include neighboring areas, oceans, and so on. The ground should fall lower than the figure in the visual hierarchy.

harmony: The property by which the elements of a map work together to create a balanced aesthetic whole.

HSI: A system for color, as values for hue, saturation, and intensity, respectively.

hue: A color as defined by the wavelength of the light reflected or emitted from the map surface.

hypsometric map: A map of topography involving a color sequence filling the spaces between successive contours, usually varying from green through yellow to brown.

image map: Using an image as a backdrop to which other map elements are added to function as a map. Images can be air photos, satellites images, or scanned images. Map elements often appear on image maps including grids, symbols, scale, projection, and so on.

inset: A map within a map, either at a smaller scale to show relative location, or a larger scale to show detail. An inset may have its own set of cartographic elements, such as a scale and graticule.

intensity: The amount of light emitted or reflected per unit area. A map that has high intensity appears bright.

isoline map: A map containing continuous lines joining all points of identical value.

label: Any text cartographic element that adds information to the symbol for a feature, such as the height number label on a contour line.

label placement rules: The set of rules that cartographers use when adding map text, placenames, and labels to features. Some rules are generic to the map as a whole, while others relate to point, line, and area features specifically. Well-designed maps follow the label placement rules and use them to resolve conflicts between the labels, as labels should never be plotted over each other.

legend: The map element that allows the map user to translate graphic map symbols into ideas, usually by the use of text.

line thickness: The thickness, in millimeters, inches, or other units, of a line as it appears on a map.

map: A graphic depiction of all or part of a geographic realm where the real-world features have been replaced with symbols in their correct spatial location at a reduced scale.

map design: The set of choices relating to how a map's elements are laid out, how symbols such as colors are selected, and how the map is produced as a finished tangible product. The process of applying cartographic knowledge and experience to improve the effectiveness of a map.

map title: Text that identifies the coverage and content of a map. This is usually a major map element and can be worded to show the map theme or the map's content.

map type: One of the set of cartographic methods or representation techniques used by cartographers to make maps of particular types of data. Data, by their attributes and dimensions, usually determine which map types are suitable in a map context.

neat line: A solid bounding line forming the frame for the visually active part of a map.

network map: A map that shows as its theme primarily connections within a network, such as roads, subway lines, pipelines, or airport connections.

orthophoto map: An image map that is an air photo, corrected for topographic and other effects. A specific type of mapping program, at 1:12,000, by the USGS.

page coordinates: The set of coordinate reference values used to place the map elements on the map and within the map's own geometry rather than the geometry of the ground that the map represents. Often, page coordinates are in inches or millimeters from the lower left corner of a standard-size sheet of paper, such as A4 or 8-1/2-by-11 inches.

permanent map: A map designed for use as a permanent end product in the GIS process.

picture symbol map: A map type that uses a simplified picture or geometric diagram at a point to show a feature type. For example, on a reference map, airports could be shown with a small airplane stick diagram, or picnic areas by a picnic table diagram.

place-name: A text cartographic element that links text for a geographic place to a feature by placing it close to the symbol to which it corresponds, such as a city name as text next to a filled circle.

realistic perspective map: A map of a three-dimensional surface showing a colored or shaded image draped over a topographic surface and viewed in perspective.

reference map: A highly generalized map type designed to show general spatial properties of features. Examples are world maps, road maps, atlas maps, and sketch maps. Sometimes used in navigation, often with a limited set of symbols and few data. A cartographic base reference map is often the base layer or framework in a GIS.

RGB: The system of specifying colors by their red, green, and blue saturations.

saturation: The amount of color applied per unit area. Perceptually, saturated colors appear rich or solid, whereas low-saturation colors look washed out or pastel-like.

scale: The part of the map display that shows the scale of the map figure as either an expression of values (the representative fraction as a number) or as a graphic, usually a line on the map labeled with an equivalent and whole-number length on the ground, such as 1 kilometer or 1 mile.

simulated hillshaded map: A map in which an apparent shading effect of raised topography is produced by computer (or manually) so that the land surface appears differentially illuminated, as it would in low sun angles naturally.

simultaneous contrast: The tendency for colors at the opposite ends of the primary scale to perceptually "jump" when placed together; for example, red and green.

stepped statistical surface: A map type in which the outlines of areas are "raised" to a height proportional to a numerical value and viewed in apparent perspective. The areas then appear as columns, with a column height proportional to value.

symbol: An abstract graphic representation of a geographic feature for representation on a map.

symbolization: The full set of methods used to convert cartographic information into a visual representation.

temporary map: A map designed for use as an intermediate product in the GIS process and not usually subjected to the normal map design sequence.

topographic map: A map type showing a limited set of features but including at the minimum information about elevations or landforms. Example: contour maps. Topographic maps are common for navigation and for use as reference maps.

visual center: A location on a rectangular map, about 5% of the height above the geometric center, to which the eye is drawn perceptually.

visual hierarchy: The perceptual organization of cartographic elements such that they appear visually to lie in a set of layers of increasing importance as they approach the viewer.

8.8   PEOPLE IN GIS

Mamata Akella, ESRI

KC: How did you get interested in Geography and GIS?

MA: I first got interested in Geography and GIS after hearing about some work that my sister was doing at Conservation International in Washington, DC—she told me about it, had me read some material, and that's how I became initially interested.

KC: Did you also work at Conservation International? What was your role there?

MA: I was sending out daily fire alerts with GIS to seven different countries where there were hot spots and then also working with satellite imagery to monitor cloud forests in the Andes Mountains.

KC: What did you gain by taking GIS courses in college?

MA: I gained a complete knowledge of GIS. And I also gained a lot of different perspectives, and I realized that we were taught to look outside of the box and not just to take what you see in front of you as being gospel but to question it and to understand how the data were actually derived in the first place.

KC: Is GIS more about theory or applications?

MA: I was glad to cover theory because it makes you understand better, about the data organization and so on. When I worked at the water district—they didn't plan it out in the ways that we were taught—things were usually learned in the process.

KC: Tell us briefly what your position was at the Water District.

MA: I was a GIS intern for 6 months working on their Geo Database. They had just recently converted all of their legacy paper maps, and all that information was imported into GIS and they now have the whole water network in GIS. But there were still a lot of things that weren't right, so I did editing and correcting. Later I worked on making the maps more useable for the service people who use them out in the field.

KC: Was there a particular project you did while you were a student that brought GIS in focus for you?

MA: I think it's the final project that we just did in class, detecting potential outbreak sites of disease in the city of LA for tuberculosis and for heart attacks using different demographic and environmental factors and map overlay. Then we compared it to actual cases, and we did a really pretty good job of predicting the sites. That was really interesting, because we had to assemble all of the data and put it into the GIS and manipulate it all to match up. There are over 800 Census tracts in the city of LA.

KC: What classes or what experiences in your background do you think helped prepare you to do well in GIS?

MA: GIS is a tool, but to understand how you can apply the tool and how you can use the tool requires you to understand the human part of Geography and the environment. I learned a lot in the environmental hazards class that I took and saw how I could relate ideas from there and apply GIS to them. It makes you think about it more than just the tool itself—more about actually what your map is showing.

KC: What are you going to do when you graduate and will it involve GIS?

MA: I'm going to go to grad school, and it will involve GIS. I'm planning on doing some cartography and geovisualization, things that all seem to revolve around GIS.

KC: What advice would you give to a new student just taking their first class in GIS in the fall?

MA: Don't let the labs scare you away. It's all worth it, and it's really a powerful thing to know how to use this type of software—you see about the possibilities and the kinds of things you can do with it seem to be endless, and that it can be applied in all sorts of aspects of the world.

KC: Thanks Mamata.

Author's note: Mamata completed her MS in Geography in 2008 at the Pennsylvania State University working on the Pennsylvania Cancer Atlas and with her adviser, Dr. Cynthia Brewer, on a redesign of the USGS topographic maps. Her thesis research focused on testing the comprehension of emergency map symbols by first responders. Mamata joined ESRI in October of 2008.

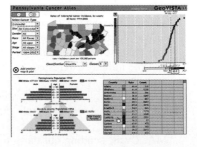

CHAPTER 9

How to Pick a GIS

If you limit your choices only to what seems possible or reasonable, you disconnect yourself from what you truly want, and all that is left is compromise.

Robert Fritz

9.1 THE EVOLUTION OF GIS SOFTWARE

One of the first tasks a GIS user often faces is deciding which GIS software to use. Even if a GIS has already been purchased, installed, and placed right in front of your nose, it is very natural to wonder whether some other GIS system might be better, faster, cheaper, easier to use, have clearer documentation, or be better suited to the actual task you are working on. This chapter gives some of the background necessary to make an intelligent choice among the many GIS software programs available. There is quite a history to learn from, including some excellent accounts of spectacular failures, but also many examples of clear statements of how things went right. Examples from the early days of GIS are the papers by Tomlinson and Boyle (1981) and Day (1981). Recent years have brought a host of new free, open source GIS software packages that promise an easier start-up experience. The philosophy taken here is that the educated consumer is the best GIS user, and an effective user soon becomes an advocate and sometimes a GIS evangelist. This chapter is not intended to tell you which GIS to buy or use. Rather, it is hoped that it will help you to decide this for yourself.

As is often the case, a good education begins with a little history. Chapter 1 introduced overall GIS development in terms of the distant origins of geographic information science as a whole. This was difficult to do without mentioning specific GIS software packages. Now it is appropriate to discuss the development of software in more detail.

9.1.1 The DNA of GIS

GIS software did not suddenly appear fully formed as if by magic. There was a lengthy period leading up to the first real GISs during which the breed evolved rather rapidly. As we saw in Chapter 1, the intellectual ancestry included the creation of a spatial analysis tradition in geography, the quantitative revolution in geography, and dramatic technological and conceptual improvements in the discipline of cartography.

An early GIS landmark was an international survey of software conducted by the International Geographical Congress in 1979 (Marble, 1980). The survey's three volumes encapsulated the state of geographic data processing in the 1970s, when the first generation of software evolved (Brassel, 1977). Most entries were single-purpose FORTRAN programs that performed individual GIS operations such as digitizing, format conversion, plotting on a specific hardware device such as a pen plotter, map projection transformations, or statistical analysis of data. None of these packages were integrated; a typical use would be to apply a series of one-at-a-time geographic operations to arrive at a final result or map.

Some of the early computer mapping systems had already devised many GIS functions by this time, however. Among these were SURFACE II by the Kansas Geological Survey, which could do point-to-grid conversions, interpolation, surface subtraction, and surface and contour mapping; CALFORM, a package that could produce thematic maps; SYMAP, a sophisticated analytical package from the Harvard Laboratory for Computer Graphics and Spatial Analysis that nevertheless ran only on mainframe computers and gave line-printer plots; and the Central Intelligence Agency's CAM, which made plots from the World Data Bank outline maps with different map projections and features.

By 1980 the first computer spreadsheet programs had arrived, led by VisiCalc, a very early microcomputer software "killer app." VisiCalc contained only a hint of the capabilities of today's equivalent packages, yet for the first time gave the ability to store, manage, and manipulate numbers in a simple manner. Above all, data could be seen as active in a spreadsheet rather than as a static "report" that consisted of a pile of computer printout. The links to statistical graphics, now common in packages such as SPSS, SAS, and R, were a natural extension of this capability.

The DNA of GIS is crossbred with the first advances in database management systems. Early systems for database management were based on the less sophisticated data models of the hierarchical and early relational data models. A landmark was the beginning of the relational database managers in the early 1970s. Relational database managers quickly became the industry standard, first in the commercial world of records management and later in the microcomputer world (Samet, 1990).

9.1.2   The Early GISs

By the late 1970s all of the necessary parts of a GIS existed as isolated software programs. The largest gap to be filled was between the relational database manager and the programs that dealt with plotting maps. The specific demands of hardware devices from particular manufacturers kept this a constantly evolving field, with frequent rewrites and updates as systems and hardware changed. Later, the device independence attributable to common operating systems such as Unix and computer graphics programming standards such as GKS, PHIGS, and X-windows led to a narrowing of this chasm, to the point where today it remains as barely a discernible dip in the GIS ground. The scene was set for the arrival of the first true GISs.

As we saw in Chapter 1, one of the earliest civilian systems to posess all the capabilities of a true GIS was the CGIS (Canadian Geographical Information System), which evolved from what was initially an inventory system to doing analyses and then management. Essential to the emergence were the georeferencing and geocoding of the data, database management capability, a single integrated software package without separate standalone elements, and a single user interface.

At first, GIS packages had unsophisticated user interfaces, and many actually forced the user to write short computer program-like scripts or to type highly structured formatted commands one at a time into the computer in response to prompts. As the GIS software evolved, the need for upward compatibility—that is, the need for existing users to be satisfied with a new version because things still work in much the same way as before—meant that many systems preserved elements of these older user interfaces long after they had been replaced by better tools.

The second generation of GIS software included graphical user interfaces, usually involving the use of windows, icons, menus, and pointers, the so-called WIMP interfaces. In the typical configuration today, the windows are standardized by the operating system and function in the same way that it does, "inheriting" its characteristics. A first generation of GIS software used windows custom-built by the vendor. Later, after the broad distribution of windowing systems such as X-Windows and Microsoft Windows, the Graphical User Interface (GUI) tools that are part of the operating system became accessible to software designers and programmers. This required the opening up of the Applications Programming Interface (API) to GIS software developers.

A typical GIS has pop-up, pull-down, and pull-right menus for selecting choices. Choices and locations are indicated with a mouse, although some systems use trackballs or light pens. Similarly, the typical GIS can support multiple windows—for example, one for the database and one to display a map—and the tasks can be opened and closed as needed. While closed, they function in the background while they are graphically represented on the screen as an icon or small picture.

9.2   GIS AND OPERATING SYSTEMS

Early GIS was heavily influenced by the types of operating systems in use. Early operating systems were quite unsophisticated but were used with GIS nevertheless. Among these were IBM's mainframe operating systems, MSDOS by Microsoft, and DEC's VMS. These were rapidly replaced as the various GUI-based operating systems came into operation and as the microcomputer and workstation took over from the minicomputer and mainframe.

In the microcomputer environment, the GUI-based operating systems include Mac OS-X, Windows-XP, Windows Vista, and Windows 7. The unified user interface, revolutionized by the Apple Macintosh's GUI and desktop metaphor, quickly took over as the dominant microcomputer operating environment, although others, particularly those based on the X-Windows standard, have remained popular also. These operating environments added three critical elements to the microcomputer's capabilities: multi-tasking (allowing many simultaneous work sessions); device independence, meaning that plotters and printers could be taken out and assigned to the operating system instead of the GIS package, in somewhat the way that printing and screen fonts are handled centrally, rather than duplicated in every Windows package; and direct connection to a network such as the Internet to extend the reach of the computing environment.

One system that had encompassed these capabilities since its inception, and that swept the workstation environment, was Unix. Unix is a very small and efficient central operating system that is highly portable across computer systems. It has been the dominant workstation environment for two reasons: first, because it has complete integrated network support, and second, because several full GUIs exist for Unix in the public domain, the most important being the X-Windows system. X-Windows implementations of most leading GUIs exist, including Mac OS-X and Linux with X. In many Unix systems, the user can switch the GUI to suit particular needs or applications. A full GUI programming tool kit, including such tools as Xt, Xview, and the X-Windows libraries Xlib, is part of the X-Windows release.

As a final benefit, several versions of Unix and all of the GUI systems run extremely efficiently on microcomputers, including freeware Unix releases such as Linux, not only outperforming the Windows-type GUIs, but also being available free or as shareware on the Internet. A key element here has been the Free Software Foundation's releases, including GNU (GNU is NOT UNIX) versions of virtually every key element of Unix. The Open Systems approach has been largely embraced by GIS, and shareware and freeware systems and browsers are now commonplace, as also are Java-based interactive front ends to distributed geographic data stored on the Internet.

Ultimately, a GIS is a collection of computer programs. Accordingly, GIS evolved its DNA along with changes in how computer programming could be conducted. During the 1960s the first structured programming languages, such as Pascal, Algol, and to some extent FORTRAN, encouraged programmers to use a "divide and conquer" approach, that is to break tasks down into units that could be coded efficiently and then reused. The first tools for code reuse were subroutine libraries. Many of these libraries are around today; indeed much of the open source revolution in GIS would not be possible without them. However, they are primarily of use to skilled computer programmers, especially working together in groups. Starting in the 1970s and '80s, software was developed using libraries that could be controlled by a skilled user via a command-line interface. UNIX in particular was favorable to developing these software tools, and GRASS is a good example of a GIS that reflects this philosophy. A command-line interface in the hands of a skilled and experienced user can almost always outperform a graphical user interface, and even some contemporary GIS and CAD software allows use of the command-line should the user request it. This approach is particularly suited to scripting, and is reflected by the incorporation of programming tools directly into GIS (e.g., ESRI's use of Visual Basic). GIS was highly influenced by the object-oriented programming paradigm, and much of the language and organization of layers, features, and geodatabases reflects this technology. Most recently GIS software has been dominated by the graphical user interface, and the operating system tools that allow its use; Microsoft's Windows and Google's Google Maps API are examples. Currently, with the rise of open source GIS and GIS mashups, the scripting approach is going through something of a revival.

9.3   GIS FUNCTIONAL CAPABILITIES

A GIS is often defined not for what it is but for what it can do. This functional definition of GIS is very revealing about GIS use, because it shows us the set of capabilities that a GIS is expected to have. A minimal set of capabilities can be outlined and each GIS package held up to see whether it qualifies. A thorough examination of GIS capabilities is the critical step in how to pick a GIS, because if the GIS does not match the requirements for a problem, no GIS solution will be forthcoming. In contrast, if the GIS has a large number of functions, the system may be too sophisticated or elaborate for the problem at hand—a sledgehammer to crack a nut.

In earlier chapters, we discussed Albrecht's division of basic GIS operations into a limited set of tasks. This minimal set has been extended in Figure 9.1 to include input and output functions. Many of these operations can, in turn, be created by combining other simpler operations. When further reduced, the remaining classes of operations closely resemble those summarized in Dueker's GIS definition.

The functional capabilities can be grouped by the categories we have used in this book, closely following the Dueker definition of GIS. These are capabilities for data capture, data storage, data management, data retrieval, data analysis, and data display. These "critical six" functions must always be present for the software to qualify as a GIS. We examine each of these in turn.

9.3.1   Data Capture

As we saw in Chapter 4, getting the map into the computer is a critical first step in GIS. Geocoding must include at least the input of scanned or digitized maps in some appropriate format. The system should be able to ingest data in a variety of formats, not just in the native format of the particular GIS. For example, an outline map may be available as an AutoCAD DXF format file. The GIS should at a minimum be capable of absorbing the DXF file without further modification. Similarly, attributes may already be stored in standard database (DBF) format, and should be absorbable either directly or through the generic ASCII format. Strangely, some GISs have rather limited input format capabilities. Work-around solutions are often employed, for example using the Global

Basic Operation Type	Task	Task2	Task3	Task4
Ingest data	Across formats	Convert	Reproject	
Measurement	Location	Distance/angle	Length/area	
Search	Interpolation	Thematic search	Spatial Search	(Re-)Classification
Location analysis	Buffer	Corridor	Overlay	Theissen/Voronoi
Terrain Analysis	Slope/Aspect	Catchments	Drainage/Networks	Viewshed
Distribution/Neighborhood	Cost/Diffusion/Spread	Proximity	Nearest Neighbor	
Spatial Analysis	Multivariate	Pattern/Dispersion	Centrality/Connectedness	Shape
Display maps	Map types	Data types		

Figure 9.1: A minimum set of GIS operations, with modification by the author. Original source: Albrecht, 1996.

Mapper software (www.globalmapper.com), which also allows conversion of map projections and coordinate systems.

Before a map can be digitized, however, it needs to be prepared. Different GIS packages handle the amount of preparation required in quite different ways. If the package supports scanning, the map needs to be clean, fold-free, free of handwritten annotation and marks, and on a stable base such as Mylar. If the map is digitized by hand it may need to be cut and spliced if the package does not support mosaicing (Figure 9.2), and control points with known locations and coordinates need to be marked for registering the map onto the digitizing tablet. Some GIS packages have extensive support for digitizing and sophisticated editing systems for detecting and eliminating digitizing errors. Others have few or none.

We also saw in Chapter 4 how essential it was to edit the maps after they have been captured. This requires the software to have an editing package or module of some kind. For a vector data set, at the minimum we should be able to delete and reenter a point or line. Snapping edited points to vertices and controlling or dissolving sliver polygons can be important. For a raster, we should be able to modify the grid by selecting subsets, changing the grid spacing, or changing a specific erroneous grid value.

Functions typical of an editor are: (1) node snapping, in which points that are close to each other and that should indeed be the same point—such as the endpoints of a line segment—are automatically placed into the graphic database with the identical coordinates; (2) dissolving, when duplicate boundaries or unnecessary lines (e.g., the digitized edges of adjacent category-type maps) are eliminated automatically or manually (Figure 9.3); and (3) mosaicing or "zipping," in which adjacent map sheets scanned or digitized separately are merged into a seamless database without the unnecessary discontinuities caused by the lack of edge matching of the paper maps (Figure 9.2). For example, a major road that crosses two map sheets does not need to be represented as two separate and independent features in the final GIS database.

Another important editing function is the ability to deal with map generalization. Many digitizing modules of GIS systems, and certainly scanning, generate far more points

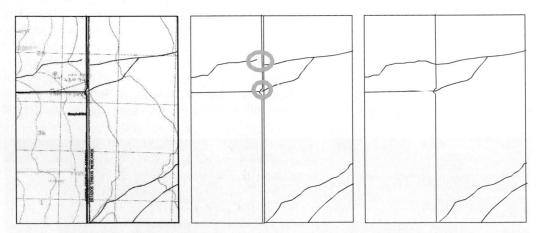

Figure 9.2: Steps in mosaicing. Left: Two maps show one feature, a dry stream, but there is a gap at the political border. Center: Map edge is merged; nodes are snapped to "zip" feature. Right: Mosaiced map with continuous feature and dissolved map edge. Example from NM–Mexico Border on Campbell Well USGS Quadrange.

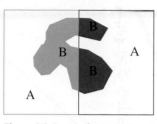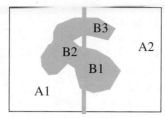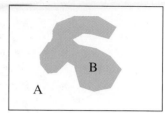

Figure 9.3: Steps in the *dissolve* operation. Left: Two maps show one feature, split across a map edge. Center: Attribute and graphic database have three records for type "B." Right: After dissolve, edge lines are removed and the three type "B" records are amalgamated into a single feature, dissolving the border.

than are necessary for the use of the GIS. This extra detail can complicate data reformatting and display, slow the analysis process, and lead to memory problems on the computer. Many GIS packages allow the user to select how much detail to retain in a feature. Most will retain points that have a minimum separation and snap together all points within a fuzzy tolerance (Figure 9.4).

For point data sets, most GIS packages will eliminate or average duplicate points with the same coordinates. Some will allow line generalization, using any one of many algorithms that reduce the number of points in a line. Common methods include extracting every nth point along the line (where n can be 2, 3, etc.) according to the amount of generalization required, and Douglas-Peucker point elimination, which uses a displacement orthogonal to the line to decide whether a point should be retained (Figure 9.5). Area features can be eliminated if they become too small, or can be grouped together, a process many GIS packages call clumping. It is also possible to generalize in the attributes, joining classes together, for example. A useful website that provides a map generalization service via the Internet is www.mapshaper.org. Figure 9.6 shows the AIMS Afghanistan rivers layer, simplified from 253,480 points to just 3% of that total using Douglas-Peucker

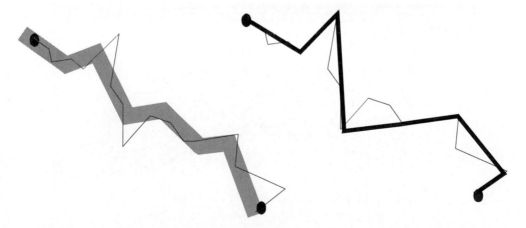

Figure 9.4: Map *generalization* includes eliminating all points along a line falling within a band or buffer around a simplification of the line, with a width sometimes called the "fuzzy tolerance." This way, minor errors such as sliver polygons can be eliminated when data have been captured from maps at different scales.

Figure 9.5: *Line generalization* alternatives. The line (left) can be resampled by retaining every *n*th point (center), or by repeatedly selecting the most distant point from a line between end nodes (right) and redividing the line until a minimum distance is reached, the Douglas–Peucker method.

simplification (Figure 9.6). Such operations are essential when merging or even viewing maps across geographic scales.

To be useful, a GIS must provide tools above and beyond the editor to check the characteristics and validity of the graphical and the attribute databases. Checking the attributes is the responsibility of the database manager. The database system should enforce the restrictions on the GIS that are specified during the data definition phase of database construction and stored in the data dictionary. Most of this checking is done at data-entry time. It checks to determine that values fall within the correct type and range (a percentage numerical attribute, for example, should not contain a text string and should have a record of less than or equal to 100).

More intricate and demanding are checks on the map data. Some GIS packages, which do not support topological structuring, do not enforce any restrictions on the map. Some simply check ranges; for example, every grid cell should have a data value between

Figure 9.6: Zoom using Map Window GIS of part of the Afghanistan rivers shape file, before (blue) and after (red), eliminating 97% of the points using Douglas-Peucker simplification.

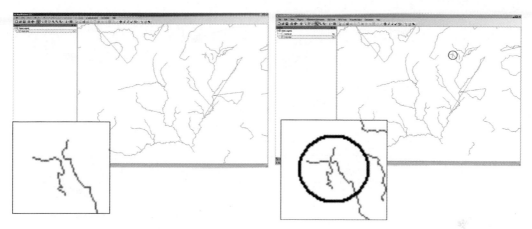

Figure 9.7: Use of Map Window GIS's "Check and Clean Up" function to simplify stream vectors for part of the Afghanistan AIMS rivers layer. Right: Original data. Left: After eliminating all points within 0.0005 degrees of a neighbor. Straight segments are drainage channels. Note loss of connectivity at circled point on zoom.

0 and 255 in an image map. These systems run the risk of lacking a match between the attributes and the space they represent. No part of the map, for example, should fall into two separate areas—that is, the areas on a polygon map should not overlap or leave gaps. This happens when maps are captured at different scales or from sources with difference levels of accuracy.

Topological GIS systems can check automatically to ensure that the lines meet at nodes and that the entire map area is covered by polygons without gaps or overlaps. Beyond simply checking, many GIS packages allow automatic cleaning of topology, snapping nodes, eliminating duplicate lines, closing polygons, and eliminating slivers. Some systems simply point out the errors and ask the user to eliminate them with the editor. Some go ahead and make the corrections without user intervention. Map Window GIS, for example, in its "Check and Clean Up" option, eliminates all points closer together than a specified distance, but aborts processing if this will eliminate any one line feature entirely (Figure 9.7). The GIS user should be careful when using automatic cleaning, for the tolerances may eliminate important small features or move the features around in geographic space by averaging their coordinates without accountability. A specific GIS package may or may not be able to deal directly with GPS data conversion, with survey-type data from COGO (coordinate geometry) systems, or with remotely sensed imagery. Some GIS packages have both functions—that is, they serve as both GIS and image processing systems. Among these are Idrisi, GRASS, and ERDAS. Most GISs will handle raster data, but only these are capable of processing imagery, for example to enhance features.

Essential to geocoding capabilities, because GIS allows maps from many sources to be brought into a common reference frame and to be overlain, is the geocoding software's ability to move between coordinate systems, datums, and map projections. Ideally, data are converted to geographic coordinates in some known datum, and map projection equations (both forward and inverse) are applied exactly, point by point. This is not possible for raster data. When projections are necessary, but not possible by projection formula, most GIS packages use affine transformations. Affine operations are plane geometry; they manipulate the coordinates themselves by scaling the axes, rotating the map, and moving

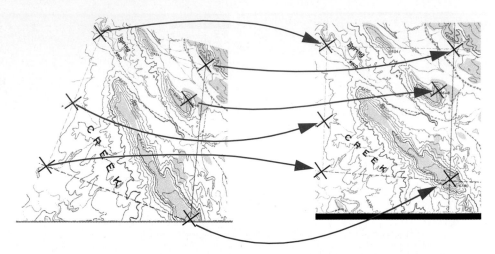

Figure 9.8: The *rubber sheeting method*. A map with unknown geometry (say an air photo taken or scanned map) can be distorted so that its geometry matches that of another map. Pairs of points must be available both on the image and on the map showing the same place or feature location, called control points. Within the GIS, rubber sheeting warps the geometry statistically into that of the map, so that the two geometries match.

the coordinate system's origin. In some cases, when no good control is available, maps must be statistically registered together, especially when one layer is a map and one an image or photograph. The statistical method known as rubber sheeting or warping is used for this and is a function inside many GIS packages (Figure 9.8).

9.3.2   Data Storage

Data storage within a GIS has historically been an issue of both space—usually how much disk space the system requires—and access, or how flexible a GIS is in terms of making the data available for use. The massive reductions in the cost of disk storage, new high-density storage media such as flash drives, and the integration of compression methods into common operating systems have made the former less critical and the latter more so. Current emphasis, therefore, is upon factors that improve data access. This has been a consequence also of the rise of distributed processing, the Internet, and the World Wide Web. In distributed processing, data files can remain remote on the Internet, and either be downloaded on demand, or simply locally displayed on a client. As a result, many GIS packages are now capable of using metadata, or data about data, in an integrated manner. Metadata support might include a system for managing a single project as a separate entity, to managing many projects with multiple versions, to full support for exchangeable metadata stored in common formats and searchable through online "clearinghouses." The USGS's Global Explorer, NASA's Global Change Master Directory, and the U.S.'s Geospatial One Stop are all examples (Figure 9.9). Participation in the common library entails both standardizing the metadata to make it searchable and agreeing to make the data available either on or offline. This effort, collectively called the National Spatial Data Infrastructure, has been emulated nation by nation in a collaborative effort known as the Global Spatial Data Infrastructure. Similar efforts are supported both as commerical efforts (e.g., ESRI's Geography Network, www.geographynetwork.com) and as open collaborative efforts (e.g., OpenStreetMap, www.openstreetmap.org).

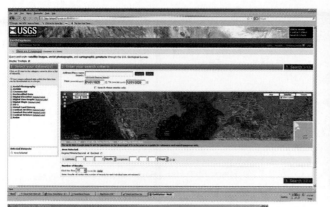

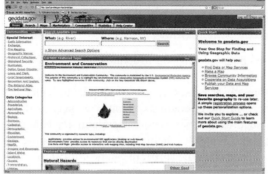

Figure 9.9: US Goverment web portals designed for easy data discovery and access.
Top left: USGS Global Explorer (http://edcsns17.cr.usgs.gov/EarthExplorer/).
Above: NASA's Global Change Master Directory (http://gcmd.nasa.gov).
Left: The e-government initiative generated Geospatial One Stop (http://gos2.geodata.gov/wps/portal/gos).

Other larger issues around GIS use, most essential to the degree of user-friendliness of the system, concern the mechanism for user interaction with the software's functionality. Virtually all GIS software allows user interaction via command lines and/or windows within a graphical user interface (GUI). The GUI interface is tedious, however, without some way of "batching" commands so that they can be executed either at another time, as a background task while the user gets on with another job, or for design-loop editing to change minor aspects of the process. Most systems, therefore, also contain a "language" for the user to communicate with or program the system's functions. This allows users to add their own custom functions, automate repetitive tasks, and add features to existing modules. These languages are usually command-line programs or macros, but they can also be enhancements of existing programming languages such as Visual Basic and C++. These have been supplemented by visual graphical languages that allow a procedure workflow to be both diagrammed as a flow diagram and then implemented directly from the diagram. An example is ESRI's Model Builder in ArcGIS (Figure 9.10).

Although disk storage is less critical than in the past, it can still be a constraint. GIS software on a microcomputer can occupy tens of megabytes even without data, and on a workstation perhaps hundreds of megabytes. As data become higher resolution, as more raster layers are used, and as finer and finer detail becomes available, many GIS data sets can easily move into the gigabyte range in size. This implies that not only is supporting multiple resolutions important—for example, using coarse browse images as samples of the real thing—but also that data compression should be supported. This can vary all the way from partitioning data sets to meet constraints (such as a maximum number of

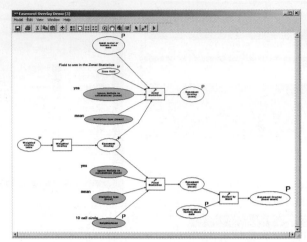

Figure 9.10: A GIS Macro applied visually—ESRI's model builder. Example from the USGS study "Acadia National Park Landscape Scale Conservation and Easement Planning", PI Jason J. Rohweder. (Source: http://www.umesc.usgs.gov/management/dss/anp_easement.html)

polygons) to supporting compressed data formats and structures such as JPEG, run-length encoding, or quadtrees.

Also of great importance from a user perspective is the degree to which the system itself provides help to users, either via the operating system or as part of the software. Integration with online manuals, such as in Unix versions, support for context-sensitive hypertext help systems, such as the Windows help feature, and, ideally, an online interactive hypertext help system connected to the software can be critical for the new (and the experienced) user. These help systems can be used only when needed rather than encumbering the advanced user with unnecessary basic information.

Support for data formats is important to a GIS when data are to be brought in from outside (e.g., public-domain data from the Internet). Ideally, the GIS software should be able to read common data formats for both raster (DEM, GIF, TIFF, JPEG, Encapsulated PostScript) and vector (TIGER, HPGL, DXF, PostScript, KML). Some GIS packages have import functions only into a single data structure, usually either an entity-by-entity structure or a topological structure. Increasingly, geographical data are embedded in ordinary XML-based web pages, as GML or GeoPDF. Interactivity with this increasingly popular and highly searchable data source seems inevitable.

For three-dimensional data, many systems support only the triangular irregular network. Others support only raster structures based on the grid, including the quadtree, and convert all data into this structure. Some GIS packages continue to support only data in a proprietary format, available only at cost from the software vendor. A rather critical GIS function is the ability to convert between raster and vector data, an absolutely essential feature for the integration of multiple data sources such as GPS data and satellite images. Vector to raster conversion is relatively straightforward. The opposite, however, is both complex and error-prone.

GIS functions often support data in standard exchange formats. At the national and international levels, several data transfer standards have been developed, such as the Spatial Data Transfer Standard and DIGEST. These standards are mandated for data exchange among federal agencies, and as the role of data exchange has increased most GIS systems support the input and output of data in these standard formats. The broad advance of Open GIS standards from the Open Geospatial Consortium (www.opengeospatial.org) has greatly improved GIS interoperability in recent years. In addition, many GIS software packages support direct linkage to files served across the Internet via server connections, such as QGIS's link to PostGIS and web-mapping services. Unfortunately, the seamless import and immediate rectification of content type, scale, datum, and projection differences across multiple data sources remains a major challenge. This is leading to research on ontologies of geographical objects, for example exact database specifications of what types and variants can be associated with a feature. Such work is often termed "data fusion," and a major goal of the research is to make such integration automatic, accurate, fast, and transparent to the GIS user.

9.3.3   Data Management

Much of the power of GIS software comes from the ability to manage not just map data but also attribute data. Every GIS is built around the software capabilities of a database management system (DBMS), a suite of software capable of storing, retrieving selectively, and reorganizing attribute information. The database manager allows us to think that all the data are available, that the data are structured in a simple flat-file "table" format, and that they constitute a single entity. In fact, the database manager may have partitioned the data between files and memory locations and may have structured it in any one of several formats and physical data models.

A database manager is capable of many functions. Typically, a DBMS allows data entry and data editing, and it supports tabular and other list types of output, sometimes independent of the GIS. Retrieval functions always include the ability to select certain attributes and records based on their values. For example, we can start with a U.S. database and select out all records for states containing cities with over 1 million inhabitants, forming a new database that is wholly enclosed by the original and that duplicates part of it. We can also perform functions such as sorting data by value, and retrieving a selected record by its identification, such as a name or a number. This can be done by clicking on the map, by clicking on the attribute rows in tables, or by using a query builder or language such as SQL.

Address matching involves taking a listed street address, such as "123 Main Street," and using the GIS's existing data to match the address with a geographic region in the GIS. The key to this capability is usually the TIGER files from the U.S. Census Bureau, which contain a topologically connected street and block network, referenced to house numbers. The address match finds the street and then moves along the street's individual blocks until the house number lies within the block and on the correct side of the street. This "geocoding" feature is present in most GIS software, but is done with different algorithms and so varying accuracy. On my own block in Santa Barbara, for example, the house numbers fall short of ending in the number 99 because the road is a cul-de-sac. As a result, when geocoding algorithms divide the length of the block by the ratio of the house number location to the assumed address at the end of the block, addresses come out far short of their actual distance along the block. Geocoding actual house locations, as is being done for the 2010 census, will correct this problem, but not the algorithm.

Many operations on data are very important from a mapping perspective. For example, very often maps captured from different sheets must be merged together, or sometimes a mask must be placed over the data to exclude features entirely from the GIS.

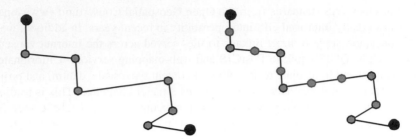

Figure 9.11: Using dynamic segmentation, the GIS can create as many segments along a feature as are necessary for analysis, merging, or display by adding new nodes (shown in magenta). Each new segment can have its own attributes. For example, it may be necessary to establish a new point to mark every mile measured along a river, and to attach river flow, or pollution data, to the points.

Examples of masks are private lands within national parks, water bodies, or military bases. Often all land surrounding a county, satellite image, or watershed must be excluded and masked. Similarly, sometimes data must be assembled in one way, by topographic quadrangle, and then cookie-cut into another region such as a state or a city boundary. Even more complex, sometimes line features such as the latitude/longitude grid, a river, or a political boundary must be sectioned up or have points added as new features or layers are introduced. This feature, called dynamic segmentation, can be done automatically by the GIS (Figure 9.11).

9.3.4 Data Retrieval

Another major area of GIS functionality is that of data retrieval. As we saw in Chapter 5, a GIS supports the retrieval of features by both their attributes and their spatial characteristics. All GISs allow users to retrieve data—they wouldn't qualify as a GIS if they did not! Nevertheless, among systems some major differences exist between the type and sophistication of GIS functionality for data retrieval.

We have seen that a GIS has the critical capability of allowing the retrieval of features from the database using the map as the query vehicle. One way, indeed the most basic way, of doing this is to support the ability to point at a feature, using a device such as a mouse or a digitizer cursor, to see a list of attributes for that feature (Figure 9.12). Again, the ability to select by pointing to a location virtually defines a GIS. If it cannot do this, the system is probably a computer mapping system, not a GIS. Just as critical is the database manager select-by-attribute capability. This is normally a command to the database query language that generates a subset of the original data set. For example, we could find all houses in a real estate GIS that had been listed on the market in the last year. Similarly, we could find all houses built after 1990. All GIS systems and all database managers support this capability.

The most basic act of data retrieval for a GIS is to show the position of a single feature. This can be by retrieving coordinates as though they were attributes, or more commonly by displaying a feature in its spatial context on a map with respect to a grid or other features. For line features, the same goes, with the exception that line features have the attribute of length, and polygon features have the attribute of area. In Figure 9.12, for example, attributes of a single land-use polygon are extracted and include the polygon type

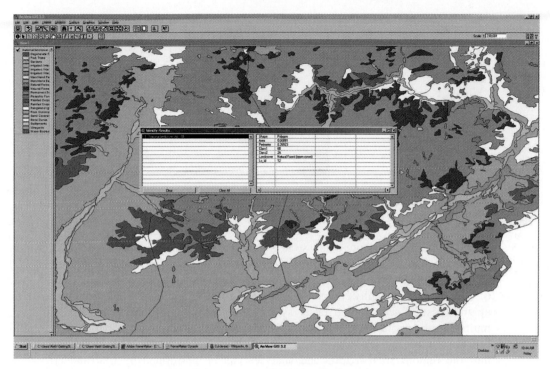

Figure 9.12: Simple information retreival by clicking (using the "I") for the AIMS Afghanistan land use coverage. Note that attributes of the polygon include the type (natural forest), the perimeter length, and the area. Example using ArcView 3.2 GIS.

and area The GIS should be able to calculate and store these important basic properties as new attributes in the database. For example, for a set of counties we may want to take a polygon attribute such as an area of forest and divide it by the county area to make a percentage density of forest cover. Another common measurement we may want is to count features. For example, with the same database we could count the number of fire stations within the same counties by doing a point-in-polygon count from a separate database of municipal utilities and then relate the forest cover to the fire-prevention capabilities.

As we saw in Chapter 5, GISs allow a set of retrieval operations based on using one or more map features as handles to select attributes of those features. Although some of them are very simple, these operations are also a real litmus test for establishing whether or not a software package is a GIS. A GIS should allow the user to select a feature by its proximity to a point, a line, or an area. For a point, this means selecting all features within a certain radius. For a line or a polygon, we have used the term buffering. Buffering allows the GIS users to retrieve features that lie within perhaps 1 mile of an address, within 1 kilometer of a river, or within 500 meters of a lake (Figure 9.13). Simple buffers can be used as a query vehicle by themselves, either finding zones within or beyond a buffer (Figure 9.14). Similarly, weighted buffering allows us to choose a nonuniform weighting of features within the buffer, favoring close-by instead of distant points, for example.

The next form of spatial retrieval is map overlay, when sets of irregular, nonoverlapping regions are merged to form a new set of geographic regions that the two initial sets

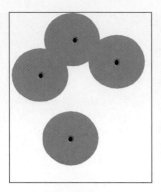

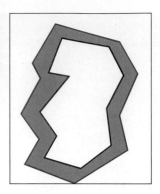

Figure 9.13: Buffers can be created around points (left), lines (center), or areas (right). Buffers can be set to a specific distance, such as 1 kilometer, or made a continuous layer of distances from each point on the map to the feature or features in question.

share. In the new attribute database it is possible to search by either set of units. A GIS should be able to perform overlay as a retrieval operation to support the many spatial analyses based on map combination and weighted layer solutions, as discussed in Chapter 6. Vector systems usually compute a new set of polygons by adding points to and breaking up the existing sets, and in raster systems we allow map algebra, direct addition, or multiplication of attributes stored in cells. Map overlay is an important part of a major GIS function, that of redistricting, in which new districts can be drawn and the data restructured into the regions so that tests and analyses can be performed by trial and error—for example, to see whether the new districts conform to the federal Voting Rights Act.

When two vector layers are overlain, two major operations are necessary. First, the graphics are intersected and new features formed. This can include points in polygons, lines in polygons, and most common polygons in polygons. A new set of least-common geographic areas must be formed. Secondly, the new attribute tables must be formed for these polygons, including attributes from both of the prior layers. When complete, almost any new data reorganization or query becomes possible. An example is shown in Figure 9.15.

Another important set of retrieval options, especially in facilities mapping and hydrological systems, are those that allow networks to be constructed and queried. Typical networks are subway systems, pipes, power lines, and river systems. Retrieval operations

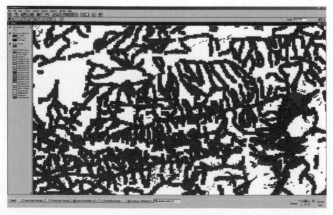

Figure 9.14: 900m wide buffer around rivers in part of Afghanistan from the AIMS data set. Purple areas are arable land more than 900m from a stream.

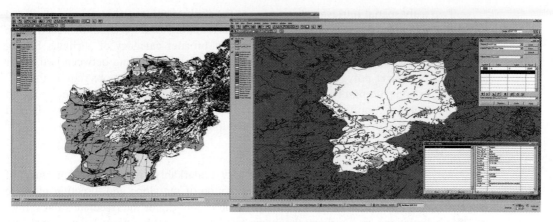

Figure 9.15: Map overlay. Left: Afghanistan land use with district boundaries drawn on top. Right: Results of overlay. One district is selected (Yakawlang in Bamyan Province), and land use specific to that district is now a new set of polygons and attributes, ending at the district boundary. Example using ArcView 3.2 GIS with the Geoprocessing extension.

involve searching for segments or nodes, adding or deleting nodes, redirecting flows, and routing. Not all GIS systems need these functions, but if the purpose is to manage a system usually abstracted as a network, such as a highway or rail system, a power supply system, or a service delivery system, obviously the GIS should then have this feature. For example, a stream network as a set of connected lines can be used to model pollution flow, or a road network used for calculating shortest routes. This function is very common in on-line map websites that give navigation directions, such as GoogleMaps and MapQuest.

Dana Tomlin (1990) has elegantly classified the operations that a raster GIS can perform into a structure called map algebra. In map algebra, the retrieval operations used are Boolean, multiply, recode, and algebra. Boolean operations are binary combinations. For example, we can take two maps, each divided into two attribute codes "good" and "bad" and find a binary AND solutions layer where both layers are "good" (Figure 9.16). Multiply allows two layers to be multiplied together—for example, two sets of weights to be combined. In recode operations a range of computed attribute codes can be reorganized. An example is taking percentages and converting them to a binary layer by making all values greater than 70% a "1" and all else a "0." Map algebra allows compute operations, such as map-to-map multiplication for a binary AND over the space of a grid.

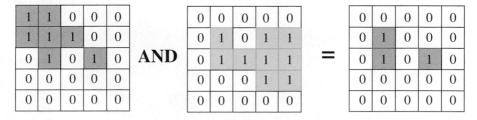

Figure 9.16: Map algebra in its simplest form: Two binary images are ANDed together to give a common area of overlap. Many other operations are possible, such as add, multiply, divide, select maximum, eliminate isolated values, etc.

Two truly spatial retrieval operations are the ability to clump or aggregate areas and to sift. For example, all areas of saturated soils surrounding swamps could be added to the swamps and recoded as wetlands, making a new, broader category of attribute. Sifting simply eliminates all areas that are too small, individual cells falling between two larger areas, or a tiny sliver polygon. Finally, some complex retrieval operations require the GIS to be able to compute numbers that describe shape. Common shape values are the length of the perimeter of a polygon squared, divided by its area, or the length of a line divided by the straight-line distance between the two endpoints.

9.3.5   Data Analysis

The analysis capabilities of GIS systems vary remarkably. Among the multitude of features that GIS systems offer are the computation of the slope and direction of slope (aspect) on a surface such as terrain; interpolation of missing or intermediate values; line-of-sight calculations on a surface; the incorporation of special break or skeleton lines into a surface; finding the optimal path through a network or a landscape; and the computations necessary to calculate the amount of material that must be moved during cut-and-fill operations such as road construction. These were discussed in Chapter 7.

Almost unique to GIS, and entirely absent in other types of information systems, are geometric tests. These can be absolutely fundamental to building a GIS in the first case. These are described by their dimensions, point-in-polygon, line-in-polygon, and point-to-line distance. The first, point-in-polygon, is how a point database such as a geocoded set of point samples is referenced into regions (Figure 9.17). Thus a set of locations for soil samples, generated at random, could be point-in-polygon merged with a digitized set of district boundaries so that a sample list can be sent to each soil district manager. Other more complex analytical operations include partitioning a surface into regions, perhaps using the locations of known points to form proximal regions or Voronoi polygons, or by dividing a surface into automatically delineated drainage basins.

Some of the most critical analytical operations are often the simplest. A GIS should be able to do spreadsheet and database tasks, compute a new attribute, generate a printed report or summarize a statistical description, and do at least simple statistical operations such as computing means and variance, performing significant testing, and plotting residuals, as in Chapter 6.

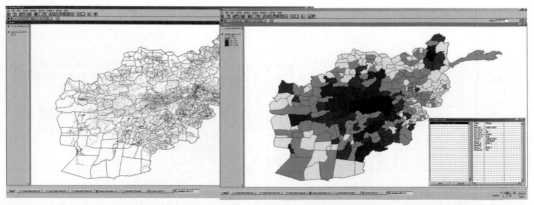

Figure 9.17: Point-in-Polygon analysis. Left: Map of settlements and districts in Afghanistan from the AIMS data. Right: Use of the "Count Points in Polygon" extension, showing raw numbers of settlements in districts added as a new attribute. Made using ArcView 3.2.

9.3.6   Data Display

Most of the display capabilities of GISs have been covered in Chapter 8. GIS systems need to be able to perform what has become called desktop mapping, generating geographical and thematic maps so that they can be integrated with other functions. GISs typically can create several types of thematic mapping, including choropleth and proportional symbol maps, and they can draw isoline and cross-sectional diagrams when the data are three-dimensional.

Almost all GIS packages now either allow interactive modification of map elements—moving and resizing titles and legends—or allow their output to be exported into a package that has these capabilities, such as Adobe Illustrator or CorelDraw. A few GIS packages include cartographic design help in their editing of graphics, defaulting to suitable color schemes, or notifying the user if an inappropriate map type is being used for the data. This would be a desirable feature for many of the GISs on today's market and could avoid many poorly designed or erroneous maps before they were created.

Some of the capabilities for display covered in Chapter 8 are barely possible in GIS. In particular, 3D visualization and animation are not well covered. In many cases, web applications and stand-alone alternatives exist. In a few cases, such as with ESRI's ArcScene and ArcExplorer, the add-on is almost a GIS by itself. Many GIS software packages now use third party software for such interactive visualization, such as producing KML versions of layers for browsing in Google Earth, or creating GeoVRML files for browsing with a standard web viewer. Often other tools, such as Adobe Flash, are used with GIS, simply playing the role of generating animation frames.

9.4   GIS SOFTWARE AND DATA STRUCTURES

In the discussion above, the focus was on what functional capabilities the typical GIS offers. It should not be forgotten that many GIS features are predetermined by the GIS's particular data structure. As we saw in Chapter 4, at the very least the underlying data structure that the GIS uses, typically raster or vector but potentially also TIN, quadtree, or another model, such as object-based, determines what the GIS can and cannot do, how operations take place, and what level of error is involved. The driving force for the choice of structure should be not only what type of system can be afforded, but more critically, what model is most suitable to a particular application, what retrieval and analysis functions will be used most, and what is the acceptable level of resolution and error.

Some examples where particular structures are favored include extensive land characterization applications such as forestry, where detailed data are not required (favors raster); applications involving irregular polygons and boundary lines, such as political units or census tracts (favors vector); applications that require the ability to register all features accurately to ground locations (favors vector); applications making extensive use of satellite or terrain data (favors raster); or applications where image processing functions and analyses such as slope and drainage analysis are to be conducted (favors raster). In many cases, the raster to vector conversion is done outside of thie GIS in specialist conversion software, so that care can be taken to avoid the most common types of error, and so that the user can be brought in to resolve cases where the software is unable to solve a rasterization problem. Geobrowsers that allow viewing the whole earth (such as Google Earth and World Wind) must rely on hierarchical data structures that only load data at the current level of detail. These are often based on hierarchical triangular networks, that use quad trees shaped as triangles for storage.

Of course, most GISs allow the user to input and keep data in both raster and vector form. The GIS user should realize, however, that virtually all cross-structure retrieval and analysis requires one (or both) of the layers to change structure, and that this transformation often stamps itself irretrievably on the data's form, accuracy, and suitability for further use.

9.5 CHOOSING THE "BEST" GIS

The term best is extremely subjective where GIS is concerned. Some systems have extremely loyal followings who advocate their system over others. A "best" system implies that one solution is best for all problems, which is of course largely meaningless. Often more than one system needs to be used, sometimes in combination with stand-alone software suited to other purposes. The following subset of GIS systems, some available commercially, is intended to illustrate the breadth and depth of systems available today and some of the major and minor differences among these systems.

No endorsement is intended, and the list is provided to further the GIS "consumer's" education. Research has shown that these packages account for a large majority of those used in educational, and many professional, settings. A major trend in the last few years has been for geobrowsers to gain some GIS functionality, and for open source software and freeware to become more available and capable. This means that while once there were only a few GIS choices, now there are a very great many.

9.5.1 Open Source GIS Software

Open source GIS software packages have added a new dimension to GIS accessibility. These are usually free, have a support system such as a Wiki, have documentation, and add new versions periodically. Most are completely open source, and follow the Free Software Foundation guidelines. Two useful information clearinghouses for information about these tools are: http://opensourcegis.org, which lists over 250 tools, not all of them GISs; and http://www.freegis.org, or FreeGIS, a clearinghouse for free mashup GIS components under the terms of the GNU General Public License as published by the Free Software Foundation.

A huge advantage of open source GIS software is that it is free. Another is that extensions and additions are commonplace, and if a capability that is desired is not available, a solution is usually forthcoming from the vast pool of users. A third plus is that many of the tools are built on common software frameworks, so their functions are somewhat interchangeable. For example, almost all of GRASS can be loaded inside Quantum GIS.

The following table is derived from Steiniger and Bocher (2008), but note that the list continues to expand rapidly. In addition to these rather whole systems, there are sets of software building blocks on which they are based. These can be accessed and used directly if the user has some programming expertise. They include: GeoTools (Open source GIS toolkit written in Java, using OGC specifications), GDAL/OGR, Proj.4, OpenMap, MapFish, OpenLayers, Geomajas, GeoDjango, GeoNetwork opensource, FIST (Flexible Internet Spatial Template), Chameleon, MapPoint, OpenMap, Xastir, and Gisgraphy.

Table 9.1: Open Source GIS software

Software	About	Information
gvSIG 1.0	Developed by the Regional Council for Infrastructures and Transportation in Spain. Open source GIS written in Java.	http://www.gvsig.gva.es/
GRASS GIS 6.4	Originally developed by the U.S. Army Corps of Engineers, open source a complete GIS.	http://grass.itc.it/
SAGA GIS	System for Automated Geoscientific Analyses- a hybrid GIS software. SAGA has a unique Application Programming Interface (API) and a fast-growing set of geoscientifc methods, bundled in exchangeable Module Libraries.	http://www.saga-gis.org/en/index.html
QGIS	Quantum GIS–QGIS is a user friendly Open Source GIS that runs on Linux, Unix, Mac OS X, and Windows.	http://www.qgis.org/
MapWindow GIS	Free, open source GIS desktop application and programming component.	http://www.mapwindow.org/
ILWIS	Integrated Land and Water Information System. Integrates image, vector and thematic data.	http://www.itc.nl/ilwis/
uDig	uDig is an open source desktop application framework, built with Eclipse Rich Client technology.	http://udig.refractions.net/
JUMP GIS / OpenJUMP– (Open)	Java Unified Mapping Platform. OpenJUMP, SkyJUMP, deeJUMP, and Kosmo emerged from JUMP.	http://www.jump-project.org/
Capaware rc1 0.1	General purpose virtual worlds 3D viewer. A free software project started in 2007 to promote the development of free software in the Canary Islands by its government.	http://www.capaware.org/
Kalypso	An Open Source GIS (Java, GML3) that focuses on water management. Supports modeling and simulation.	http://www.ohloh.net/p/kalypso
TerraView	Desktop GIS that handles vector and raster data stored in a relational or geo-relational database, a frontend for TerraLib.	http://www.dpi.inpe.br/terra-view/index.php
GeoServer	GeoServer is an open source software server written in Java that allows users to share and edit geospatial data. Designed for interoperability.	http://geoserver.org/display/GEOS/Welcome
WebMap Server	Open source protocol and tools for serving GIS data over the Internet.	http://terraserver-usa.com/ogcwms.aspx
MapGuide Open Source	Web-based platform that enables users to quickly develop and deploy web mapping applications and geospatial web services.	http://mapguide.osgeo.org/
MapServer	Web-based mapping server, developed by the University of Minnesota.	http://mapserver.org/
PostGIS	Spatial extensions for the open source PostgreSQL database, allowing geospatial queries.	http://postgis.refractions.net/
H2Spatial for	Spatial extension for an open source DBMS H2_(DBMS).	http://geosysin.iict.ch/irstv-trac/wiki/H2spatial/Download
SpatialLite for SQLite	SpatiaLite extension enables SQLite to support spatial data in a way conformant to OpenGis specifications.	http://www.gaia-gis.it/spatialite-2.0/index.html
MySQL Spatial	MySQL spatial extensions following the specification of the Open Geospatial Consortium.	http://dev.mysql.com/doc/refman/5.0/en/spatial-extensions.html

9.5.2 Commercial GIS Software

The world of GIS has been well served by a great many GIS software companies over the years. As the industry matured, the larger vendors emerged and companies known for other fields, aerospace for example, became interested in building GISs. The following list is derived also from Steiniger and Bocher (2008).

Table 9.2: Commercial GIS software

Software	About	Information
Autodesk	Map 3D, Topobase, MapGuide and other products that interface with its AutoCAD CAD package.	usa.autodesk.com/
Bentley Systems	Products include Bentley Map, Bentley PowerMap, and other products that interface with its MicroStation software.	www.bentley.com/en-US/
Intergraph	GeoMedia, GeoMedia Professional, GeoMedia WebMap, and add-on products for industry sectors, as well as photogrammetry.	www.intergraph.com/
ERDAS	Leica Geosystems subunit encompassing GIS, Photogrammetry, and Remote Sensing. Main software is Imagine.	www.erdas.com
ESRI	ArcView 3.x, ArcGIS, ArcSDE, ArcIMS, ArcWeb services, and ArcServer.	www.esri.com
ENVI	From ITT. Image analysis, exploitation, and hyperspectral analysis.	www.itt.com.
MapInfo	From Pitney Bowes. Includes MapInfo Professional and MapXtreme. Integrates GIS software, data and services.	www.mapinfo.com
Manifold	Full capability GIS software package.	www.manifold.net
Smallworld	Developed in Cambridge, England; now owned by General Electric and used primarily for public utilities.	http://www.gepower.com/prod_serv/products/gis_software/en/smallworld4.htm
Cadcorp	Cadcorp SIS (desktop), GeognoSIS (web), mSIS (mobile), and developer kits.	www.cadcorp.com
Caliper	Maptitude, TransCAD, and TransModeler. Develops GIS and the only GIS for transportation.	www.caliper.com
GeoConcept	GeoMap 3D,Topobase, GC Standard, GC eterprise, Sales & Marketing, routing, Geo optimization, Geo Server and other products.	www.geoconcept.com/en
IDRISI	Taiga GIS product developed by Clark Labs.	www.idrisi.com
TatukGIS	TatukGIS Developer Kernel (SDK), GIS Internet Server, GIS Editor, and free GIS Viewer software products.	www.tatukgis.com
SuperGeo	SuperGIS Desktop, SuperPad Suite, SuperWebGIS, SuperGIS Engine, SuperGIS Mobile Engineg, SuperGIS Image Server, SuperGIS Server, and other desktop extensions.	www.supergeotek.com

There are other GIS software packages with markets primarily in China, Korea, and Japan. Some other international packages include: Axpand (Germany/Switzerland), Clarity by 1Spatial (UK), SavGIS (France), VISION MapMaker (India), and Elshayal Smart (Egypt).

There are some GIS proprietary software programs that are built into database management systems. These include Boeing's Spatial Query Server for Sybase ASE, Oracle Spatial for Oracle, ESRI's ArcSDE, IBMs DB/2, and SDL Server 2008.

9.5.3   Selecting Software: Issues

Selecting the best GIS for use involves many other aspects than simply the technical capabilities of the software package. It could be argued that very little difference actually exists between desktop GIS packages other than their user interfaces and their data structures. Conversely, many of the issues that determine how satisfied we are with the GIS we choose relate to how we acquire the software, how easily it installs itself on our computer, whether or not it is flexible enough to run on a given computer system, and how satisfied we are when the software is up and running.

Obviously, cost is an important factor. Although the cost of basic GIS packages has fallen remarkably in recent years, to zero in the case of open source GIS, cost can still be significant, especially when the hidden costs are taken into account. For example, GIS companies may charge not only a software purchase fee, but also include a maintenance fee, a fee for upgrades, a per call support cost, and sometimes other fees. Maintenance fees for workstation licenses, the sorts of licenses that would be used in a local area network configuration, can be a major proportion of the software cost. In addition, there is constant pressure to upgrade to new versions, usually by discontinuing support for older versions of the software. Especially if a large project is to be undertaken by the GIS user, this fact should be budgeted into the GIS software costs. Shareware and freeware, by contrast, may have less support infrastructure, but the software and update costs are zero. The hidden costs are familiarizing oneself with another software package, assembling the information necessary for add-on capabilties, the overhead of following support networks, and the need to track changes.

Training is another important factor. Few GIS packages can be used by a novice right out of the box. The user may need help from a systems expert, may have special installation requirements, and may require some formal GIS training. Of course, this book can go a long way toward helping the user to understand GIS, but there is a great deal of other straightforward technical information available too. Many GIS users take technical training from one of the GIS vendors or from other sources. These vary from one- or two-day workshops to entire college semester classes. They can also be rather expensive and time-consuming. Many GIS implementations, although well thought out and organized, fail for the lack of one or two people with the right technical expertise at the right time. Some technical problems are hardware and even version specific.

Once technical training ends, the real GIS use begins. At this stage, with 5pm Friday approaching or a project deadline looming, the usual sole self-help mechanism is the GIS system printed or on-line manual. Again, these vary considerably in readability, comprehensiveness, and user-friendliness. Some are excellent, others poor. FAQ lists and blog entries often have answers to obscure problems but lack answers to simple ones. The user should examine the documentation carefully before making a major GIS purchase, as users will spend many hours poring over these pages. Best of all are online manuals, which can be searched, may be context-sensitive, have hypertext links, and are available on a computer while the software is running in another window. This feature is worth any extra expense, since it can speed learning and still serve as a reference later. Most shareware GIS have excellent on-line documentation.

Regardless of the GIS's self-help capabilities, sooner or later almost all GIS users will eventually call a help line, e-mail a help desk, or interact in some way with the GIS creator or vendor's technical support staff. In some cases this is done by telephone, but e-mail, blogs, and network conference groups all act as help facilities. Telephone help lines can involve being placed on hold for long periods, confusion over GIS versions, or worse, waiting to be called back after leaving a phone number. E-mail is better and gets around the time-zone problem of phone lines. On-line conference groups and Wikis are best. When seeking help, a concise statement of the problem and a full set of information will greatly speed up your response. E-mailed screen snapshots showing error messages are particularly useful. In general, using the reference manual or user guide until there is no other means of finding information is far preferable to calling a help line. Remember, if all else fails, read the manual!.

Software maintenance can be another major consideration. For example, most software is updated by complete version upgrades, which require a new installation, or by "patches," a self-contained fix for a specific problem in the software. Maintenance is more of a consideration for large and networked systems, but every user needs to be concerned about too many large files and about how critical data are to be backed up in case of emergency. A GIS should also not be seen as a static entity, but rather one that will grow and evolve. A system that is big or powerful enough for a small prototype project today may not be able to deal with the follow-up project. Fortunately, as time passes the hardware becomes faster and faster, the disks get bigger and bigger, and the cost actually remains the same or falls. Conversely, the expertise required to install, maintain, and use the system is also important and should be planned for. GIS technicians typically get experienced enough to compete for better jobs very quickly. This should also be a part of the GIS cost plan.

Picking a GIS is obviously a complex and potentially confusing process. The most productive approach to the problem is to adopt the attitude of someone about to purchase a new car. First, the GIS user should assemble all the available details about the system requirements, the functional capabilities, the system constraints, and so on. The car buyer could, for example, determine a need for four doors, power steering, adequate luggage space, and front-wheel drive. Next, these should be matched against the systems available. Perhaps a trade-off is necessary between capabilities? Next comes the visit to the car dealership, followed by a test drive. Many demonstration or shareware versions of GIS packages are available to give a flavor of the system use before the purchase. Some demo versions can be downloaded free over the Internet or are given away at GIS conferences.

Finally, "You pays your money and you takes your choice." After the fact, however, the car will need to be maintained and perhaps repaired. One day it may be traded in for a new car. Every one of these issues should be considered. Although every vehicle will probably allow you to get home from the dealership, fundamental differences exist between a sports car and a sports utility vehicle. Just so with GIS. To summarize: Before you choose—research, select, test, and question. Fortunately for the new GIS user, the early days of GIS failures are now over. Technically, today a GIS is much like a reliable automobile. Where and how you drive, however, is still entirely up to you!

9.6   STUDY GUIDE

9.6.1   Quick Study

Chapter 9: How to Pick a GIS

O *GIS users need to be aware of different GIS software products during system selection and beyond* O *Informed choice is the best way to select the best GIS* O *GIS software has evolved very rapidly over its brief history* O *A historical GIS "snapshot" was the IGC survey conducted in 1979* O *In the 1979 survey, most GISs were sets of loosely linked FORTRAN programs performing spatial operations* O *Many early computer mapping programs had evolved GIS functionality by 1979* O *In the early 1980s, the spreadsheet was ported to the microcomputer, allowing "active" data* O *In the early 1980s, the relational DBMS evolved as the leading means for database management* O *Addition of a single integrated user interface and a degree of device independence led to the first true GISs* O *The second generation of GIS software used graphical user interfaces (GUIs) and the desktop/WIMP model* O *Unix workstations integrated GIS with the X-Windows GUI* O *As GUIs became part of the operating system, GISs began to use the operating system's GUI instead of their own* O *PCs integrated GIS with the variants of Windows and other OSs* O *GIS features are known as functional capabilities* O *Functional capabilities fall into the "critical six" categories* O *The critical six functional capabilities are data capture, storage, management, retrieval, analysis, and display* O *Some data capture functions are digitizing, scanning, mosaicing, editing, generalization, and topological cleaning* O *Storage functions are compression, metadata handling, control via macros or languages, and format support* O *Some data management functions are physical model support, the DBMS, address matching, masking, and cookie cutting* O *Some data retrieval functions are locating, selecting by attributes, buffering, map overlay, and map algebra* O *Some data analysis functions are interpolation, optimal path selection, geometric tests, and slope calculation* O *Some data display functions are desktop mapping, interactive modification of cartographic elements, and graphic file export* O *Many GIS functional capabilities are by-products of their particular data structure* O *Raster systems work best in forestry, photogrammetry, remote sensing, terrain analysis, and hydrology* O *Vector systems work best for land parcels, census data, precise positional data, and networks* O *GIS software can be divided into open source and proprietary systems, with many to choose among* O *A variety of issues should be considered in system selection: cost, upgrades, network configuration support, training needs, ease of installation, maintenance, documentation and manuals, help-line and vendor support, means of making patches, workforce* O *Selecting a GIS can be a complex and confusing process* O *The intelligent GIS consumer should research, select, test, and question systems before purchase* O

9.6.2 Study Questions and Activities

The Evolution of GIS Software

1. Make a timeline from about 1960 to today, using as your basis one of the on-line GIS timelines such as http://www.casa.ucl.ac.uk/gistimeline/, http://www.gisdevelopment.net/history/index.htm, or any other. Place on the time line a subset of the packages mentioned in this chapter. How does the sequence of software packages relate to the history of GIS discussed in Chapter 1?

GIS and Operating Systems

2. Make a list of all the operating systems, mainframes, workstations, and microcomputers that can run the GIS packages in Tables 9.1 and 9.2. Which are mentioned most frequently? Why?

3. If you have the ability, install another operating system on your workstation or microcomputer, such as Linux and Windows, or find two computers with them already installed. Do a few simple tasks—say, enter 50 numbers into a spreadsheet file—in each of the two operating systems. Time each process and make a chart showing how much total time you spent on each task. How much did the operating system help or hinder the task? How much system help was available in each system?

GIS Functional Capabilities

4. Make a word list of key functional capabilities structured under the headings of the "critical six." Score the functional capabilities by how essential they are for a GIS to qualify as a "true" GIS. Match the capabilities of a particular GIS against the list.

5. Examine the manuals for two different GIS packages. Read the same section—for example, the section on digitizing lines—in each manual. Which is the better explanation? Why? Make a list of the features that you consider desirable in GIS documentation.

GIS Software and Data Structures

6. Review the Chapter 4 coverage of the different data structures for GIS. Classify a subset of the software for GIS in this chapter by which data structure(s) they support and whether or not they support data structure conversions. List some operational reasons why you might need to convert between data structures.

Choosing the Best GIS

7. Go through the "People in GIS" sections in this book and tally the mentions of specific GIS software packages. How does your list match up with the characteristics of the software packages covered in this chapter?

8. If you have access to more than one GIS, download and ingest the AIMS Afghanistan data and perform a simple retrieval or analysis operation such as a buffer or overlay. Take careful note of how long each step took, how many steps were necessary, and how useful the manuals and help systems were in troubleshooting.

Place the two output maps together at the same size and scale. Are they identical? What might be the factors contributing to the differences?

9.7 REFERENCES

Albrecht, J. (1996) Universal GIS Operations for Environmental Modeling *Proceedings, Third International Conference/Workshop on Integrating GIS and Environmental Modeling.* http://www.ncgia.ucsb.edu/conf/SANTA_FE_CD-ROM/sf_papers/jochen_albrecht/jochen.santafe.html.

Buckley, A. and Hardy, P. (2006) "Cartographic Software Capabilities and Data Requirements: Current Status and a Look toward the Future,"*Cartography and Geographic Information Science*, vol. 34, no. 2, pp. 155–157.

Brassel, K. E. (1977) "A survey of cartographic display software," *International Yearbook of Cartography*, vol. 17, pp. 60–76.

Day, D. L. (1981) "Geographic information systems: all that glitters is not gold." *Proceedings, Autocarto IV*, vol. 1, pp. 541–545.

Marble, D. F. (ed.) (1980) *Computer Software for Spatial Data Handling.* Ottawa: International Geographical Union, Commission on Geographical Data Sensing and Processing.

Moreno-Sanchez, R. Anderson, G., Cruz, J., and Hayden, M. (2007) "The potential for the use of Open Source Software and Open Specifications in creating Web-based cross-border health spatial information systems." *International Journal of Geographical Information Science*, vol. 21, no. 10, pp. 1135–1163

Samet, H. (1990) *The Design and Analysis of Spatial Data Structures.* Addison-Wesley, Reading, MA.

Steiniger, S. and Bocher, E. "An Overview on Current Free and Open Source Desktop GIS Developments." http://www.spatialserver.net/osgis.

Steiniger, S. and Weibel, R. (2009) "GIS Software—A description in 1000 words" http://www.geo.unizh.ch/publications/sstein/gissoftware_steiniger2008.pdf.

Tomlin, D. (1990) *Geographic Information Systems and Cartographic Modelling.* Upper Saddle River, NJ: Prentice Hall.

Tomlinson, R. F. and Boyle, A. R. (1981) "The state of development of systems for handling natural resources inventory data," *Cartographica*, vol. 18, no. 4, pp. 65–95.

9.8   KEY TERMS AND DEFINITIONS

active data: Data that can be reconfigured and recomputed in place. Spreadsheet term for data for attributes or records created by formulas within a spreadsheet.

address matching: Address matching means using a street address such as 123 Main Street in conjunction with a digital map to place the street address onto the map in a known location. Address matching a mailing list, for example, would convert the mailing list to a map and allow the mapping of characteristics of the places on the list.

affine transformation: Any set of translation, rotation, and scaling operations in the two spatial directions of the plane. Affine transformations allow maps with different scales, orientations, and origins to be coregistered.

batch: Submission of a set of commands to the computer from a file rather than directly from the user as an interactive exchange.

buffer: A zone around a point, line, or area feature that is assumed to be spatially related to the feature.

CALFORM: An early computer mapping package for thematic mapping.

CAM (Computer-Assisted Mapping): A map projection and outline plotting program for mainframe computers dating from the 1960s.

CGIS (Canadian Geographic Information System): An early national land inventory system in Canada that evolved into a full GIS.

clump: To aggregate spatially; to join together features with similar characteristics into a single feature.

compression: Any technique that reduces the physical file size of data in a spatial or other data format.

cookie-cut: A spatial operation to exclude areas outside a specific zone of interest. For example, a state outline map can be used to cut out pixels from a satellite image.

critical six: The GIS functional capabilities included in Dueker's GIS definition: map input, storage, management, retrieval, analysis, and display.

data exchange format: The specific physical data format in which exchange of data between similar GIS packages takes place.

data structure: The logical and physical means by which a map feature or an attribute is digitally encoded.

DBMS (Database Management System): Part of a GIS; the set of tools that allow the manipulation and use of files containing attribute data.

desktop mapping: The ability to generate easily a variety of map types, symbolization methods, and displays by manipulating the cartographic elements directly.

desktop metaphor: For a GUI, the physical analogy for the elements with which the user will interact. Many computer GUIs use the desktop as a metaphor, with the elements of a calendar, clock, files and file cabinets, and so on.

device independence: The ability of software to run with little difference from a user's perspective on any computer or on any specialized device, such as a printer or plotter.

dissolve: Eliminating a boundary formed by the edge or boundary of a feature that becomes unnecessary after data have been captured; for example, the edges of sheet maps.

DXF: Autocad's digital file exchange format, a vector-mode industry standard format for graphic file exchange.

dynamic segmentation: GIS function that breaks a line into points at locations that have significance, and that can have their own attributes. For example, the line representing a highway can have a new node added every mile as a mile marker that can hold attributes about the traffic flow at that place.

edge matching: The GIS or digital map equivalent of matching paper maps along their edges. Features that continue over the edge must be "zipped" together and the edge dissolved. To edge-match, maps must be on the same projection, datum, ellipsoid, and scale and show features captured at the same equivalent scale.

entity by entity: Any data structure that specifies features one at a time, rather than as an entire layer.

FORTRAN: An early computer programming language, initially for converting mathematical formulas into computer instructions.

functional capability: One of the distinctive processes that a GIS is able to perform as a separate operation or as part of another operation.

functional definition: Definition of a system by what it does rather than what it is.

fuzzy tolerance: Linear distance within which points should be snapped together.

generalization: The process of moving from one map scale to a smaller (less detailed) scale, changing the form of features by simplification, and so on.

geometric test: A test to establish the spatial relationship between features. For example, a point feature can be given a point-in-polygon test to find if it is "contained" by an area.

GNU: Free Software Foundation organization that distributes software over the Internet.

GUI (Graphical User Interface): The set of visual and mechanical tools through which a user interacts with a computer, usually consisting of windows, menus, icons, and pointers.

help line: A telephone service available to software users for verbal help from an expert.

import: The capability of a GIS to bring data in an external file and in a nonnative format for use within the GIS.

installation: The step necessary between delivery of GIS software and its first use, consisting of copying and decompressing files, data, registering licenses, and so on.

integrated software: Software that works together as part of a common user interface rather than software that consists of separate programs to be used in sequence.

local area network: An arrangement of computers into a cluster, with network linkages between computers but no external link. Usually, this allows sharing data and software licenses, or the use of a file server.

macro: A command language interface allowing a "program" to be written, edited, and then submitted to the GIS user interface.

map algebra: Tomlin's terminology for the arithmetic of map combination for coregistered layers with rasters of identical size and resolution.

map overlay: Placing multiple thematic maps in precise registration, with the same scale, projections, and extent, so that a compound view is possible.

mask: A map layer intended to eliminate or exclude areas not needed for mapping and analysis.

metadata: Data about data. Index-type information pertaining to the entire data set rather than the objects within the data set. Metadata usually include the date, source, map projection, scale, resolution, accuracy, and reliability of the information, as well as data about the format and structure of the data set.

mosaic: The GIS or digital map equivalent of matching paper maps along their edges. Features that continue over the edge must be "zipped" together and the edge dissolved. To edge-match, maps must be on the same projection, datum, ellipsoid, and scale, and show features captured at the same equivalent scale. See also *edge matching*.

multitask: The ability of a computer's operating system or a GIS to handle more than one process at once; for example, editing and running a command sequence while extracting data from the database and displaying a map.

node snap: Instructing the GIS software to make multiple nodes or points in a single node so that the features connected to the nodes match precisely, say at a boundary.

online manual: A digital version of a computer application manual available for searching and examination as required.

patch: A fix to a program or data set involving a sequence of data that are to be overwritten onto an older version.

proprietary format: A data format whose specification is a copyrighted property rather than public knowledge.

relational DBMS: A database management system based on the relational data model.

renumbering: Use of the DBMS to change the ordering or ranges of attributes. Also, especially in raster GISs, to change the numbers within grid cells into categories.

rubber sheeting: A statistical distortion of two map layers so that spatial coregistration is accomplished, usually at a set of common points.

sift: To eliminate features that are smaller than a minimum feature size.

spatial data transfer standard (SDTS): The formal standard specifying the organization and mechanism for the transfer of GIS data between dissimilar computer systems.

spreadsheet: A computer program that allows the user to enter numbers and text into a table with rows and columns, and then maintain and manipulate those numbers using the table structure.

SURFACE II: An early computer mapping package from the Kansas Geological Survey.

SYMAP: An early multipurpose computer mapping package.

topologically clean: The status of a digital vector map when all arcs that should be connected are connected at nodes with identical coordinates and the polygons formed by connected arcs have no duplicate, disconnected, or missing arcs.

Unix: A computer operating system that has been made workable on virtually every possible computer and has become the operating system of choice for workstations and science and engineering applications.

upward compatibility: The ability of software to move on to a new version with complete support for the data, scripts, functions, and so on of earlier versions.

user interface: The physical means of communication between a person and a software program or operating system. At its most basic, this is the exchange of typed statements in English or a program-like set of commands.

vector: A map data structure using the point or node and the connecting segment as the basic building block for representing geographic features.

version: An update of software. Complete rewrites are usually assigned entirely new version numbers (e.g., Version 3), while fixes and minor improvements are given decimal increments (e.g., Version 3.1).

VisiCalc: A spreadsheet package for first-generation microcomputers. Supported data tables in flat files.

warping: See rubber sheeting.

WIMP: A GUI term reflecting the primary user interface tools available: windows, icons, menus, and pointers.

X-Windows: A public-domain GUI built on the Unix operating system and computer graphics capabilities, written and supported by the Massachusetts Institute of Technology and the basis of most workstation shareware on the Internet.

zip: See *mosaic*.

9.9   PEOPLE IN GIS

Jonathan Raper, Professor of Information Science, giCentre, City University, London

KC: We're at the Royal Geographical Society, London in the old President's Office. Jonathan, what got you interested in Geography and/or mapping science?

JR: I trained as a geographer at Cambridge in the UK. I did my PhD at the University of London at Queen Mary College then went to Birkbeck College, where I began specializing in Geographic Information Systems and Science. What got me interested? When I was a teenager, I was asked by an uncle what I wanted to do when I grew up. I said well I want to work in the outdoors, and he said, "Foolish boy. Just get a desk by the window if you want to work in the outdoors." This so aggravated and annoyed me that its been a sort of personal driver ever since. And so, it's been fantastic to be working and researching

as a geographer and to be able to do outdoor geomorphological-type work.

KC: Why is a geographer in a school of informatics?

JR: Geography is one of those universal subjects; the geographical perspective on human society ranges from understanding the very specific ways in which human society is organized to the way the environment works. There are geographic perspectives to the way we use information technology, and the group that we've founded at City University deals with the geographical dimensions of social informatics—with the interface between people's geographic knowledge and their experience of technology. I think that you could say that geographic information science doesn't

necessarily intrinsically belong in Geography. Geography is ambivalent about representation, but Information Science sees representation as its point of existence, and so it's a comfortable home.

KC: Tell us about some of the research projects your group is conducting.

JR: We've been working with the interface between technology and wayfinding and awareness of your surroundings. So we are looking at how people use information provided by mobile devices to help them find out about their surroundings and find their way. For tourists and traveler's needs we did a project called WebPark which explored the way in which geographic information could help people enjoy their experience of visiting a national park, and we created a spin-off company which is selling a system that people rent in a variety of European national parks. We've moved on to look at accessibility and the way that a variety of other communities use geographic information, principally those with mobility challenges. The "Location Based Services for all" project is currently looking at the design of information support for navigational wayfinding for older people, visual impaired, and blind. Here at the Royal Geographical Society, we've been running tests with an older subject tester who's been trying out our technology and giving us feedback

KC: Are these glimpses of the next generation of systems that we might expect to find as consumer products in the future of GIS?

JR: Well, at this stage of the development of geographic information science things have been dominated professionally by planners, utilities managers, foresters, and so on—all with different professional contexts. The great challenge that we yet face is to put geographic information and geographic information systems into the hands of ordinary people. Everybody has wrist watches—so everyone knows the time. But not everybody has such an easy, simple, universal way to tell you where they are and where they're going and where they've been. We don't have a universal location and place finding tool. One of the challenges which I think GIS faces, is making GIS personal. Now what would a personal GIS look like in the future? It clearly has to be very easy to use. It has to work with naïve geography concepts, where naïve means a native and inbuilt sense of direction and things of that kind. And that means that we have to understand egocentric geographic information.

KC: Thanks, we look forward to the results of your research.

CHAPTER 10

GIS in Action

A computer with a bullet in it is just a paperweight. A map with a bullet in it is still a map.

— Major Keith Hauk, US Commander in Afghanistan

10.1 GIS STUDY BY EXAMPLE

In earlier chapters we met the theory and principles surrounding the basic understanding of GIS. It is impossible, however, to truly learn GIS without using the software to solve geographical problems in the real world. Many of the lessons about GIS, indeed some of the most important ones, can only be learned by seeing this powerful and versatile tool in action. To get an appreciation of GIS in action, in this chapter we adopt the case study method of instruction. Four case studies are presented, just as they were reported in the GIS news literature. The case studies are from the GeoPlace GeoReport archives, available by free registration on-line at www.geoplace.com under "archives." These reports are used with permission of the copyright owner.

In each case, the GIS application report is given, followed by a summary of the research problem, the solution, and the GIS software employed. In accordance with our educational goal, we will also draw conclusions and lessons learned from the case studies that might be useful for someone getting started with GIS. The four case studies cover two types of natural disasters, hurricanes and wildfires, and two industries, agribusiness and electrical power. As you will see, the focus (and often the data) are local, but the broader implications are very general. Global climate change, for example, increases the likelihood of hurricanes and wildfires, changes what agricultural crops can be grown in particular regions, and also directly impacts how much power is needed to heat and cool cities. We will return to this theme at the end of the chapter.

10.2 GIS IN POST-HURRICANE DISASTER RESPONSE AND RECOVERY

GeoReport Issue Date: March 2009, Posted On: 4/1/2009
By Gregory S. Fleming, Frank C. Veldhuis and Jason D. Drost (Gregory S. Fleming is president, Frank C. Veldhuis is vice president and Jason D. Drost is a GIS systems analyst, NorthStar Geomatics)

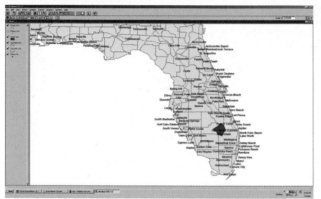

Figure 10.1: ArcView 3.2 map showing Martin County, Florida.

In 2004, Martin County, Fla., endured one of the most devastating storm seasons in recorded history, with two major hurricanes (Frances and Jeanne) making landfall in the county within a three-week period. NorthStar Geomatics, a Florida firm specializing in surveying/mapping, GIS, and asset inventory, played a major role in collecting, verifying, and charting data on the massive debris collection that followed these storms.

The company's efforts would prove to be instrumental in helping the county obtain more than $17 million in cost reimbursement for debris removal. NorthStar's documentation would later become critical in successfully resolving an issue involving a denial from FEMA for $3.45 million of this cost reimbursement. The process used has since become a benchmark for documenting and mapping debris removal in similar emergency-response situations.

Figure 10.2: Following hurricanes Frances and Jeanne, massive debris piles were picked up throughout Martin County, Fla., and taken to several dump sites in the area.

Figure 10.3: By the end of the debris-removal process, information on more than 32,000 loads was documented.

The Storms

The battering that hurricanes Frances and Jeanne delivered to Martin County in September 2004 delivered an unprecedented one-two punch, destroying homes, businesses, and public infrastructure. Damages from the storm created an avalanche of refuse, including construction debris (e.g., wood siding, roofing materials, screen rooms, etc.) and large amounts of trees and vegetation. Ultimately, more than 1 million cubic yards of material would be collected, presenting a major logistical and administrative challenge for the county's engineering staff.

Martin County's disaster-response plan included a process for debris management, and disaster recovery contracts were already in place with area hauling contractors. Just as critical, however, was the need to accurately document the massive collection of debris, per Federal Emergency Management Agency (FEMA) guidelines. County officials knew they had to collect and store data on the debris collection and report it to FEMA in an efficient, timely, and credible manner. Accuracy was crucial, as was the ability for the supporting documentation to withstand a FEMA audit.

Fortunately, in 2001 Martin County had instituted an asset-management program for the county's infrastructure. NorthStar, in its consultant role as GIS coordinator for the county, had assisted in maintaining and expanding the database during the ensuing years. NorthStar's knowledge allowed its team to work seamlessly with county engineering staff to swiftly develop a process for documenting debris removal. Also key to the success of the documentation collection was the prompt involvement of NorthStar in the process. County officials contacted the company immediately following the storms to assist with data collection and tracking.

The Process

Following the hurricanes, contractors were dispatched throughout Martin County to pick up debris and haul it to one of three previously determined dump sites. The specifications for each load were documented on "load tickets" that provided detailed data on the contractor, crew, vehicle, load size and pick-up location, pick-up date and time, and dump site. At the dump sites, three representatives—one from the county, one from FEMA, and one an impartial third-party contractor—would reach agreement on the cubic yards hauled in each load. Load tickets were gathered and reconciled daily, and the information was entered on a spreadsheet by the hauling contractor. This was no small task—by the end of the removal process, information on more than 32,000 loads was collected.

Figure 10.4: NorthStar used road centerline data from Martin County's asset-management system to link ticket information to the appropriate road segment.

The first step in the documentation process was a comprehensive audit of each load ticket. NorthStar manually verified every ticket to ensure the accuracy of the information and that the ticket data matched the information provided by the contractor in spreadsheet form. This process was necessary to determine the correct payment amounts to the contractor and identify tickets with missing or incorrect data that would not pass a FEMA audit.

Mapping the Progress

Taking the process one step further, NorthStar recommended that all ticket information be geocoded against the specific street locations where each load of debris was collected. This innovative concept became extremely important, as FEMA later required extensive documentation on the location and amount of debris removed from each type of roadway (local, state, and federal).

Roads are color-coded to denote maintenance responsibility. NorthStar geocoded the information using ESRI ArcGIS software. To map the collection points, the digital version of the load-ticket spreadsheets was converted to a database format. Road centerline data from Martin County's asset-management system then was used to link each ticket's information to the appropriate roadway segment.

Charting the debris-removal process via road centerlines in the county's GIS dataset allowed NorthStar and the county to analyze the debris-removal process by road, quantities per debris zone, quantities per truck, and removal cost. It also provided a means to verify that debris removed from city and state roads (which weren't eligible for federal funding) was excluded from the FEMA reimbursement request.

Figure 10.5: The progress of debris pick-up was mapped by zone and made available via Martin County's website.

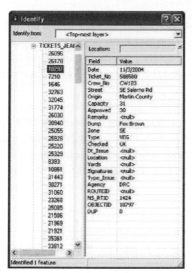

Figure 10.6: Each load ticket included information on the driver, crew, vehicle, pick-up location, pick-up time and date, and dump site.

Integrating the collection data with the county's existing asset-management system eliminated confusion, improved the accuracy of the documentation process and allowed county personnel to access the information in a wide variety of formats.

NorthStar also used the county's ArcIMS Web services to map the condition of area roadways, which had been severely damaged by the hurricanes from downed trees and power lines and flooding. Roads were designated as open, closed (impassable), and emergency passage only. The operational status of traffic lights also was charted. Finally, the progress on debris pick-up was mapped by zone. This information was updated continually and available live on the Internet through the county's website, for access by a variety of emergency-response crews, news organizations, and the general public.

Dispute and Resolution

Although the accumulation of debris collection data was necessary for documentation for FEMA, the mapping aspect became critically important when Martin County faced an unexpected hurdle: a denial for reimbursement of $3.45 million in expenditures for debris removal. At issue in FEMA's reimbursement denial was debris collection in private, or "gated," communities, for which FEMA claimed Martin County wasn't responsible. Data compiled and mapped by NorthStar showed the portion of debris removal under dispute to be 174,000 cubic yards, representing $3.45 million in expenses fronted by the county.

Although the basis for FEMA's decision centered primarily on a legal issue (i.e., whether Martin County had the legal authority and responsibility to remove the debris), the county relied on the GIS data compiled by NorthStar to document and map the collections where eligibility was in question. Without this detailed supporting information, it would have been difficult to resolve the reimbursement issue.

After three years and two unsuccessful appeal processes, county officials turned to their legislative delegation in Washington, D.C., for assistance in securing an audience with key FEMA officials, including R. David Paulison, FEMA administrator. Following their meeting on September 18, 2007, Paulison agreed to revisit the issue, ultimately resulting in an approval of the $3.45 million in reimbursement.

Lessons Learned

The unprecedented challenges faced by Martin County in 2004 underscore the importance of establishing and maintaining comprehensive enterprise GIS, ArcIMS Web services and asset-management systems. During and after a disaster, these tools can be used to provide continual updates to emergency-response crews and the general public on essential safety and recovery issues.

Equally vital is creating a methodology within an emergency-response plan that provides for the accurate collection, organization, and mapping of recovery data. Such pre-planning can pay major dividends in the long run. GIS use by NorthStar Geomatics and Martin County's commitment to building a roadway asset dataset were major components of the county's successful debris-removal program. Effective use of this technology enabled the county to present critical debris information to FEMA in a credible format that satisfied the agency's needs.

Case Study Summary

Component	Item
Where	Martin County, Florida
Who	Northsat Corp., County Government
Problem	Manage debris removal after hurricanes
Customer	Residents of the County, FEMA
Software used	ArcGIS, ArcIMS Web services
Data	Existing road centerlines. Collected and georeferenced new load ticket information.
Maps created and used	Load amounts by street segment. Road condition. Traffic lights. Progress by zone. Maps served over Web.
Issues	Gated communities removal disputed with FEMA
Lessons learned	Value of comprehensive GIS and prior planning for tracking methodology.

10.3 AGRIBUSINESS GROWS WITH CROP-SPECIFIC MAPS

Issue Date: January 2009, Posted On: 2/1/2009
By Jessica Wyland (writer for ESRI).

Figure 10.7: U.S. Department of Agriculture National Agricultural Statistics Service Research and Development Division website at: http://www.nass.usda.gov/research/Cropland/SARS1a.htm.

Crop-specific maps, created by combining survey data and satellite images, provide a literal "lay of the land" for farmers and agribusinesses such as seed and fertilizer companies. Corn, soybean, rice and cotton crops grown in the Corn Belt and Mississippi River Delta areas of the United States are mapped extensively in the Cropland Data Layer (CDL) now available for download or on DVD from the U.S. Department of Agriculture (USDA)/National Agricultural Statistics Service (NASS). GIS software from ESRI is used to prepare and manage agricultural data and build geospatial snapshots of cropland.

"There are many possible uses for the Cropland Data Layer inside and outside the farming community," says Rick Mueller, a GIS expert with NASS. "CDL can be leveraged in a GIS to perform spatial queries against other enterprise GIS data layers. It can be extracted and masked out so public or private entities can focus solely on their own interests."

GIS and Agribusiness

Enhancing a GIS with land-cover data layers has proved helpful to crop growers associations, crop insurance companies, seed and fertilizer companies, farm chemical companies, libraries, universities, federal and state governments, and value-added remote-sensing/GIS companies. Agribusinesses refer to the data to site new facilities for retail supplies and equipment, route transportation of crops and goods, and forecast harvests and

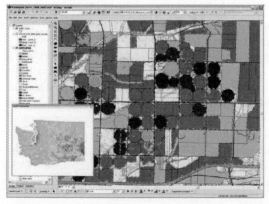

Figure 10.8: Areas of Washington are shown with the NASS 2007 Cropland Data Layer with USDA/ Farm Service Agency with Common Land Unit data overlay.

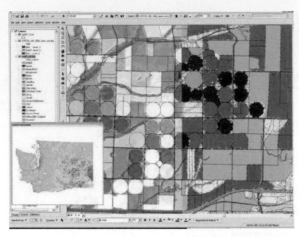

Figure 10.9: A combination of the Washington 2007 Cropland Data Layer with raw AWiFS data taken on July 25, 2007, with a Common Land Unit overlay using the swipe function. The band combination displayed is 3,4,2.

sales. A fertilizer company, for example, can use the CDL to better anticipate how much fertilizer will be needed in specific regions.

The data also are used by pesticide companies to study pest-migration trends and pesticide applications. They're used by farmers and conservationists to perform risk assessment in terms of wildlife habitat, crop stress, and blight locations. Educators determine research locations based on crop density distribution and develop ecosystem models with CDL figures and images.

For each state in the Corn Belt and Mississippi River Delta areas, CDL provides the categorized raster data along with accuracy statistics and metadata by state. CDL is a unique product that provides annual updates of the agricultural landscape. The entire inventory of CDL products is available for download from the Geospatial Data Gateway.

"ArcGIS Desktop from ESRI makes it possible for us to create resourceful maps to identify the spatial extent and associated acreage of the crops grown in these specific states," notes Mueller.

ESRI's ArcMap application also is used to create finished products—detailed, informative maps of U.S. cropland for agricultural stakeholders. GIS specialists use ArcMap to create maps that are distributed to NASS field offices, where they are used at trade shows and distributed to customers.

CDL Web Atlas

ArcMap also is used to create the CDL Web Atlas, where each county within a state is plotted with the location of acreage planted with corn, oats, winter wheat, peas, and other crops, and encapsulated into a single PDF file. Each year, the CDL program focuses on highly intensive agricultural regions to produce digital, categorized, georeferenced output products. NASS uses ESRI's ArcGIS Desktop to manage and edit administrative ground-reference data such as the Common Land Unit (CLU) from the USDA/Farm Service Agency. CLU data are survey-based records of where specific crops are grown. That information is combined with satellite-based remote-sensing imagery to produce super-vised classifications of each field within the state. Satellite imagery is provided by the Resourcesat-1 AWiFS sensor, launched in 2003 by the India Space Research Organization.

The CDL Program was created in 1997 as an offshoot from the NASS Acreage Estimation Program, established to sync satellite images with farmer-reported surveys.

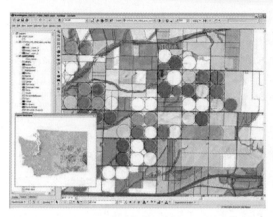

Figure 10.10: A raw AWiFS image dated July 25, 2007, is shown with a Common Land Unit data overlay. The band combination displayed is 3,4,2.

Research and development have been ongoing since the mid-1970s to deliver real-time estimates of acreage at state and county levels using remote-sensing science. Acreage estimates are used for legislation and government programs pertaining to agriculture. The CDL program produced real-time acreage estimates over the Midwest and Mississippi River Delta areas for crop-year 2008 and delivered a unique geospatial product to the GIS and remote-sensing user community.

Author's Note: For more information or to download the Cropland Data Layer, visit www.nass.usda.gov/research/Cropland/SARS1a.htm. For more information on GIS for agriculture, visit www.esri.com/industries/agriculture.

Case Study Summary

Component	Item
Where	US Corn Belt and Mississippi Delta
Who	US Department of Agriculture, National Agricultural Statistics Service
Problem	Provide timely agricultural crop information
Customer	Agricultural businesses, farmers
Software used	ArcGIS Desktop, ArcMap
Data	Agricultural survey by Common Land Units. Harvest forecasts. AWifS-ResourceSat-1 satellite imagery. Accuracy statistics and metadata.
Maps created and used	Crop type. Acreage statistics. CLU survey results.
Issues	Integrate data with other layers. Provide data to agricultural businesses and farmers
Lessons learned	Value of information dissemination via GIS and the Web.

10.4 FIGHTING FIRES WITH COMPUTERS

Issue Date: August -- 2008, Posted On: 8/22/2008

Figure 10.11: Western US Fire information products compiled at the USDA Forest Service (USFS) Remote Sensing Applications Center in cooperation with NASA Goddard Space Flight Center, the University of Maryland, the National Interagency Fire Center, and the USFS Missoula Fire Sciences Lab. Source: US Forest Service.

Fighting Fires with Computers

When you're battling a dangerous, wind-driven wildfire, and people's lives are at stake, you want access to the toughest, most innovative tools and technology available. That's precisely what local, state and federal emergency-response crews got throughout the summer and fall of 2007 when Southern California was ravaged by a series of wildfires that blackened hundreds of thousands of acres across more than half a dozen counties.

Fire and emergency officials knew they needed to deploy crews, equipment, water, and fire retardant to key strategic locations as quickly as possible and constantly monitor the fire's status. To develop a plan of attack against the firestorm, emergency responders called on wildfire strategists armed with GETAC rugged notebook PCs and GIS software from ESRI.

Figure 10.12: GETAC's M230 rugged notebook PC helped the San Bernardino Sheriff's Department search for a hiker lost in the San Bernardino Mountains. Deep snow and extremely cold conditions made the search difficult, but authorities credit GIS technology and hard work with the recovery.

Using rugged, military-grade GETAC M230 rugged laptop PCs with daylight-viewable screens and built-in GPS and wireless radio-transmission capabilities, ESRI wildfire specialists operating behind the fire lines with local, state, and federal emergency responders employed the company's ArcGIS software to integrate, manage, and analyze large amounts of real-time geographic and other fire and terrain conditions and instantly generate and share detailed maps, charts, and other information.

Mapping Out a Firefighting Strategy

ESRI Wildland Fire Specialist Tom Patterson remembers when mapping out a strategy for fighting a wildfire meant unfolding a topographic map on the hood of a truck, covering it with a sheet of Mylar and penciling in fire boundaries and other information. These days, Patterson opens his GETAC rugged M230 notebook PC with sunlight viewable screen and uses ESRI's ArcGIS software to help fire commanders generate 2-D and 3-D maps of fire perimeters and progression, analyze vegetation and other physical features, allocate resources and equipment, and perform property and community damage assessments. The maps and other information then are distributed via the M230's wireless radio connection to other command locations and centralized emergency-response centers.

"It's so much faster, and you make smarter decisions," says Patterson, a retired National Park Service fire-management officer and former deputy chief of the U.S. Bureau of Land Management's California Desert District.

"For the first time in our county's history, we were able to deploy maps that allowed all of the agencies involved to see the problem from the same point of view," says Zacarias Hunt, Santa Barbara County geographic information officer, who spent more than two months tracking the huge Zaca fire, which burned almost 250,000 acres. As the fire moved from wilderness toward populated areas, Hunt and two other technicians used the ESRI-powered GETAC notebooks to generate detailed plans that mapped out evacuation routes, Red Cross shelters, historic sites, schools, and other infrastructure that might have to be defended—even the location of people with disabilities who might need help during an evacuation. Officials knew that if the fire destroyed major power lines, evacuating nearby Santa Barbara and surrounding communities would be more difficult.

"Using the maps, we were able to determine decision points where evacuation warnings and then full evacuation plans would be put into action," says Hunt. "We were even prepared to go out [to the community] with a series of 11-by-17 maps to make it happen."

Figure 10.13: GIS technology, along with the GETAC M230's stability in extreme conditions, helped the rescue team map out a search area based on a 911 call from a lost hiker in the San Bernardino Mountains in early January.

Figure 10.14: A member of the San Bernardino Sheriff's Department views a detailed area of the San Bernardino Mountains during a search and rescue mission. Area details are sent directly from the GETAC M230 to the printer to create a hardcopy.

Solving an Old Problem

When it comes to working in rugged environments, it's vitally important to have a computer that offers various degrees of protection against physical and environmental conditions.

"I used to have a Toshiba Tecra that had one of those glare guards you attached with Velcro, and even then I had to shade it with my hand to be able to read it, and I had to replace the display two or three times because some of the circuitry wasn't heat-resistant," recalls Patterson. "Protection from heat is critical out in the desert, where it can get up to 140 degrees Fahrenheit inside a vehicle if you leave the windows up." Patterson upgraded to a Panasonic CF29, which had some, but not all, of the rugged features he wanted—like a display he could read while working outdoors or in a vehicle.

That changed when he tested a fully rugged GETAC M230 notebook outfitted with a sunlight-readable touch-screen display that delivers image quality in bright sunlight. GETAC designs and builds its M230 and other rugged PCs with extreme outdoor computing in mind. The M230 is MIL-STD-810F and IP54 compliant, which means it has undergone a series of rigorous tests designed by the U.S. Army as the standard for rugged performance. GETAC is the only rugged PC manufacturer with its own inhouse testing facility, where units are subjected to the kinds of abuse users may experience in performing their jobs: shocks to the hard drive, drops, repeated blows from heavy objects, vibration, and water and dust.

Power is another important feature for emergency services personnel in the field. "We typically get three-and-a-half to four hours on a battery, depending on whether or not we are using GPS," states Patterson. "And I like the fact that the battery has a power indicator that tells me how much power I have left so I can decide whether to throw an extra one on a helicopter before I head out on a recon flight."

Other features are designed to allow users to work the way they need to work. For example, operating ESRI's ArcGIS software requires a sentinel key that plugs into a parallel port, which negates the need for a USB key that can get lost, stolen or broken off from the port. The GETAC M230's serial port allows Patterson to plug in a digital radio for real-time mapping of wildfires or search-and-rescue operations from the backseat of a helicopter. The built-in GPS and antenna at the top center of the display eliminate the need for an external antenna or an external GPS receiver. Security is a common concern for anyone working with government agencies. The GETAC M230 includes a swappable hard drive bay.

"We can go into any command post, log into any agency's net, and swap out the hard drive later, which allows us to do our job and also comply with government security policies," he adds. "It's like having two separate computers."

Case Study Summary

Component	Item
Where	Southern California
Who	Federal, state and local emergency response crews. Wildfire specialists.
Problem	Plan crew, equipment, water, and fire retardant deployment. Support fire protection, evacuation planning, and fire mapping.
Customer	Residents in fire-prone areas.
Software used	ArcGIS running on a ruggedized PC with GPS and digital radio.
Data	Real time fire boundaries from GPS. Terrain characteristics. Vegetation. Damage assessments. Emergency services and evacuation routes.
Maps created and used	Red cross evacuation centers. Historical sites. Schools. Infrastructure. Accessibility. 2d and 3D fire perimeters and progress.
Issues	Need hardware suitable for hot and hard conditions, long battery life, and backup.
Lessons learned	GIS provides unified point of view.

10.5 MUSIC CITY'S POWER COMPANY PLUGGED INTO GIS

Issue Date: January 2008, Posted On: 2/1/2008
By Matt Freeman (writer for ESRI)

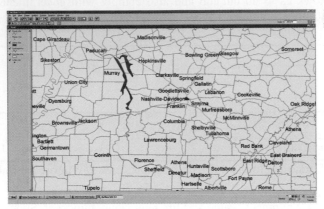

Figure 10.15: Nashville, Tennessee, and region shown in ArcView 3.2.

Nashville, America's Music City, is the home of many country stars whose songs "warm the soul." Unfortunately, the city also is plagued by inclement weather that wreaks havoc on the local power grid. Nashville Electric Service (NES) is working hard to keep its customers from "singing the blues" about power losses.

The electric company is using GIS software to assist in the planning, design, operation and maintenance of its complex electric network. GIS has been instrumental in improving NES' operations management and decreasing storm-outage recovery time. Since a decade ago, when the company first implemented ArcGIS software, NES has added many more updates, and now most processes at NES incorporate some form of GIS applications. The company's 87,538 distribution transformers, 253 distribution substations, 5,619 miles of pole lines and approximately 200,000 utility poles all can be viewed via the NES intranet, complete with company-wide access to GIS data and up-to-date information on its electrical system and basemap data.

The entire enterprise now relies on a single GIS-based mapping system with applications for multiple assets such as circuit maps, property boundaries, pole attachments, street lighting, private lighting, an underground network, circuit profile, and communication infrastructure.

Real-Time Outage Maps and AVL

A component of NES' GIS that's used to manage normal and emergency work on a daily basis is the company's automated vehicle-location (AVL) system. It helps dispatch and service crews better serve the company's 351,000 customers within NES' 700-square-mile service territory, which includes all of Davidson County and portions of six surrounding counties in central Tennessee.

AVL gives dispatchers the ability to view work-crew locations in real time. The company's management reports that the system works well for dispatchers, because it makes their jobs easier and allows them to work more efficiently. Since NES began using AVL, its hazardous-energy control procedures are less complicated and more rapid. As a result, safety has been enhanced and reliability has been improved.

The use of NES real-time maps isn't limited to dispatchers. The "Estimated Customers without Power" Map Web page on www.nespower.com provides customers with their own view of NES work-crew activities and locations. The real-time map highlights regions that have reported an outage and gives details of the time the outage was reported, the number of customers affected, a work-crew assignment, and a street map of the affected area.

Reducing Tree-Caused Outages

NES' vegetation-management workers rely on GIS to prioritize tree-trimming cycles. In 2002, a worldwide study of 110 utilities, conducted by Environmental Consultants Inc., found that NES had the highest number of tree-caused outages per 100 miles of power lines. The outages were 10 times that of best-practice utilities included in the study.

NES set out to improve its power-reliability record by managing its tree-trimming cycles using GIS data. In June 2005, NES completed its initial three-year trimming cycle, and, as a result, power outages were reduced 19 percent. NES has made trimming-cycle information available to its customers via the interactive "Tree-Trimming Crew Activity Map" Web page.

Accuracy for Outage-Management Planning

By integrating data from its network-analysis and outage-management systems (OMSs) into a single database, NES created a transmission and distribution (T&D) planning forecast, which is a 20-year forecast that includes five-year increments of estimated calculations.

Previously, the utility had outsourced this planning effort, costing the company approximately $1 million. Using its GIS, NES has been able to perform T&D planning inhouse with the same amount of effort, but at a much–reduced cost. The results will help determine NES' projected land use for the year 2020 and estimate new customer numbers. The projected land-use report is proving to be a helpful tool for NES operations teams as they plan future work cycles and pole-replacement needs.

NES is responsible for nearly 200,000 poles. By using an NES facilities map, staff can generate a needs assessment for each pole based on variables such as height, age, location, and number of customers the pole serves. NES work crews place or replace approximately 600 poles a year—sometimes many more. GIS generates a risk-based analysis of the variables that's used to prioritize the annual pole replacements.

Success of Disaster Preparedness Plan

Response time has improved since the days before NES implemented GIS. The service area has a history of being ravaged by weather such as blizzards and ice storms, which have caused as many as two-thirds of the utility's customers to be without electricity. Although NES crews would work around the clock following an ice storm or tornado, customers were sometimes in the dark for days and even weeks. In the mid-1990s, NES implemented a GIS-based solution that allowed the company to track reported outages and isolate probable causes so crews could be dispatched where they were most needed. The GIS integration with OMS gave NES the ability to track reliability statistics at the circuit level. After the company migrated its CAD drawings into ArcGIS, technicians were able

Figure 10.16: A 3-D flood map of a local river-system flood zone provides a visual model useful for creating a disaster plan.

to access this database and use applications for maintaining data and producing cost-efficient maps.

The implementation of GIS into the utility's OMS came just in time. In 1998, a tornado ripped through Nashville, east of downtown, taking out more than 500 poles and leaving 75,000 customers without power. With GIS in place, NES was able to restore power to all of its customers within a week—most within three days—a feat much hailed by customers and the media.

A second test of the utility's ability to respond came in 2006, when a tornado leveled homes and knocked out power to more than 16,000 NES customers. A significant portion of the territory's electrical system was destroyed in the storm. GIS was used in restoration operations that entailed the mobilization of work crews and 70 bucket trucks to quickly replace more than 100 poles. Due to the company's aggressive efforts in developing disaster preparedness, the majority of NES customers had their power restored within 48 hours.

NES preparedness plans now include the possibility of flood-caused outages. Reports of leaks springing from the nearby Wolf Creek and Center Hill dams on the Cumberland River system have raised concerns for many of the utility's customers. They're inquiring about NES' disaster plan should the problem reach disastrous proportions before the dams are fixed by the U.S. Army Corps of Engineers.

NES used its GIS expertise to implement flood maps of its service territory in the company system. By doing so, NES discovered that it has some power substations that were potentially in danger. More detailed investigation using an integration of its GIS flood map and a 3-D map video confirmed that an NES substation is in the flood zone.

"Sometimes it's not enough to just tell somebody that there is a problem," says Paul Allen, vice president of operations at NES, about generating the flood map and 3-D video. "You have to show them, and our GIS is a good tool for presenting that story. Based on our GIS analysis, our staff is in the process of writing an intelligent flood-disaster plan."

Case Study Summary

Component	Item
Where	Nashville, Tennessee
Who	Nashville Electric Services and Nashville residents
Problem	Planning growth, protecting against natural disasters, minimizing downtime.
Customer	Nashville power users.
Software used	ArcGIS.
Data	Electrical system—transformers, substations, utility poles. Vehicle locations. Base map. Tree trimming and work assignments. CAD drawings. Demand forecasts.
Maps created and used	Circuits, property, pole attachments, street lights, underground utilities, customers without power.
Issues	Need coordination during tornadoes and floods. Need to plan for future demand.
Lessons learned	GIS has measurably improved and reduced time of recovery after disasters.

10.6 SUMMARY OF THE CASE STUDIES

It should be apparent from the case studies in this chapter that, as Chrisman noted in his GIS definition featured in chapter 1, GIS is always associated within the context of its users and their organizational settings. The staffing, software, experience, leadership, and other characteristics of a GIS group or enterprise have a great deal to do with how GIS is used as a tool to craft a solution to a geographical problem. Also of great importance is the problem itself and its complexity, the resources available for a solution, and the goal of the task. Whether something needs to just get fixed for now or whether a long term solution is needed for the future has a major impact on how problems get solved in the real world. A GIS is just as effective at damaging the environment for short term profit as it is at sustaining it for future generations.

GIS is in many ways not unlike many other technological solutions in business or organizations. In terms of planning for a GIS problem solution, important steps are often written in a formal document called a "Statement of Work." This document should contain a task-by-task "Scope of Work," the location of where work will be performed, who will do each step, dates, and a schedule when items will be forthcoming (called "deliverables"), a list of standards that must be applied or followed, what criteria are necessary for the deliverables to be accepted, and any special requirements. Often an organization when deciding what tasks are necessary will perform a needs assessment, in which the most critical processes to be created are delineated and prioritized, perhaps in the light of the interests of the customers or stakeholders. This is often followed by a problem definition step, goal setting, and creating the statement of work. Lastly risks are assessed, including identifying failure and decision points, and alternative plans compiled. Often it is felt important to name specific measurable outcomes (termed performance metrics), so that after the fact it can be assessed whether these were accomplished and the project goals met.

Many of GISs most challenging problems relate to the environment and how it is used. Increasingly, GIS must be linked with models to explore the consequences of future actions and decisions. A whole branch of GIS research overflows into decision-support systems and theory. GIS is likely to play an increasing role in managing the world's environmental systems as global changes wrought by human-induced change impact our planet. Effective GIS managers, as well as analysts, therefore have a very important role to play in the future.

10.7 STUDY GUIDE

10.7.1   Study Questions and Activities

CHAPTER 10: GIS IN ACTION

GIS Study by Example

1. What can be learned using the case study approach that can't be learned from studying GIS theory and software?

2. Research what is meant by a "Scope of Work." What does it contain? Write a scope of work for one of the case study GISs in this chapter.

3. Make of list of the software used in the case studies, and what it was used to do. Why do the applications seem to use the same software?

4. Check the website for GeoReports used in this chapter. Choosing a story about a new application, make up the same table used for the Case Study Summaries.

5. In any of the case studies, imagine that you are a consultant hired to improve the systems as described. What would you recommend?

6. Choosing any one of the case studies, make a list of all of the employees that you think might be necessary for the tasks described. Now assign GIS tasks to each person on the list, and suggest how much GIS education and training each needs.

7. Go to the website for your own county, university, or community. What GIS data is available? What administrative units use GIS in their work? How many people do they employ? Put your results together in a brief report that assesses GIS use in your community.

10.8 FURTHER READING

Croswell, P. L. (2009) *The GIS Management Handbook.* URISA: Kessey Dewitt Publications.

Grimshaw, D. J. (1999) *Bringing Geographical Information Systems into Business.* New York: Wiley.

Huxhold, W. E. and Levinsohn, A G. (1995) *Managing Geographic Information System Projects.* New York: Oxford University Press.

Longley, P. and Clarke, G. (Eds) (1995) *GIS for Business and Service Planning.* Cambridge: Geoinformation International.

McGuire, D., Kouyoumijan, V. and Smith, R. (2008) *The Business Benefits of GIS: An ROI Approach.* Redlands, CA: ESRI Press.

Pick, J. B. (2008) *Geo-Business: GIS in the Digital Organization.* New York: Wiley.

Pinto, J. K. and Obermeyer, N. J. (2007) *Managing Geographic Information Systems.* 2ed. New York: Guilford Press.

10.9 KEY TERMS AND DEFINITIONS

agribusiness: The various businesses involved in food production, including farming and contract farming, seed supply, agrichemicals, farm machinery, wholesale and distribution, processing, marketing, and retail sales.

ArcIMS web service: An ESRI software package that enables GIS users to distribute mapping services via the Internet.

asset-management: A management system that produces efficient use of organizational resources and equipment such as vehicles, measuring equipment, etc.

basemap data: Reference data of general rather than specific purpose, for example coastlines, placenames, roads, and rivers.

benchmark: A standard by which something can be measured or judged.

best-practice: A well defined procedure that is known to produce near-optimum results.

CAD drawing: The digital equivalent of a drawing, figure, schematic, or architectural plan created using a computer-aided design system.

case study: One of several ways of doing research, an intensive study of a single group, incident, activity, or community.

centerline data: Digital street map data consisting of a set of single connected vectors that represent each street link that runs down the center of the actual road.

Common Land Unit (CLU): A geographic division used by the US Department of Agriculture that is the smallest unit of land that has a permanent, contiguous boundary, uses common land cover and land management, and has a common owner or a common producer association.

consultant: A person hired to give expert or professional advice.

crop-specific map: Map of the geographic distribution of a single agricultural product, such as soybeans.

damage assessment: A quantification or estimate of the extent, type or cost of destruction due to natural or man-made causes. Usually a summation from individual-level data.

daylight-viewable screen: A screen capable of displaying its contents under direct sunlight. Often by being very bright, or by avoiding reflected light.

decision point: Latest time at which a planned course of action should be or is initiated.

emergency response crew: A trained group of individuals equipped to deal with a critical situation involving risk.

enterprise GIS: A geographic information system that is integrated through an entire organization so that many users can share and use geospatial data to address a variety of needs, such as data creation, modification, visualization, analysis, and dissemination.

geographic information officer (GIO): Geospatial equivalent of the CIO, or Computer Information Officer, a widely recognized executive management position in Information Technology based organizations.

GeoReport: A weekly electronic newsletter for the Geospatial industry. A service of the Geoplace.com GIS information website.

incorrect data: Data that are in a database, but that are known to be wrong.

infrastructure: The underlying foundation especially for an organization or system and necessary for functioning.

intranet: A private computer network that uses Internet technologies to securely share any part of an organization's information or operational systems with its employees.

load ticket: In the Martin Co. case study, a paper form filled out with information about each truck delivery. The data were entered into a spreadsheet, checked, then imported into the GIS.

mask out: In the GIS, invoke a layer used to exclude a zone from mapping or processing.

methodology: A body of practices, procedures, and rules used by those who work in a field or engage in inquiry.

missing data: Records for which an attribute is not present when it should be, or records that are not in the database when they should be.

needs assessment: A process for determining and addressing gaps between current and desired conditions, often used for improvement projects in education/training, organizations, or communities.

operations management: An area of business concerned with the production of goods and services, and the responsibility of ensuring that business operations are efficient and effective in terms of meeting customer requirements.

PDF file: The Portable Document Format created by Adobe Systems in 1993 for independent and cross-platform document exchange.

power grid: An interconnected network for delivering electricity from suppliers to consumers.

real-time map: Map in which the data are being updated and redisplayed as events actually happen.

roadway segment: A single block-to-block street face, or subset of such links in a digital street network map.

risk assessment: Part of the risk management process involving determination of quantitative or qualitative value of risk related to a situation and a threat or hazard. Quantitative risk assessment requires calculations of the potential loss and the probability that the loss will occur.

sentinel key: A hardware device that monitors or controls use of GIS software on a particular computer.

supervised classification: Conversion of raw imagery bands into a set of land use or other classes, using patches of known land use identified by the analyst.

USB key: A small memory chip with a built-in USB interface, used to control the use or licensing of GIS software on a particular computer.

10.10 PEOPLE IN GIS

Brenda G. Faber, Owner, Fore Site Consulting, Loveland, Colorado

KC: What got you started in GIS?

BF: My graduate degree was in Electrical Engineering and Image Processing. I went to work for IBM after graduate school to develop Robotic Vision techniques. Later, a job transfer within IBM led me to a group doing volumetric GIS research. I found that my background in image processing was a natural fit for scientific and raster GIS work.

KC: What is your role in the GIS industry?

BF: I own a small consulting company that develops Planning Support Systems for municipalities and federal land manage-

ment agencies. Planning Support Systems are a relatively new class of planning tools that extend the capabilities of traditional GIS to include impact, simulation, and visualization options. These systems are primarily concerned with exploring the implications of land use alternatives.

KC: What is CommunityViz?

BF: CommunityViz is a Planning Support System. It is an integrated suite of GIS extensions that provide a customizable framework for evaluation of land use proposals, including 3D exploration, impact analysis, and predictive forecast modeling. CommunityViz is unique among Planning

Support Systems, because it integrates three unique planning perspectives into one multi-dimensional environment. The development of CommunityViz has been sponsored by the Orton Family Foundation to enhance collaboration and participation among planners, officials, and citizens. I have worked as a consultant to the Foundation for several years now as a primary developer of CommunityViz.

KC: How is visualization integrated into ArcView in the CommunityViz software?

BF: CommunityViz 3D Visualization allows you to "experience" proposed land scenarios by taking a virtual stroll through a 3D GIS scene. Realistic scenes can be quickly created directly from standard GIS datasets including terrain data (such as a DEM or TIN), orthophotographs, and 3D representations of common GIS features such as buildings, tree canopies, roads, rivers, and fences. You can move about in this virtual scene to explore not only existing landscapes, but the visual impacts of proposed land changes as well.

KC: What is your experience working with GIS with planning groups around the country?

BF: I love the natural balance that my work allows between technical development and direct involvement with planning groups. I think one of my greatest strengths is the ability to act as a liaison between the very technical world of computer modeling and real folks who want to make informed decisions about the future of their communities. So many important decisions today are made by "gut feel." By making GIS and analysis techniques more accessible to citizens and policy makers, we will have fewer unanticipated consequences down the road.

KC: What would you advise someone to do who would like to follow in your career footsteps?

BF: Exposure to a wide array of technical disciplines and experience with real-world project implementation are both very valuable. I believe that innovative concepts are rarely "new," but most often evolve from unique combinations of existing concepts from diverse fields of study or experience. A GIS background combined with additional talents in programming, public speaking, engineering, biology, psychology, etc. can make for a dynamite career in GIS research and development.

KC: Brenda, thanks for the information.

Used with permission. Original interview for 4th edition in 2003

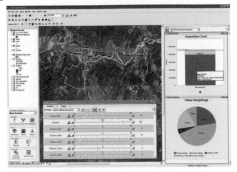

CHAPTER 11

The Future of GIS

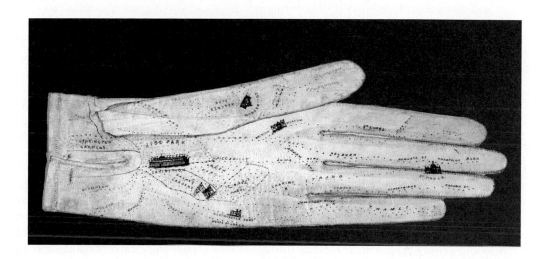

After growing wildly for years, the field of computing appears to be reaching its infancy.

John Pierce

11.1 FUTURE SHOCK

The theme of this book has been an examination of the value that GIS brings as a tool for understanding geographic distributions, and for describing and predicting what will happen to these distributions in the real world. History has shown just how powerful GIS can be as a new mechanism for managing information. From humble origins, a set of simple ideas, and some rather inefficient software, GIS has grown into a sophisticated, multibillion-dollar industry in only half the length of a human career. GIS's dual role as a mainstream technology for the management of geographic information and as an effective tool for the use of resources is no longer a promise, but a reality. Many refer to GIS as a "mature industry."

So why bother to discuss the future of GIS? Quite simply, why speculate? The future always seems to catch up with the present at an alarming rate, what Alvin Toffler called "future shock." In only a few years, for example, GIS and GPS technologies have met and merged in a seamless way without the slightest hitch. Why not just wait and let the technology deliver our dreams and speculations if they are realistic enough to come into being?

There are three good reasons to speculate. First, GIS use involves planning for the purchase or acquisition of hardware and software. If a new product that is just around the corner can do more for less money, the GIS user stands to lose money and perhaps build obsolete systems if attention is not paid to market trends. Second, GIS has created a science, geographic information science, that has built itself as a field for study and research and which now is designing the logic of the systems of tomorrow. As Jonathan Raper states in his interview in Chapter 9, the next generation GIS needs to be much easier to use. If GIS users, even new users, do not follow this research, they will miss information about what has been accomplished. Finally, GIS is expanding continuously into new fields and new areas of application, discovering new uses, and solving new problems.

In doing so, GIS is bringing fresh ideas into parallel sciences—oceanography, epidemiology, facilities management, disaster planning and relief, environmental management, archaeology, forestry, geology, real estate—and the list goes on and on. Each new application brings GIS concepts and techniques to a new set of individuals and in turn is influenced by their own expert knowledge and needs. Many of these second-generation GIS users are not just scientists and professionals, but ordinary people. GIS can be used to defend or develop a community, to plan basic services, to educate, to win an election, to plan a major event, to avoid traffic jams—and again the list goes on. The inexpensive but powerful microcomputer has placed the basic tools of high technology into the hands of the educated citizen. Quite simply, things can never be the same again.

If speculation about the future is indeed valid, we should recognize two types. First, present trends and ideas already at the cutting edge of research can be extended into the future. This is the most certain form of speculation, for most of these predictions will actually come about, especially if the items are new products coming to market. The second type of speculation is really gazing into the crystal ball. Many of these ideas will be wrong, but what the heck, some will probably be right! In this chapter, we examine the future of data for GIS, and the future of computing. The book then concludes with a few of the broader issues and problems that GIS has brought to light and that we will have to grapple with in the future. If there is a GIS in your future, at some stage you will most likely be dealing with one or more of these issues. Just remember, you read it here first!

11.2 FUTURE DATA

11.2.1 Data Are No Longer the Problem

The blood of a GIS is the digital map data that runs through its software veins and hardware body. The future holds immense promise for new types of data, more complete data, higher-resolution data, and more timely data. Once the major obstacle to GIS development, data have now become GIS's greatest opportunity. Some of the types and sources of GIS data have already been described in detail in this book. The years ahead will bring us even more new types of data, and vast revisions of the existing types. As such, this summary of future data can be only a glimpse of what is still to come.

First, it should be stressed yet again that the entire mechanism for GIS data delivery has been revolutionized by the Internet and by the search tools built upon the structure of the World Wide Web. Most public-domain data, most shareware and freeware, and an increasingly large proportion of commercially produced GIS data use the Internet in place of computer tapes, diskettes, the U.S. Postal Service, and the so-called sneaker-net (i.e., hand-delivery). This single trend has had, and will continue to have, the most impact on the field of GIS. Rarely does a new GIS project have to begin by digitizing or scanning

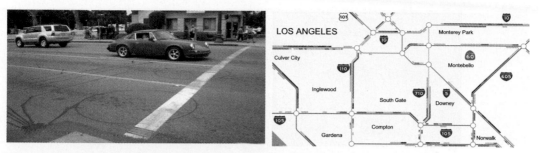

Figure 11.1: Left: Embedded inertial coil traffic sensor providing traffic flow data via the Internet. Right: Experimental design for web-based traffic flow map. Source: Goldsberry, 2007. Used with permission.

geographic base maps. Instead, the majority of GIS work now involves bringing into the system a base layer of public-domain data and enriching it by capturing new layers pertinent to a particular GIS problem.

Within the last few years, entirely new sources of data have arisen in the guise of sensor networks. A sensor network is a wireless network consisting of spatially distributed autonomous devices using sensor technology to cooperatively monitor physical or environmental conditions. An example is the network of traffic sensors in the form of inertial coils embedded in highways all over America. These send traffic flow information via the Internet to servers, which then redistribute it automatically with updates every five minutes, allowing services such as Google Maps and SigAlert to provide real time traffic information for major cities (Goldsberry, 2007: Figure 11.1). Wireless sensor networks were originally created for military applications such as outdoor surveillance. Today, wireless sensor networks are being used in many industrial and civilian applications, including industrial process monitoring and control, machine health monitoring, environment and habitat monitoring, healthcare applications, oceanography, and traffic control. Invariably, GIS is used to pull together the data and to do analysis and visualization.

It seems likely that sensors will become far more ubiquitous in the future. Mobile telephones have been equipped with GPS, making them essentially spatial sensors that can be linked to many other activities. Supermarket checkouts and credit card scanners record data, and many appliances and cars now use sensor technologies for a host of applications. Web-cams are increasingly accessible via on-line map-based systems. As more data from sensors become georegistered, GIS will be ever more important for its role as an information integrator.

Another major trend in GIS has been the appearance of numerous Web portals, entry points into vast reserves of geospatial data. One vision for this is to link all data sources together into a digital database containing all geospatial data in the world, what has been called a "digital earth" (Grossner et al. 2008). The International Society for Digital Earth calls digital earth "a visionary concept, popularized by former US Vice President Al Gore, for the virtual and 3-D representation of the Earth that is spatially referenced and interconnected with digital knowledge archives from around the planet with vast amounts of scientific, natural, and cultural information to describe and understand the Earth, its systems, and human activities." Annual international conferences since 1999 and an international journal support this vision. While a single worldwide integrated GIS data source has yet to appear, there are now a large number of geobrowsers, on-line search engines that use geographical location and the visualization tools of a virtual globe, to link

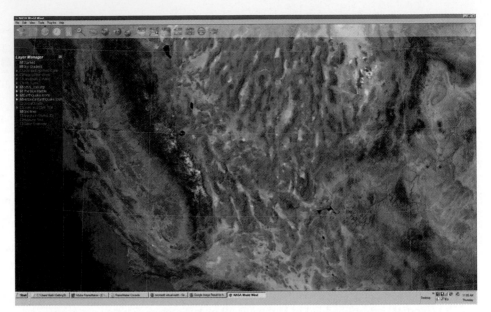

Figure 11.2: NASA's Worldwind geobrowser, capable of browsing most NASA and other public domain geospatial data worldwide. Downloadable from: worldwind.arc.nasa.gov.

other Web-based information. Examples are NASA's Worldwind, Google's Google Earth, and Microsoft's Bing Maps 3D. Given what already exists, a concerted effort could realize the digital earth vision with relative ease.

11.2.2   GIS and GPS Data

Another critical step in data provision has been the ability, using the global positioning system (GPS), to go directly to the field to collect data rather than relying completely on maps. The GPS has also improved mapping significantly, because the geodetic control once only marginally available to mapping projects is now as easy as pushing a GPS receiver button and doing a differential correction to sub-meter accuracy. So precise is this new mechanism for data collection that existing GIS maps of cities, parks, and other areas will have to be revisited for field verification. The ability to register a map quickly to a given map geometry (projection, ellipsoid, and datum) means that GIS layers can quickly and efficiently be brought into registration for overlay and comparative analysis. The field of GIS has greatly benefited, and the GIS-to-GPS link is now such that many GPS receivers and their data loggers can write data directly into GIS formats or include satellite images, air photos, or regular photographs directly in the field.

The flexibility of this system, when integrated with in-vehicle navigation systems that also use inertial navigation and stored digital street maps, has evolved a technology that is becoming standard equipment in public and private vehicles. Never again will the driver have to stop to ask the way to a destination! Now available as a car order option, or as the popular add-on in-vehicle GPS, the cost of these systems is such that they pay for themselves in saved time almost immediately (Figure 11.3). Second generation in-vehicle systems offer real-time traffic updates, tourist and other sites of interest, and links to business information. The rapid generation of street, highway, and city maps resulting

Figure 11.3: In-vehicle GPS-based navigation system in use (TomTom One).

from the growth of these systems—data that are by definition of great locational accuracy—is greatly benefitting GIS. Although the data have so far been digitized almost exclusively by private companies, competition has led to a data price war in recent years, and costs have fallen remarkably. Many systems offer direct supplement of the signal via the Wide Area Augmentation System (WAAS), which gives basic receivers meter level accuracy. The WAAS is an air navigation aid developed by the Federal Aviation Administration to augment GPS, with the goal of improving its accuracy, integrity, and availability. Hand-held receivers with map displays can now be purchased for under a hundred dollars, and are finding uses in hunting, hiking, travel, and even sports such as orienteering and geocaching (Figure 11.4). In almost all cases, data from these units can be downloaded for use in GIS. GPS has also found use in fleet vehicles such as the trucking and moving industries, and in the delivery business. In each case, the common element is the need for moving around a street network efficiently.

GPS is one example of a global navigation satellite system or GNSS. Even while GPS is itself in the process of upgrading to a next generation system, other systems are arising as competitors, and often allow redundancy in computing a position. The Russian GLONASS system, which lapsed after the fall of the Soviet Union, is now being revised. The European Space Agency's Galileo system hopes to begin launching satellites in 2010. China is building its COMPASS navigation system. Other nations also have plans to develop GNSSs. It is not unimaginable that within a decade, cross system receivers will be able to offer extraordinary levels of positioning accuracy, and more ubiquitous coverage. Experiments are also under way to make GPS signals readable indoors, and by tiny

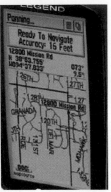

Figure 11.4: Inexpensive GPS receivers for outdoor recreational use.

receivers (Figure 11.5). Lastly, almost all GISs now support direct input from GPS receivers, making the use of GPS data in GIS straightforward.

11.2.3 GIS and Image Data

The popularity of geobrowsers and acceptance of oblique and overhead imagery as valuable sources of geospatial information imply that the standard symbolized cartographic map is now only one of the ways that people perceive the world. Another significant new data source has existed since the arrival of the digital orthophotoquad (DOQ). Digital orthophotoquads are geometrically corrected air photos with some cartographic annotation. Their historical use has been as sources of information for the U.S. Department of Agriculture. Recently, however, these data have been made available by the USGS in digital format in quarter-quadrangles; that is, one-fourth of a 1:24,000 7.5-minute quadrangle as one data set, with an equivalent scale of 1:12,000 and a ground resolution of 1 meter (hence the abbreviation DOQQ for quarter-quad). These images have been supplemented with even more detailed representation of the major U.S. cities, and with other DOQs supplied by state and other programs.

Figure 11.5: World's smallest complete RF-Front-End GPS receiver chip from New Zealand's Rakon Corp (www.rakon.com). Used with permission.

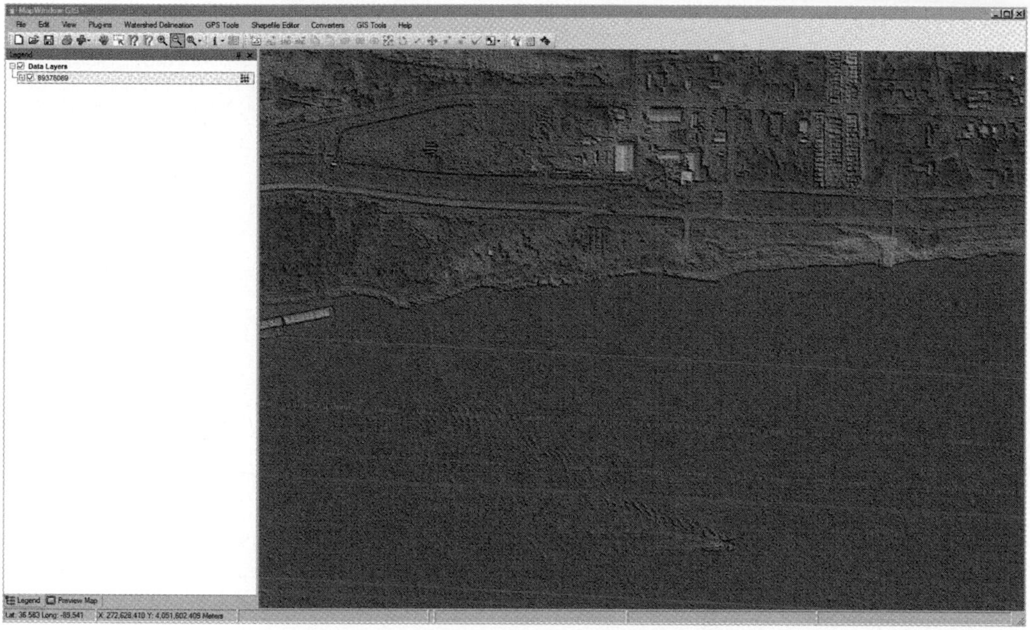

Figure 11.6: Part of the USGS's nationwide DOQ imagery coverage. Shown is New Madrid, Missouri on the Mississippi River, using MapWindow GIS.

These astonishingly accurate data have already become a new national base layer, eclipsing the role of the current digital line graph data. Rather than being vector data, though, the raster nature of this layer and the fact that it is partly monochrome have resulted in its use as a background image for GIS, over which field and existing geocoded data are assembled. The primary function of the orthophoto is to assure the same type of layer-to-layer registration discussed in the case of GPS above. The coverage has expanded to cover the entire United States, and regular revisit assures that maps are updated as required. This imagery is accessible through the National Map viewer and seamless serverThe digital raster graphic (DRG) is a scanned image of a U.S. Geological Survey (USGS) topographic map, including all the information on the map edge, or "collar." The image inside the map neat line is georeferenced to the surface of the earth. The USGS has produced DRGs of the 1:24,000-, 1:24,000/1:25,000-, 1:63,360-(Alaska), 1:100,000-, and 1:250,000-scale topographic map series. These maps make excellent starting points for GIS projects, and they often contain many features that can be extracted for use, such as contour lines and building footprints. Their release as GeoPDF files means that some basic GIS functionality is available just in viewing the files. With these projects now complete, the major mapping agencies, led by the USGS, are supporting the National Map, a tiled coverage built from both contributed and existing data and imagery, served seamlessly. The National Map viewer has been covered in earlier chapters, and is adding both higher resolution imagery and new imagery sources. Unlike the commercial geobrowsers that show proprietary data, National Map information can be downloaded for free and imported directly into a GIS. Figure 11.6 shows a DOQ imported directly as a GeoTIFF file into MapWindow GIS..

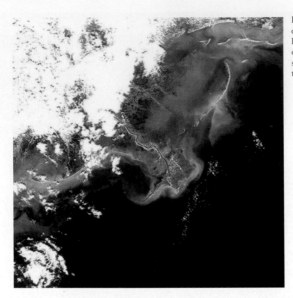

Figure 11.7: Image of the Mississippi Delta acquired on February 24, 2000, by the Moderate Resolution Imaging Spectroradiometer (MODIS) on the *EOS-Terra* spacecraft. Source: MODIS Instrument Team, NASA Goddard Space Flight Center.

11.2.4 GIS and Remote Sensing

GIS is increasingly a tool for continental and global scale analysis and display. A major source of map data is that coming from aircraft and spacecraft in the form of remote sensing data. New spacecraft with the next generation of space instruments will provide an extremely rich set of both new and existing forms of data. Among the new programs are NASA's Earth Observation System (EOS), consisting of a huge variety of new instruments for mapping that will continue the NOAA polar orbiting programs and Landsat type data flows. NASA's *Terra* satellite, launched in 1999, and including ASTER and MODIS (Figure 11.7), has already established the flow of Earth Science Enterprise data into the NASA databases. The ASTER-based Global Digital Elevation Model is an example of derived data, but the amount of imagery is extreme. *IKONOS* and other commercial satellites such as DigitalGlobe's *Quickbird* return high resolution data at about a 1-meter ground resolution and better. Several new commercial satellites, such as a new generation of French SPOT satellites, ensure that the diversity of instruments will increase.

Similarly, the shuttle-carried radar mapping capabilities of SIR (shuttle imaging radar), as well as the Canadian RADARSAT-2 and the European Space agency's Envisat, all provide nighttime and weather-invariant terrain mapping capabilities. The Shuttle Radar Topographic Mapping mission of Spring 2000 returned highly detailed topographic data and radar images for much of the world. This was the source of the SRTM data featured in earlier chapters.

Finally, the release of previously top-secret government spy satellite data, from the CORONA, LANYARD, and ARGON programs during the 1960s and 1970s, has allowed a significant amount of historical high-resolution imagery, much of it covering the United States, to be used for new mapping purposes (Figure 11.8). Evident after the release from the "black" world of intelligence is the fact that this program and its successors have contributed significantly to the U.S. national mapping program, perhaps implying a higher degree of fidelity in these data than might have been imagined. As a historical record, these

Figure 11.8: A GIS view backwards in time. KH-4B camera image from the CORONA spy satellite of Moscow, taken in 1970. Source: National Reconnaissance Office.

data are often able to show the "before" image necessary to understand the "after" of the present-day information. It can only be hoped that more of this historical imagery can be made available as its original purpose fades in time, allowing snapshots of land use change at different time periods in the last decades.

Clearly, remotely sensed data are highly structured around the raster data format. As much more data become available in this format, the demand upon intelligent software for correcting brightness differences and boundaries that come from pixel-based images will increase. If this software becomes powerful and inexpensive, the possibility of having it work directly on the spacecraft data stream becomes attractive. Over the last few years, private companies that generate and supply data from satellites have increasingly post-processed the data to automatically georectify, correct, and even mosaic the data. As a result, GIS based coverages of imagery can be acquired by geographical units, such as counties, rather than by coverage tile, a vast improvement in usability.

11.2.5   GIS and Location-based Services

Location-based services (LBS) are computer-based services that exploit information about where a user is located in geographic space. Location-based services take advantage of GPS, but also may rely on E911, an initiative of the Federal Communications Commission that requires wireless telephone carriers to pinpoint a caller's telephone number to emergency dispatchers. This can use the location of the telephone itself with respect to the nearest cellular transmitters, solved by signal triangulation. E911 is the most widely used location-based service in the U.S., although manufacturers of cellular telephones now incorporate GPS chips into them. LBS already accounts for billions of dollars in revenue and is growing worldwide. The power of LBS means that the Internet can also be made location-oriented. For example, Internet search engines such as Google Maps or MapQuest return "hits" based on the distance that the feature is from the user's current location. Many such Web-based services exist, often using Google Maps for Mobile and attached to cellular phone services, for example the iPhone (Figure 11.9).

Users of LBS so far seem to be either vehicle-based, where the GPS and computer are resident in the car and are used to query geographically ordered information, or mobile. Mobile users are usually either working on a personal digital assistant that contains a cellular phone connection to the Internet and a GPS card, or they are using an Internet-enabled cellular telephone. Early uses of the systems have included automotive roadside assistance, emergency and collision notification, stolen vehicle tracking, on-demand navigation assistance, traffic alerts, and vehicle diagnostics. For example, On-Star LBS is available in many models of cars, and the TeleAid system is standard on all Mercedes cars sold in the United States. Other systems include tracking devices for children and house-arrest prisoners, pet location devices, and others.

In essence, LBS uses selected subsets of GIS functionality, but delivers them to the user on demand. Most future applications will be in navigation, route finding, and space-constrained search. For example, one could search for nearby French restaurants in Dallas, Texas (Figure 11.10). One unresolved issue with LBS is how "open" the geographic information will be. For example, consumer goods can be made to report back their location and condition to their manufacturers or sellers. While this has some positive implications for avoiding breakdowns, the privacy issues involved are obvious. Sales calls

Figure 11.9: Google Maps running on Apple's iPhone. Photo by Matt Clarke-Lauer.

Figure 11.10: A spatially-constrained search. Google Maps search results for French restaurants in Dallas, Texas. (Use follows Google Guidelines at: http://www.google.com/permissions/geoguidelines.html).
Source: © 2009 Google, Map Data © 2009 Tele Atlas.

to cellular telephone owners who walk close to a particular store or restaurant may be informative, but are likely to be very intrusive. Similarly, cellular calls could be used to locate people in time and space. This may be relevant for criminals or terrorists, but if the information were broadly available some significant abuses would be possible. Nevertheless, interest in LBS by some very mainstream businesses is high, and we are likely to see them in far more common use in the near future.

11.3 FUTURE COMPUTING

11.3.1 Geocomputation

Geocomputation has been defined as the art and science of solving complex spatial problems with computers. GIS was in its early years constrained by the limitations of early computers. Several factors have freed geographical problem-solving in this regard. First, ordinary PCs and workstations are now very much more cheap and capable, disk storage is also cheap, and processors fast. Secondly, software has become far more interoperable, and open source solutions have opened up GIS to those without the ability to buy into expensive software suites. Thirdly, after many years at the extremes, high performance computing, by the so called supercomputers, has migrated slowly downward, to the extent that many desktop PCs now have multiple CPUs. This new power means that the old restrictions on spatial analysis, limited size samples, coarse resolutions, aggregated regions, and overly small areas of study have given way to analysis using entire populations, high resolutions, and disaggregated data.

With these new capabilities, the methods of computer science, such as advanced programming, parallelization, information-theoretic approaches, and formal theory, have found their way into GIS. For example, the Association for Computing Machinery (ACM) now has an annual conference and special interest group in GIS (www.sigspatial.org). GIS can learn a great deal from advanced computation, high performance computing, large scale simulation and modeling, scientific visualization, visual analytics, data mining, image processing, machine vision, distributed computing, cyberinfrastructure, and formal theory, for example on spatial ontologies.

Geocomputation faces new intellectual challenges with powerful implications for GIS. Some problems are methodological. Gahegan (2000) has set forth the following list of issues that Geocomputation needs to solve:

1. the inclusion of geographical "domain knowledge" within the tools to improve performance and reliability
2. the design of suitable geographic operators for data mining and knowledge discovery
3. the development of robust clustering algorithms able to operate across a range of spatio-temporal scales
4. obtaining computability on geographical analysis problems too complex for current hardware and software
5. visualization and virtual reality paradigms that support a visual approach to exploring, understanding and communicating geographical phenomena

Recent developments in computing have raised the idea of a computing cyberinfrastructure, "the new research environments that support advanced data acquisition, data storage, data management, data integration, data mining, data visualization and other computing and information processing services over the Internet." Cyberinfrastructure has also been called grid computing, in which the user connects via a network to a system containing data, programs, information, and means to communicate with other researchers. Another term used is "cloud computing," where it is assumed that the network can deliver every component of computing (data, computers, software, knowledge) out of the "ether." Key to this has been the migration of software toward open source, interoperable component-based systems; the use of standard protocols for communication and query such as the extensible markup language (XML) and the simple markup language (SML); and the movement of spatial data to portals that are ubiquitously available. Software solutions are moving toward "services," that is, instead of installing GIS software on a desktop computer, all of the functions of GIS can be engaged remotely, and the results visualized by the end user, who is now relieved of much of the detailed work necessary to get results. Research on exactly how this will happen is now under way, and the GIS analysis of the future may be very different from the examples given in Chapter 6. Cloud computing is already moving into practice, with many on-line software applications such as spreadsheets, becoming available. One positive outcome will be that more of an analyst's time will be spent on generating solutions to the problem itself, and less on the details of projections, datums, and buffers. Another advantage is that data and results can be saved on the server side, for possible use in a digital earth.

11.3.2 Geovisualization

A critical issue for the future of GIS is the degree to which the systems become integrated with those new parts of computer graphics and cartography most suitable for GIS applications. The entire field of scientific visualization is an example. Scientific visualization seeks to use the processing power of the human mind, coupled with the imaging and display capabilities of sophisticated computer graphics systems, to seek out empirical patterns and relationships visible in data but beyond the powers of detection using standard statistical and descriptive methods.

Key to the issue of visualization is the ability to model very large and complex data sets and to seek the inherent interrelationships by visual processing alone or with the assistance of standard empirical and modeling methods. Obviously, GIS is the provider of such data sets. GIS data are complex, and the use of maps to begin with already implies that a visual processing mechanism is being used. GIS should move toward full integration with the tools and techniques of scientific visualization and has much to gain by doing so. This would greatly enhance the analysis and modeling component of GIS use, and in a way that is inherently compatible with a GIS and the tools in the GIS toolbox.

Many GIS data are also inherently three-dimensional, such as atmospheric and ocean concentrations of chemicals, topography, or abstract statistical distributions such as crime rates and population densities over space. New software allows the user of a GIS not only to map and analyze three-dimensional distributions, but also to model and display them in new ways. Among the cartographic methods now familiar to GIS and to automated cartographic system users are simulated hillshading, illuminated contours, gridded perspective and realistic perspective views, and stepped statistical surfaces. On-line geobrowsers have made the exploration of 3D geospatial data commonplace.

Even simple maps, such as weather maps, now use sophisticated hypsometric coloring with interwoven hillshading. In addition, new types of display, such as stereo screens with shutters and head-mounted displays, along with the new types of three-dimensional input devices, gloves, trackballs, and three-dimensional digitizers, have expanded the suite of interaction means for the GIS user remarkably. Many people who deal with image registration and digitizing work with anaglyphic (red and green) stereo and use soft-copy or computer screen photogrammetry to take measurements. Animation has added another dimension to display and is now commonplace. What was once highly innovative is now commonplace during the weather forecast of the evening television news. Usually, weather satellite data such as GOES is animated and the perspective changed to simulate a flyby. These data can be viewed within geobrowsers, such as Google Earth (Figure 11.11).

The possibilities of animated and interactive cartography, the sort we now see as interactive kiosk-type displays at hotels, airports, and supermarkets, are remarkable, and will strongly influence the future of GIS, especially as the computing power and tools necessary for animation become cheaper and more widespread. Animation has a particular role to play in showing time sequences in GIS applications. Just as it is hard to see exactly what happened during a particular play in a sports contest without slow-motion viewing of

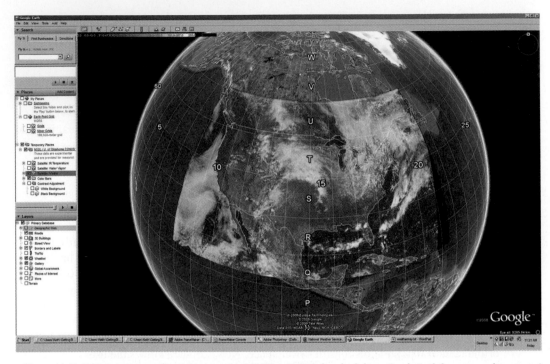

Figure 11.11: NOAA weather satellite visible image for 7/3/09 at 19:00Z, downloaded into Google Earth from www.srh.noaa.gov/gis/kml.

Source: © 2009 Google, Map Data © 2009 Tele Atlas.

film or video, so GIS users can compress long time sequences or view short time sequences to reveal geographic patterns that were not visible in other ways.

11.3.3   Object-Oriented Computing

Another major development in the software world has been languages, and now databases, that support "objects," called object-oriented systems. Geographic features map very closely onto objects. Object-oriented programming systems (OOPSs) allow the definition of standard "classes" that contain all the properties of an object. As a simple example, an object class could be a point containing the latitude and longitude of the point, a feature code for the point such as "Radar Beacon," and any necessary text describing the object. If we wish to create another point feature, this can be done simply by cloning the original with all its class information, a process called inheritance. This has given GIS terms like feature class and geodatabase.

In addition, we can encode the fact that points often have data conversion or analysis constraints. For example, the centroid of a set of points is itself a point and can inherit a point's properties. This approach has allowed the development of entire GIS packages, and is seen as a way of building far more intelligent GIS systems in the future. While the OOPS is not the tool for all GIS operations or systems, it is indeed a powerful way of modeling data and will influence the future of GIS software significantly. The first GIS based entirely on the object model was the Smallworld GIS, now a brand name within General Electric.

Probably the most important aspect of object-oriented GIS is that the objects, once designed, can be used as independent components or building blocks for larger systems. Much of the open source GIS software developed in recent years, and many of the web-oriented server systems owe their success to this property of OOPs, as embedded in the Java and C++ programming languages.

11.3.4   User-Centered Computing

The computer era has seen radical changes in the very nature of both the computer and GIS user interfaces. Early systems used only the screen and the keyboard to communicate to the user. Systems now have these same functions, but also a mouse, pointing devices such as a trackball or light pen, multiple windows on the screen, sound, animation, and many other options.

Most significant has been the rise of the WIMP (windows, icons, menus, and pointers) interface. Windows are multiple simultaneous screens on a single display, usually serving different tasks and fully under user control. When inactive, windows can be closed and kept visible as icons, or icons can be attached to tasks and used to activate them. Menus can take a variety of forms. Many user interfaces place a set of menus along a bar at the top of the screen, controlling more and more specific tasks as one goes from left to right. Menus are often "nested," that is, a selection reveals another menu level and that one even more selections. Menus can "pop up" from a space or window, or can be "pulled" from other menus or messages. Pointers are devices for communicating location on the screen and in windows, and they most commonly take the form of a mouse or a trackball.

Central to recent GUIs has been a metaphor. The metaphor most commonly used has been the desktop; that is, the screen of the computer is designed to resemble the top of a desk, and the icons and other elements are allowed to rest on it, awaiting use. Some operating systems have gone beyond the constraints of this suite of interactions, and many operating systems now allow input from voice, touch screen, and direct input from GPS receivers and other recording devices, such as digital cameras and videocams.

The map itself is a useful metaphor, and a future GIS can easily be imagined in which the map and its elements, such as the scale and the legend, are used to manage and manipulate the data. This is already what a GIS does, but the user-interactive element would be a new addition to the system. Several systems already use icons as elements of a process or transformation model to track sequences of operations, for example ESRI's Model Builder for ArcGIS. A flow might consist of selecting an image, running a geometric correction routine, classifying the data, selecting a category, doing a map overlay, and then printing the result. The entire sequence of operations can then be copied, subsetted, or further manipulated just by treating the flow diagram as a graphic, with the diagram showing the data sets, GIS operations, and cartographic transformations they represent in both a virtual and a real sense. These models can be shared, but as yet they cannot be linked to advanced processing. For example, a raster routine suitable for parallel processing could check for the availability of multiple CPUs and reconfigure itself automatically to save time.

When GIS operations, even simple ones, are implemented on mobile devices such as GPS receivers, cell phones, or palm-top computers, much of the desktop metaphor becomes unsuitable. This has led to a new dimension in GIScience research, often called "cognitive engineering," concerned with GIS design and testing, using human subjects tests, and systems that work in practical settings, such as outdoor pedestrian navigation or in-vehicle routing. The success of this research will do much to lead GIS toward intuitive human-computer interfaces that work in mobile settings. Human subjects tests, in which studies are conducted on people while they are using systems, will play an increasing role in the future of GIS user interface design.

11.3.5 Open Source Computing

Open source computing has its roots in the Free Software Foundation and the UNIX-based user forums. Open source means that no-one "owns" software; instead any user can choose to contribute their expertise in improving the source code for the program, which can be downloaded and worked on, then contributed back for other users. GIS had an early pioneer in open source GIS in the GRASS GIS, which was written using the open-source approach as early as 1978 (Neteler and Mitasova, 2008). The number of modules and components contributed to the base program by the Army Corps of Engineers is huge, reflecting the variety of applications where a "fix" to one problem could be shared with a whole community. With the advent of entire open source alternatives to all aspects of computing (e.g., Linux, OpenOffice.org), it was natural that GIS would also be replaced by open systems. A major contributor to this trend has been the Open Geospatial Consortium (www.opengeospatial.org), who pioneered and helped built by consensus the standards and basic toolboxes for open source GIS.

Many of the open source GIS programs were listed in Chapter 9. Some are the equivalents of desktop GIS, built in Java or similar environments (e.g., Quantum GIS, MapWindow GIS). Some are toolsets built on standards, and so allow simpler development of additional GIS tools. For example, reading ESRI shape files is supported by a set of C language programming libraries called ShapeLib (shapelib.maptools.org). This is just one small part of a set of programing libraries called MapTools (www.maptools.org) that support everything from drawing graphics to reading file formats and building web-accessible data bases. Still other open source tools allow the construction and use of GIS Web servers, so that data can not only be automatically available to a desktop GIS across the Internet, but users' results and maps can then be displayed on the server too. This is the equivalent of commercial products such as ArcGIS Server, which normally require a systems-level programmer. MapServer (mapserver.org), by the University of Minnesota, is an example, as also is PostGIS (postgis.refractions.net), built on the Postgres database system. These entire GIS systems are largely device-independent; that is, they can run on any operating system and on almost any computer. Hardware independence is a very important part of the Open Source philosophy.

Another component of open source computing has been open source geographical data. In the U.S., federal government data has been considered public domain, the equivalent of open source. This approach has been emulated by most states and an increasing number of local authorities such as counties and tribal groups. One attempt to provide an open portal to this open data is the National Map viewer and seamless server. Much data, however, remains proprietary. The terms of use for images taken from Google Maps and Google Earth are good examples of the constraints that proprietary ownership of geographic data places on geospatial data. An alternative approach is becoming important, that of "crowd sourcing," as represented by the popular Wikipedia (Sui, 2008). This has been called both user-contributed geographic information and VGI, or Volun-

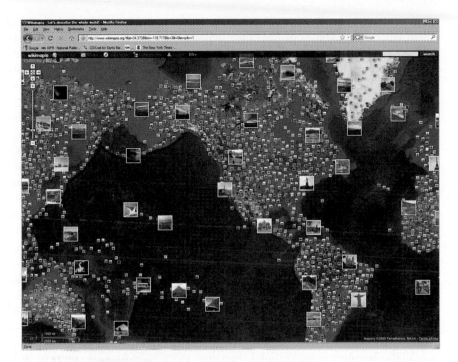

Figure 11.12: User contributed Pictoramia photographs displayed on Wikimapia.

teered Geographic Information (Goodchild, 2007). An excellent example is the open source world street map at openstreetmap.org. Individuals contribute their GPS tracks to a communal map, built by conflating all the detailed information into a whole. This is similar to users contributing their photographs to Pictoramia and Flickr (often geographically tagged), or their video to YouTube.

User contributed geographic information can be detailed, accurate, and very much demand-related. However, such systems are open to abuse and can need moderating or control. There is no control over theme, and users can contribute vastly varying information across many media. When focused, for example, on ornithology during the annual Christmas bird count, some very useful and timely data can be collected quickly. This trend will obviously continue into the future, and expand to new map layers, for example the 10.4 million photographs on Pictoramia (www.wikimapia.org; Figure 11.12).

11.3.6   GIS as a Virtual Organization

GIS users have broadened into two types: the large organization-wide projects with huge databases and often specific missions and the small, usually one-person operations run by a jack-of-all-trades. Although GIS can serve both sets of users, the specifics of hardware, software, and the computing environment mean that different GISs suit each world.

At the organizational level, labor can be divided. One staff member can take care of data maintenance and software updates, one education and training, one data analysis, and so on. In the one-person shop, all of these tasks are the responsibility of one individual, often a GIS pioneer who is the key person in championing the use of GIS to begin with and who is also the computer expert, systems administrator, hardware engineer, and coffee-maker. Small users will probably not be able to add significant amounts of new data, with the exception of field data collection with GPS. They will be more reliant on public-domain data, and the data will probably be less up-to-date and at a coarser scale.

It is at this level that the GIS use is closest to the domain expertise. Getting the GIS as far into the field as possible is often a key to the success of a system. Field operatives can use the GIS quickly to make ordinary but informed decisions about the use of resources on a day-to-day basis, where the payoff is greatest. Sophisticated analysis may not even be necessary at this level, and using the GIS as a graphic inventory and map production system is more than sufficient for success. An example of such a GIS group is a community support organization's GIS, often termed a "public participation GIS."

Large systems, by contrast, can maintain up-to-date and detailed information, and can use it in its full GIS context, performing the roles of inventory, analysis, decision making, and management. Here, also, better information means better use of resources. Clearly, the GIS industry must continue to exploit both types of environment. Often, this means taking large systems and packaging them small, or taking lessons learned by advanced users and translating them for the general user.

Finally, the GIS users themselves have become a sort of self-help facility. Most major software packages or regional-interest organizations using GIS have user groups, often with special conferences, workshops, newsletters, and Internet discussion groups. Recent years have supplemented these fora with blogs and newsfeeds. This is an excellent grass-roots level for GIS to flourish, one that GIS vendors have discovered. As GIS packages become more complex but also more user-friendly, these user groups will converge on some common principles for GIS use. These principles should be, and are, shared with all users. Often, a good idea in one software environment can lead to productive duplication in another.

11.3.7   Future GIS

Finally, what about the speculative future? Some of the trends on the edges of computer science and engineering have real prospects for GIS application. Among these are stereo and head-mounted displays; input and output devices that are worn; parallel and self-maintaining fault-tolerant computers and above all, mass storage and computing power much greater and faster than those available today. A vision of a future GIS system might be a pocket-held integrated GIS, GPS, and image-processing computer capable of real-time mapping on a display worn as a pair of stereo sunglasses. An example is the vision of Anywhere Augmentation (Wither et al., 2006). Data capture would consist of walking around and looking at objects, and speaking their names and attributes into an expert-system-based interpreter that extracts features from imagery, encodes and structures the data, and transmits them immediately to a central network-accessible storage location (Figure 11.13). This implies that a single person, or even an unmanned vehicle or pilotless aircraft, could move around gathering data while any interested person displays and analyzes the information in real time in his or her office or home. Perhaps a nationwide set of mobile data collectors could roam the countryside, constantly field-checking and updating the digital maps being used by virtually every automated system, from power supply to emergency vehicles to the U.S. postal system. Rather than being the "big brother" of fiction, universal and open access to the information generated would ensure

Figure 11.13: Anywhere Augmentation. Color panorama (top) and semi-automatically generated depth map (below). Darker regions of the depth map are closer to the user. Images generated by a user simply looking around the scene wearing display AR glasses and a laser ranging device. Source: Wither et al., 2008.

that the public good is being served. Tangible examples are the user contributed map information supporting wildfire emergencies (Pultar et al, 2009).

Another future prospect is that of the data analyst becoming a data explorer, delving into three-dimensional realistic visualizations of the data, seeking out patterns and structure instead of the use of the simple statistical analysis of today. The human mind is capable of some amazing parallel processing of its own and can easily seek out structures that computers and even some scientists often miss. Similarly, the same systems could manage the very systems they support, perhaps allowing for integrated modeling and prediction of future "what-if" scenarios.

11.4   FUTURE ISSUES AND PROBLEMS

This chapter, and indeed the entire book, will conclude with the issues and problems we are likely to face with the future of GIS. How well we as a user community react to the challenges of the issues will play a major role in the future of GIS. As a person now introduced to the possibilities, it is you, the reader, who will have to deal with these issues at a practical level.

11.4.1   Geoprivacy

An issue raised again and again as GIS databases become more and more widespread is that of personal privacy. We very often take our right to privacy for granted, yet all the time, by the use of telephones, credit cards, mail order, and the like, we are constantly revealing to other people what can be personal property. Facts we consider of the greatest privacy—our personal income, information about the family, our health record, and employment history—are all tucked away in somebody's database. GIS offers the integration of these data through their common geography. Although it is to the public benefit, for example, to build a link between environmental pollution and health, the more local and individual the link, the more the issue of personal privacy arises. Even the federal census, with its highly general information about groups of individuals, has strict restrictions on availability of information that can identify specific people, holding such data private for over 70 years before releasing it.

Whole sectors of the economy now rely upon linking data from individuals, such as magazine subscriptions and purchases over the Internet, with demographic and other information by district, such as census tract or ZIP code. A personal credit history can be amazingly revealing about an individual, and data are often bought and sold as a side benefit of computerized ordering and mailing systems. Just assembling every item of information about an individual, once an extremely difficult task, is now considerably easier. An entire field of business GIS, with its own specialist data and software, now examines these issues, named Geodemographics. This field uses small local area census and other marketing data to precisely characterize residents of neighborhoods, so that they can be selected for marketing of particular good and services—baby products for new parents, for example.

Who draws the lines? GIS is a successful technology for spying on people (Monmonier, 2002). A whole new area of sub-interest in GIS is in GIS and the law. As GIS becomes used in lawsuits, in lobbying, voting district delineation, and, as always, in mapping of property, the legal profession will come increasingly to use GIS as a tool, and then by extension to challenge the means by which data are collected and transformed, analyses are conducted, and conclusions are drawn. This is becoming more and more common, as people take developments such as Zillow.com (real estate) and Google Street View (pictures of homes) as threats to privacy. In one case, villagers in the UK actually turned away a Google Street View camera car to keep views of their neighborhood off of the map. On the other hand, people volunteer an extraordinary level of personal information on social web 2.0 sites such as Facebook and MySpace.

This will force GIS analysts to become somewhat more explicit in their methods and more accountable in their operations. GIS software, for example, should keep a log of the functions used, commands given, menu choices selected, and somehow attach this "data lineage log" to the data sets themselves, so that GIS results are more accountable. It is well known that regular statistics can be used to support many viewpoints, and even maps can be manipulated to show different points of view (Monmonier,1996). GIS offers the mapping and analysis processes full accountability, and this must be stressed in the future if GIS is not to become yet another courtroom gimmick as far as the law is concerned.

11.4.2 Data Ownership

There are two largely incompatible philosophies about GIS data ownership. At the one extreme, the federal government produces and distributes digital data in common formats at the marginal cost of distribution, the "cost of fulfilling user requests." This means that the cost of producing the data should not enter into the pricing of the data. The logic here is that, because the federal government has already created the data at the public's expense, it cannot charge a second time for data to the same people when they request copies for their own use. The Internet and map Web servers have made the dissemination cost for the geospatial user and the producer effectively zero, so that data are usually available for setting up and using a GIS free or at least for only a very modest price.

At the opposite extreme lie the groups (and nations) who believe that GIS data are a commodity, a product to be protected by copyright and patent and sold only at a profit. The argument for this view is that when the market demands a data set, the profit motive will generate the data, and the profit will draw in competitive data producers, who will eventually drive down the cost. In a few cases, this has happened, but rarely is the profit motive capable of generating data in complete, systematic, and standardized coverages that are regularly maintained. There is a great deal of motive to produce a data set that may sell many times, but little motive to map a corner of the country with little demand and poor existing digital maps. Extended to the international context, neither will there be a motive to map for GIS the poorest and most needy nations, especially in Africa and South America.

Most nations have evolved some combination of these two approaches. The United States uses the federal government data, especially the TIGER files, as a base, but adds other, more detailed, and up-to-date information by geocoding new maps or buying the data from private companies. The companies sell their data based on its timeliness, accuracy, completeness, and so on, but originally most of the data derives from one or another of the free federal data sets.

This two-way relationship between government and business has generally served GIS well, although it should be clearly noted that without the free federal data, the entire system collapses. The private mechanism can rarely produce data for every small planning office and project, and rarely can the small office afford the high cost of such data. As always, most people will continue to work with the "least cost solution." For GIS, this usually means a microcomputer, inexpensive or public-domain software, and free federal data.

11.4.3   Spatio-temporal Dynamics and GIS

For some time, scholars in GIS have been interested in the impact on GIS of supporting geographic and attribute data from many time periods. Obviously, the digital map in a GIS is "time stamped" at the time the data were created. In the real world, however, data become out of date and must be revised, or new data sets are released to replace the old.

Some data have very short duration—weather forecasts or shipping notices, for example—and revision and update quickly become a major part of the GIS maintenance. In most cases, GIS data are simply given an additional attribute of the date the data were created, even though often the date of the data and the date of entry into the GIS are not the same. The implications on the design of the GIS to facilitate use, automatic update, for instance, or automatic selection of the most up-to-date version of every feature are now being integrated into the GIS's functions.

GISs are current only weakly suited to handling data from multiple time periods. One approach to dealing with time is to add a timestamp as a feature attribute. For example, a land use database could label each land use polygon with the date it changed uses. Alternatively, the GIS could use multiple "snapshot" coverages for different times. For example, land use maps may be available for 2008, 2001, 1990, 1975, and 1940. Note that using these snapshots to analyze change is problematic, since the data are not a regular sample in time. So adding time to GIS has been seen as a data structure challenge, and many models have been suggested to deal with this problem.

However, a focus on the data misses an important fact about spatial dynamics. Return to the tornadoes example in Chapter 6. While from the static analysis point of view, one tornado is much the same as another, varying only by time and place or occurrence, yet there are broad weather and climate conditions associated with tornadoes. Given a developing weather pattern, it is possible to predict that a tornado will develop, and that many tornadoes will be linked to one weather system. Rather than a static map with tornadoes simply happening spatially, a developing weather system with a beginning, middle, and end moves as a single spatial pattern across the landscape, with individual tornadoes literally spinning off. The spatial development of the pattern as a whole and the ability to match entire systems over time are important in weather analysis and forecasting, not to mention for warning systems. For example, it is desirable to compare the developing weather conditions with the development patterns of all previous similar weather systems. Only a few studies in GIS have used this approach (e.g., Macintosh and Yuan, 2005).

11.4.4 Changing Roles for GIS

As GIS moves into the future, changes are inevitable, for GIS is a science and a technology based on change. Nevertheless, there are broad movements within science toward topics or challenges that are national or international areas of new emphasis. A few trends are already obvious; fortunately, GIS has a role to play in each of them. In addition, the imperatives of our times—war, recession, unemployment, food prices, alternative energy, global warming, universal health care, emerging epidemics, terrorism—all need the tools and methods of spatial analysis if progress is to be made.

Science has become increasingly focused on issues of global importance. Viewing the earth as a whole system is now a valid way to approach issues of global climate change such as global warming and the ozone hole; global circulation, such as the patterns and flows within the earth's oceans and atmosphere; and the global scale of the impact of people on the environment. The new global nature of the world economy, the increasingly strong efforts to solve the world's problems with global legislative bodies, such as the World Bank and the United Nations, and the coming into being of methods and tools for approaching these problems with hard data have all led toward a new global science.

GIS has an immense amount to offer this global science. Global distributions need mapping, global mapping needs map projections, and the understanding of flows and circulations are based on an understanding of spatial processes. Many global data collection efforts for GIS are now under way, and organizations use GIS to attack global problems such as crop-yield estimation and famine prediction.

Moreover, GIS has also been at the forefront of a new approach to science. More and more the traditional boundaries between disciplines in the sciences and the social sciences have disappeared, although there are many who fail to recognize it and even resist this trend. Most major research is now conducted by teams, with representatives from a host of different but interrelated sciences working together on a problem. GIS is a natural tool for this sort of work environment because it is able to integrate data from a variety of contexts and sources and seek out interrelationships based on geography, the mapping of distributions, and visualization. Just as no scientist is literate without calculus, matrix algebra, and statistics, the methods and principles of GIS are likely to become essential tools in the scientist's toolbox, at least as an integral part of one's educational background, well into the future. This new understanding has been termed spatial literacy. Spatial thinking, the reasoning behind spatial analysis, will become part of everyone's basic education. With this book, you too have gotten started with GIS, and by now, if you have followed the assignments, questions, and projects, you are already on your way toward getting a jump-start on this new scientific approach.

11.5 CONCLUSION

As a new GIS user, your first experience was probably in the "push-button" category. You carried out a program of analysis or processing already prescribed by your instructor, perhaps using the laboratory manual for this book. It is hoped that this book has helped you gain the substantial knowledge needed to understand more deeply what you are doing. A second group of those experiencing GIS for the first time are those who have been hired as a GIS expert with perhaps only a little experience, and are then thrust deep into a new project with only vendors' manuals or on-line Web pages for help. In this case, it is hardest of all to see the forest for the trees, especially when deadlines loom. Here too this text can help, by adding supplementary background to the often minimalist explanations and references in manuals. In either case, you are well on your way to mastering the basics of geographic information science, those principles necessary to build a GIS expert. Nevertheless, you will probably spend many hours perplexed and frustrated at first. If so, do not hesitate to fall back on the sources of help covered in this book. At the very least, the fact that you are not the only person in the world facing a seemingly impossible GIS problem is reassuring.

If you have reached this page, either working alone or with a class, and have grasped the concepts behind GIS, there are two paths forward. First and foremost, as stressed in Chapter 10, there is no substitute for grappling with the problems and issues raised in this book. Many GIS packages are inexpensive, are shareware or freeware, and may even be available at your public library or school. Many agencies seek out GIS interns, and volunteers with GIS skills are always well received. Dive in headfirst, and use the knowledge that *Getting Started with GIS* has given you to master the technology. If you have followed along with the assignments in this book or have used the parallel manual, you are already most of the way to being a GIS specialist.

At this point you are ready for the next step. The title of this book was chosen deliberately. It is a first guidebook, a tutorial to get you started in GIS with the necessary background information to avoid major mistakes. Quite simply put, there is much more. Most of the next steps you should take were dealt with in Chapter 1. From now on, you must be your own guide. In preparation for this step, take a look at the code of ethics for GIS professionals prepared by the Urban and Regional Information Systems Association at http://www.urisa.org/about/ethics. As a professional entrusted with spatial information, you will be confronted with both positive and negative uses of this powerful technology, and indeed, you must make your own choices. Please do so wisely.

As you move onward with GIS, or even if you have used this book to find your way out of a one-time-only GIS problem, always bear in mind the power that GIS can bring to bear. Many are the problems and ills of society and this world that GIS can help with. Most of all, GIS is a tool for a sustainable human future because it offers the promise of managing what resources we already possess with the highest level of efficiency. Indeed, GIS can help us live better with the resources we have rather than always demanding more.

Finally, GIS offers the promise of reducing waste, improving living standards, eliminating disease, helping to manage risk and disasters, understanding global change, and even advancing the principles of democracy. In the new era of the Information Age, it is you, the intelligent GIS user and analyst, who can use this tool productively, or perhaps just as easily waste its immense capabilities. To paraphrase Sir Arthur Conan Doyle, you know my methods—now go out and apply them!

11.6 STUDY GUIDE

11.6.1   Quick Study

CHAPTER 11: THE FUTURE OF GIS

O *GIS has become a multibillion dollar industry in one human career* O *Methods from GIS are being used in parallel disciplines* O *Data availability is now rarely a constraint in GIS projects* O *Sensor webs can supply huge amounts of geospatial data in real time, and will become more commonplace* O *Many web portals provide access to private and public geospatial information* O *Digital earth is a vision for a future data repository of all human knowledge about the earth* O *Geobrowsers provide a visual means to access spatial data and information* O *GIS gets much of its field data from GPS and provides GIS data for GPS navigation systems* O *GPS is one case of GNSS, with more systems now planned* O *Scanned maps and air photos are an increasingly important source of base map data for GIS* O *Satellite remote sensing systems can provide global level data for global GIS* O *Location-based services are a new industry using GIS and GPS for mobile applications* O *Geobrowsers can do spatially-constrained Web searches* O *Geocomputation is increasingly applied to GIS problems* O *Next generation computing will use cloud computing and the grid* O *Geovisualization holds potential for better exploration and analysis of spatial data* O *The object-oriented computing paradigm has been very important in GIS, especially open source GIS* O *User centered computing and cognitive engineering are improving the human-computer interface and making GIS easier to use, especially on mobile devices* O *The traditional desktop metaphor for interaction is not useful for small mobile computer devices* O *Open source GIS has become a major source of GIS software and facilitates interoperability and Web services* O *User contributed or volunteered geographic information as contributed to Web services is a new source of geospatial data and will become more valuable in the future* O *GIS user communities can be considered virtual organizations, and have new web-based support mechanisms* O *Future GIS data collection could come from autonomous vehicles or humans equipped with sensors* O *Personal privacy is threatened because GIS can integrate data about individuals using space as the unifying theme* O *Data can be proprietary or public domain, and are often both with the profit motive supporting timeliness* O *GIS is not yet well equipped to map, model, and predict phenomena that are dynamic in time and space, such as storm systems* O *GIS is increasingly able to study global scale phenomena and change* O *There are few major social or economic problems that cannot be approached using GIS* O *GIS suits the interdisciplinary approaches of twenty-first century science* O *Spatial literacy and thinking will become more important in education and society* O *GIS practitioners should consult the URISA code of ethics* O *The reader is now equipped to go forth and use GIS; please do so with care* O

11.6.2 Study Questions and Activities

Future Data

1. In studies from the 1980s, data digitizing and assembly costs were thought to account for anywhere from 60% to 80% of the total costs of setting up a GIS. Why will this change in the future? What new data sources are available now that were not available then? What data distribution systems might be used in data assembly now?

2. How do GPS data find their way into a GIS? What use would a large number of GPS differentially corrected points be to a GIS project?

3. What function might a geometrically correct air photo play in the development of a GIS project on environmental remediation? Under what circumstances might satellite remote sensing data be preferable to an air photo?

4. Investigate for your GIS, or for a GIS package about which you can find data, what the GIS license agreement says about redistribution of the GIS software and what rights you have as the creator of a GIS data set using the software.

5. Download from the Internet both a satellite image or an air photo for a district and a vector map coverage. Overlay the two digital maps, and examine carefully the discrepancies between features on the two data sources. What sources of error account for the differences? How might this and other layers be rectified so that they can be used together in overlay analysis?

Future Computing

6. Draw a generic diagram for a new user of any GIS package with which you are familiar so as to instruct that user on the characteristics of the user interface of the GIS. Suggest three ways that the user interface could be improved. How might you objectively test how effective the user interface is?

7. How has GIS interoperability assisted the GIS professional's day-to-day work?

8. How big would a GIS need to be to be used effectively to hold data for the whole world at the resolution of MODIS instruments? What layers might a small "global awareness" GIS need to serve as an effective GIS learning tool in high schools? What is an appropriate resolution for such a project?

Future Issues and Problems

9. How might GIS be used to impinge on the individual's right to privacy? What GIS applications are closest to these scenarios today?

10. What mechanisms of enforcement of ownership are available for GIS packages, and how might these make using a GIS more difficult?

11. Examine the URISA GIS code of ethics. Make a set of narrative examples that show actions on the right and wrong sides of the code.

11.7 REFERENCES

Dykes, J., MacEachren, A. M., and Kraak, M-J. (2005) *Exploring Geovisualization.* International Cartographic Association: Elsevier.

Gahegan, M. (2000) What is GeoComputation? A history and outline. http://www.geocomputation.org/what.html.

Goldsberry, K. (2007) *Real-Time Traffic Maps for the Internet.* Ph.D. Dissertation, Department of Geography, University of California, Santa Barbara.

Goodchild, M. (2007) "Citizens as Sensors: the World of Volunteered Geography." *Geo-Journal,* vol. 69, pp. 211–221

Grossner, K. E., Goodchild, M. F., and Clarke, K. C. (2008) Defining a digital earth system. *Transactions in GIS.* vol. 12, no. 1, pp. 145–160.

McIntosh, J. and Yuan, M. (2005) Assessing similarity of geographic processes and events. *Transactions in GIS.* vol. 9, no. 2, pp. 223–245.

Monmonier, M. (1996) *How to Lie with Maps* (2ed.) Chicago: University of Chicago Press.

Monmonier, M. (2002) *Spying with Maps: Surveillance Technologies and the Future of Privacy.* Chicago: University of Chicago Press.

Neteler, M. and Mitasova, H. (2008) *Open Source GIS: A GRASS GIS Approach.* 3 ed. International Series in Engineering and Computer Science: Volume 773. New York: Springer.

Pultar, E., Raubal, M. Cova, T., and Goodchild, M. (2009) Dynamic GIS Case Studies: Wildfire Evacuation and Volunteered Geographic Information. *Transactions in GIS.* vol. 13, supp. 1, 85–104.

Sui, D. Z. (2008) "The wikification of GIS and its consequences: Or Angelina Jolie's new tattoo and the future of GIS." *Computers, Environment, and Urban Systems,* vol. 32, no. 1, pp 1–5.

Wither, J., DiVerdi, S. and Höllerer, T. (2006) "Using Aerial Photographs for Improved Mobile AR Annotation", *Proceedings, International Symposium on Mixed and Augmented Reality.* Santa Barbara, CA, Oct. 22–25.

Wither, J., Coffin, C., Ventura, J., and Höllerer, T (2008) "Fast Annotation and Modeling with a Single-Point Laser Range Finder", *Proceedings, ACM/IEEE Symposium on Mixed and Augmented Reality,* Sept. 15–18.

11.8 KEY TERMS AND DEFINITIONS

ARGON: The camera code-named KH-5 carried by a series of reconnaissance satellites between February 1961 to August 1964 and used for reference mapping.

ASTER: (Advanced Spaceborne Thermal Emission and Reflection Radiometer) One of five remote sensory devices on board the *Terra* satellite collecting remotely sensed data since February 2000.

blog: A web-log, usually maintained by an individual with entries of comments, descriptions of events, or graphics or video, often listed in reverse-chronological order.

cloud computing: Computing with scalable and often virtual resources provided as services over the Internet.

cognitive engineering: An interdisciplinary approach to the development of principles, methods, tools, and techniques to guide the design of computerized systems intended to support human performance.

COMPASS: A planned Chinese GNSS of 35 satellites that will offer complete coverage of the globe, also known as Beidou-2.

CORONA: A U.S. reconnaissance satellite system operated by the CIA assisted by the U.S. Air Force, used for photographic surveillance of the Soviet Union, China, and other areas from August 1960 until May 1972.

crowd sourcing: Taking a task normally performed by an employee and outsourcing to an undefined group of people or community using an open call. The call could be to develop a new technology, design something, carry out a parallelized computation, or gather data.

cyberinfrastructure: Research environments that support advanced data acquisition, data storage, data management, data integration, data mining, data visualization, and other computing and information processing services over the Internet.

desktop metaphor: An interface design constituting a set of unifying concepts for human-computer interaction that treats the monitor of a computer as if it were a desktop, with folders and files on it.

device-independent: The ability of a software program to interact with different input, output and storage devices generically.

digital earth: A system providing access to what is known about the earth and its inhabitants' activities—currently and for any time in history—via responses to queries and exploratory tools such as virtual globes.

digital orthophotoquad (DOQ): A computer-generated image of an aerial photograph in which the image displacement caused by terrain relief and camera tilt has been removed combining the image characteristics of the photograph with the georeferenced qualities of a map.

disaggregated data: Data that have not been grouped into summary units, e.g., census tracts or counties, but kept as records of individual instances.

E911: Enhanced 9-1-1 service is a North American telecommunications based system that automatically associates a physical address with a telephone number and sends the call to the best Public Safety Answering Point. The caller's address and information is displayed to the call-taker immediately upon answering.

Envisat: (Environmental Satellite) is an Earth-observing satellite built by the European Space Agency launched in 2002 carrying nine instruments for remote sensing.

Free Software Foundation: A non-profit corporation founded in 1985 to support the free software movement, which promotes the distribution and modification of computer software without restriction.

freeware: Computer software that is available for use at no cost or for an optional fee.

future shock: A book written by sociologist Alvin Toffler in 1970 defining a state of "too much change in too short a period of time."

Galileo: A GNSS planned by the European Union and European Space Agency scheduled to be working after 2012, and expected to be compatible with the modernized GPS system.

geobrowser: Software for accessing information on the Internet by browsing content visually, typically on a map.

geocomputation: The art and science of solving complex spatial problems with computers.

geodemographics: The science of profiling people based on where they live.

geoprivacy: The degree to which the geographic location and activities of a person are shared or protected via computing or sensing devices and networks.

geovisualization: A set of tools and techniques supporting geospatial data analysis through the use of interactive visualization and mapping.

global navigation satellite system (GNSS): A generic term for satellite navigation systems that provide autonomous geo-spatial positioning with global coverage.

GLONASS: A radio-based GNSS, developed by the former Soviet Union and now operated for the Russian government by the Russian Space Forces.

Google Maps: A Web mapping service application and technology provided by Google, that powers many map-based services and maps embedded in websites via an open API.

Google Earth: A virtual globe, map and geographic information program originally created by Keyhole, Inc., and acquired by Google in 2004 that uses superimposition of images obtained from satellite imagery, aerial photography, and GIS.

GPS: A GNSS developed by the United States Department of Defense and managed by the United States Air Force 50th Space Wing the only fully functional current GNSS used extensively for navigation purposes.

grid computing: The application of several computers to a single problem at the same time.

head-mounted display: A display device, worn on the head or as part of a helmet, that has a small display optic in front of one or each eye.

Ikonos: A commercial earth observation satellite, the first to collect publicly available high-resolution imagery at 1- and 4-meter resolution.

in-vehicle navigation system: A vehicle-based system including a display and usually a GNSS receiver for applying GIS data and algorithms to assist navigation.

interoperable: The ability of GIS software and data systems to work together.

LANYARD: A short-lived series of reconnaissance satellites produced by the United States from March to July 1963 that carried the KH-6 camera developed for the Samos program.

Location-based Service (LBS): Information services accessible with mobile devices through a wireless or other network and exploiting information about the user's location.

MapTools: A suite of software tools intended as a resource for users and developers in the open source mapping community.

MODIS: (Moderate-resolution Imaging Spectroradiometer) A remote sensing instrument launched by NASA in 1999 on the *Terra* satellite, and in 2002 on the *Aqua* satellite.

National Map viewer: A Web portal and geobrowser intended to provide access to all public geospatial data for the United States, operated by the USGS.

National Map seamless server: USGS software tool for the extraction and downloading of public map data to a user's specifications.

newsfeed: A Web application that aggregates content such as news headlines, blogs, and podcasts, in a single location for easy access.

object-oriented system: A programming paradigm that uses data structures consisting of data fields and methods to design applications and computer software.

Open Geospatial Consortium: A non-profit, international, voluntary consensus standards organization that is leading the development of standards for geospatial and location-based services.

Personal Digital Assistant: A handheld computer, also known as a palmtop computer.

photogrammetry: The making of precise measurements from photographs, mostly to support mapping.

public domain: Intellectual property that is not owned or controlled by anyone, making it available for anyone to use for any purpose.

Quickbird: A high-resolution commercial earth observation satellite, owned by Digital-Globe and first launched in 2001.

RADARSAT-2: An Earth observation satellite launched in 2007 by the Canadian Space Agency.

remote sensing: The acquisition of information of an object or phenomenon, by measurements made without physical contact with the object (e.g., from an aircraft, spacecraft, satellite, buoy, or ship).

seamless: Geospatial data that have been processed to remove the effects of independent origins, matched across edges and boundaries, and given uniform characteristics.

sensor network: A wireless network of self-contained sensing devices to pool monitored information on physical or environmental conditions from spatially distributed locations.

shareware: Proprietary software provided to users without payment on a trial or demonstration basis, often with limited functionality, availability, or convenience.

SigAlert: Any unplanned event that causes the closing of one lane of traffic for 30 minutes or more. A website serving such data for Western U.S. cities (www.sigalert.com).

SIR: The Shuttle Imaging Radar, a C-band radar instrument used for terrestrial mapping from the Space Shuttle.

SML: Simple Markup Language, an approach to simplify XML to make interpretation easier for XML-consuming devices.

soft-copy: A term used to distinguish transient computer display of imagery as opposed to paper or film output.

source code: A set of statements or declarations written in a human-readable computer programming language to allow a programmer to communicate with a computer.

spatial literacy: The ability to use spatial thinking to communicate, reason, and solve problems.

spatial thinking: Analysis, problem solving, and pattern prediction involving objects and their spatial relationships.

Spatially-constrained search: A search of World Wide Websites of interest around a known point in geographic space, ordered by distance from that point.

SPOT: (Satellite Pour l'Observation de la Terre) A high-resolution, optical imaging Earth observation satellite system operating from space.

Terra: A multi-national, multi-disciplinary satellite mission involving partnerships with the U.S., Canada and Japan, and several remote sensing instruments.

user-contributed geographic information: Geographically tagged content on the World Wide Web submitted voluntarily by the owner for general distribution.

Virtual organization: An organized entity, such as a collaborative group of scientists or a social network, that does not exist at one location, but solely through the Internet.

Volunteered geographic information (VGI): The harnessing of tools to create, assemble, and disseminate geographic data provided voluntarily by individuals.

Web 2.0: A second generation of World Wide Web development and design, characterized as facilitating communication, information sharing, interoperability, user-centered design, and social collaboration.

Web portal: A way to present information from diverse Internet sources in a unified way through a single Web interface.

wide area augmentation system: An air navigation aid developed by the Federal Aviation Administration to augment the Global Positioning System, to improve its accuracy, integrity, and availability.

Wikipedia: A free, web-based multilingual encyclopedia project supported by the non-profit Wikimedia Foundation.

WIMP: Abbreviation: Windows, icons, menus, and pointers, denoting a style of human-computer interaction using these elements.

WorldWind: A free, open source virtual globe developed by NASA and the open source community for use on personal computers.

XML: (Extensible Markup Language) A general-purpose specification for creating custom markup languages, extensible because it allows the user to define the mark-up elements.

11.9 PEOPLE IN GIS

Daniel Z. Sui, Director, Center for Urban and Regional Analysis, Professor of Geography, and Distinguished Professor of Social and Behavioral Sciences, The Ohio State University

KC: Please give some sense of your educational background.

DS: I have a Bachelors degree in Geography and a Masters degree in GIScience and Remote Sensing from Peking University, and a Ph.D. in Geography from the University of Georgia.

KC: Why did you choose to go into Geography and GIScience in particular?

DS: My first choice was really mathematics and computer science, but the computer science program at Beijing in the early '80s was extremely competitive.

KC: So Geography was your second choice?

DS: Yes, but soon I discovered GIS, remote sensing, and quantitative geography. I could move back to my first love and that's how I ended up doing GIS.

KC: How important do you think it is for students in GIS to have a background in mathematics and computer science?

DS: I think critically important. Otherwise, they will be just the average GIS users instead of developing some of the cutting edge new concepts and implementing them using some of the new program tools. It's critically important to have a solid foundation in computer science and mathematics.

KC: You just received an award to support work on a new book. Correct?

DS: Yes, a Guggenheim Foundation fellowship. I will be working on the book discussing the social implications of geospatial technology, primarily on the need for geoprivacy for citizens in their ordinary lives.

KC: What is geoprivacy?

DS: Privacy originally was defined legally speaking to leave people alone so that they can enjoy their private lives, but, with the pervasive application of geospatial technologies, nowadays we can track peoples' whereabouts 24-7. There has been a huge public outcry about the potential invasion of people's privacy and even the possibility of "geoslavery." My book will investigate both sides of the issue. People are conservative about privacy. On the other hand, using Web 2.0 technology and on reality TV, people are voluntarily disclosing all sorts of excruciating details about their lives. In my book I'll discuss the complexity of this situation.

KC: What is volunteered geographic information?

DS: The term was coined by Michael Goodchild. It refers to geospatial information, primarily longitude and latitude type spatial information, contributed by citizens on a voluntary basis. Broadly speaking, it refers to all different sorts of geospatial information that could identify citizens or even commodities. It could refer to the whereabouts of people, animals, or commodities disclosed on a voluntary basis.

KC: Do you think GIS has taken away people's privacy?

DS: In a way, yes. I can give you a personal example. Last semester I hosted a party for new students at my house. Just five minutes after I gave my address to the students, I got an e-mail saying, "Dr. Sui, you did such a lovely job with the repair to your driveway." On Google Street View they could see that I'd repaired my own driveway, but that I didn't do a good job—that's how much detail they can see!

KC: How would you compare GIS developments in China to those in the United States?

DS: The U.S. is leading the world in GIS, both in terms of technology and the development of new concepts and theories. When I go to conferences in China, I am amazed by the technology they are using and the software tools. But when I talk to the GIS researchers I quickly realize that they are not so much interested in the theoretical foundation of GIScience, more the technology.

KC: What do you think will be the major differences in GIS 20 years from now?

DS: I think it will become more user friendly, it will become used more by ordinary folks— that is already happening—and I think it will be more penetrating into the details of people's lives and also more intimately involved in business operations. Geospatial will become so pervasive that it will be an integral part of the information-based society, and there will be more demand for students with geospatial skills—so GIS will be the equivalent basic skill as knowing how to use a word processor today.

KC: Thanks, and good luck with the book and your recent move to Ohio State. (Used with permission).

Index

Chapter–Opening Artwork Credits

Artwork credits: Chapter 1—Vermeer, Jan *The Geographer.* (c.1668-9) Stadelsches Kunstinstitut, Frankfurt am Main. Used with permission. Chapter 2—Hockney, David. *Garrowsby Hill.* (1998). Used with permission. Chapter 3—Clarke, Caro G. *Vermeer's Geographer. Detail as paint-by-numbers (2009).* Chapter 4—Unknown Artist. Raku Bowl. Photograph by K. Clarke (2009). Chapter 5—Clarke, Caro. *G. Geography Man.* (2009). Chapter 6—Avercamp, Hendrik. *Winter Games* (1609) Rijksmuseum, Amsterdam, The Netherlands. Used with permission. Chapter 7—Yasuda, Kim "Domestic Bathymetry (Crater Lake)", recycled flannel blankets, 72"x72"x9"(2006). Used with permission. Chapter 8—Ortellius, Abraham. Compass Rose from map sheet in Theatrum Orbis Terrarum. French ed. (1572). From the collection of the author. Chapter 9: Cézanne, Paul *Still Life with a Curtain* (1895). Hermitage Museum. Used with permission. Chapter 10—Clarke, Keith *Hammer Projection.* (2009) Used with permission. Chapter 11: George Shove. *Map of London on Glove for the Great Exhibition of 1851.* Walters Art Museum and The National Archives of the United Kingdom, Kew. Used with permission.